HPE ASE – Server Solutions Architect V3
OFFICIAL CERTIFICATION STUDY GUIDE
(EXAMS HPE0-S46 AND HPE0-S47)

First Edition

Ken Radford

HPE Press
660 4th Street, #802
San Francisco, CA 94107

HPE ASE – Server Solutions Architect V3
Official Certification Study Guide
(Exams HPE0-S46 and HPE0-S47)
Ken Radford
© 2016 Hewlett Packard Enterprise Development LP.

Published by:

Hewlett Packard Enterprise Press
660 4th Street, #802
San Francisco, CA 94107

ISBN: 978-1-942741-44-2

Printed in Mexico

WARNING AND DISCLAIMER
This book provides information about the topics covered in the HPE ASE – Architecting HPE Server Solutions (HPE0-S46 and HPE0-S47) certification exams. Every effort has been made to make this book as complete and as accurate as possible, but no warranty or fitness is implied.

The information is provided on an "as is" basis. The author and Hewlett Packard Enterprise Press shall have neither liability nor responsibility to any person or entity with respect to any loss or damages arising from the information contained in this book or from the use of the discs or programs that may accompany it.

The opinions expressed in this book belong to the author and are not necessarily those of Hewlett Packard Enterprise Press.

TRADEMARK ACKNOWLEDGEMENTS
All third-party trademarks contained herein are the property of their respective owners.

GOVERNMENT AND EDUCATION SALES
This publisher offers discounts on this book when ordered in quantity for bulk purchases, which may include electronic versions. For more information, please contact U.S. Government and Education Sales 1-855-447-2665 or email sales@hpepressbooks.com.

Feedback Information

At HPE Press, our goal is to create in-depth reference books of the best quality and value. Each book is crafted with care and precision, undergoing rigorous development that involves the expertise of members from the professional technical community.

Readers' feedback is a continuation of the process. If you have any comments regarding how we could improve the quality of this book, or otherwise alter it to better suit your needs, you can contact us through email at hpepress@epac.com. Please make sure to include the book title and ISBN in your message.

We appreciate your feedback.

Publisher: Hewlett Packard Enterprise Press

HPE Contributors: James Robinson, Chris Powell and Chris Bradley

HPE Press Program Manager: Michael Bishop

About the Author

Ken Radford is a certified HPE ASE—Cloud Architect, HPE ASE—Converged Infrastructure Architect, HPE ATP—Cloud Administrator, and HPE Certified Instructor. He was author of the HPE ASE Cloud Architect V1 and V2 study guides, and also authored the HPE ATP—Data Center and Cloud V2/HPE ASE—Data Center and Cloud Architect V3 study guide.

Introduction

This study guide helps you prepare for the HPE ASE – Architecting HPE Server Solutions V3 certification exam (HPE0-S46) and delta exam (HPE0-S47). This guide provides a solid fundamental overview of the HPE ProLiant server portfolio and supported operating systems. Topics discussed include differentiating among enterprise server solutions including HPE ProLiant ML, DL and BL servers as well as HPE Apollo and HPE Synergy solutions. The book also covers the processes, technologies, and tools involved with managing a data center, including on-system, on-premises and on-cloud management. With a focus on meeting business, financial, and technical goals, the book serves as a practical on-the-job reference tool to describe, position, and plan advanced HPE server solutions based on customer needs.

Content developed prior to the Hewlett-Packard Company separation may contain branding, logos, web page links, and other elements/information that has not been updated for each HP Inc. and Hewlett Packard Enterprise. The general knowledge and skills are still considered of value to HP Inc. and Hewlett Packard Enterprise employees (partners/customers) respectively, so these legacy materials are being made available here. Plans are in place for updating the most highly-used content for each HP Inc. and Hewlett Packard Enterprise.

QR codes and Videos

For additional learning content and practical demonstrations, you will find URLs to YouTube videos throughout the guide.

To expand your print to digital reading experience, Quick Response (QR) codes are also used throughout this book. These two-dimensional bar codes can be read by a smart phone or tablet that has a QR reader installed. Here is an example:

Note

If you do not have a smart phone or tablet with a QR reader, enter the URL beneath the QR code into a browser to access the content.

http://www.hpe.com

Scan this QR code with a smart phone app that reads QR codes.

Certification and Learning

Hewlett Packard Enterprise Partner Ready Certification and Learning provides end-to-end learning programs and professional certifications that can help you open doors and succeed in your career as an IT professional. We provide continuous learning activities and job-role based learning plans to help you keep pace with the demands of the dynamic, fast paced IT industry; professional sales and technical training and certifications to give you the critical skills needed to design, manage and implement the most sought-after IT disciplines; and training to help you navigate and seize opportunities within the top IT transformation areas that enable business advantage today.

As a Partner Ready Certification and Learning certified member, your skills, knowledge, and real-world experience are recognized and valued in the marketplace. To continue your professional and career growth, you have access to our large HPE community of world-class IT professionals, trend-makers and decision-makers. Share ideas, best practices, business insights, and challenges as you gain professional connections globally.

To learn more about HPE Partner Ready Certification and Learning certifications and continuous learning programs, please visit http://certification-learning.hpe.com

Audience

Typical candidates for this certification are presales technical specialists with at least three years' experience who plan and design HPE server and Composable Infrastructure solutions.

Assumed Knowledge

Although anyone may take the exam, it is recommended that candidates have a minimum of three years' experience with architecting HPE server solutions. Candidates are expected to have industry-standard technology knowledge from training or hands-on experience.

Minimum Qualifications

The HPE ASE Server Solutions Architect certification is targeted at the Accredited Solutions Expert (ASE) skill level and is the next progressive step after a candidate has achieved HPE ATP—Server Solutions certification.

Although anyone may take the exam, it is recommended that candidates have a minimum of three years' experience with architecting HPE server solutions. Candidates are expected to have industry-standard server technology knowledge from training or hands-on experience.

Relevant Certifications

After you pass these exams, your achievement may be applicable toward more than one certification. To determine which certifications can be credited with this achievement, log in to The Learning Center and view the certifications listed on the exam's More Details tab. You might be on your way to achieving additional HP certifications.

Preparing for Exam HPE0-S46

This self-study guide does not guarantee that you will have all the knowledge you need to pass the exam. It is expected that you will also draw on real-world experience and would benefit from completing the hands-on lab activities provided in the instructor-led training.

This exam tests candidates' knowledge and skills on architecting HPE server products and solutions. Topics covered in this exam include server architectures and associated technologies as well as their functions, features, and benefits. Additional topics include knowledge of planning, designing, and positioning HPE server solutions to customers.

Preparing for the Delta Exam HPE0-S47

Candidates who want to acquire the HPE ASE Server Solutions V3 certification and have already acquired the previous version of the server ASE certification – HP ASE Server Solutions V2 – can sit the delta exam HPE0-S47. HPE recommends that candidates read this study guide in its entirety to prepare for the delta exam.

Recommended HPE Training

Recommended training to prepare for each exam is accessible from the exam's page in The Learning Center. See the exam attachment, "Supporting courses," to view and register for the courses.

Obtain Hands-on Experience

You are not required to take the recommended, supported courses, and completion of training does not guarantee that you will pass the exams. HP strongly recommends a combination of training, thorough review of courseware and additional study references, and sufficient on-the-job experience prior to taking an exam.

Exam Registration

To register for an exam, go to http://certification-learning.hpe.com/tr/certification/learn_more_about_exams.html.

CONTENTS

1 Hewlett Packard Enterprise in the New Compute Era

WHAT IS IN THIS CHAPTER FOR YOU?

After completing this chapter, you should be able to:

- ✓ Summarize the Hewlett Packard Enterprise (HPE) Transformation Areas

- ✓ Explain how server trends present challenges and opportunities for enterprise businesses

- ✓ Provide a high-level overview of HPE ProLiant Gen9 server innovations

- ✓ Name the features of an HPE Converged Infrastructure

- ✓ Outline the HPE position and market share in the server solutions industry

- ✓ Position HPE Technology Services in the new compute era

OPENING CONSIDERATIONS

Before proceeding with this section, answer the following questions to assess your existing knowledge of the topics covered in this chapter. Record your answers in the space provided here.

1. What are some of the major marketplace trends affecting IT and servers?

_____

_____

_____

_____

_____

_____

2. How familiar are you with HPE Converged Infrastructure solutions?

_____

_____

_____

_____

_____

_____

3. How would you describe HPE compared to its market competitors? How do HPE servers stack up against third-party offerings?

_____

_____

_____

_____

_____

HPE Transformation Areas

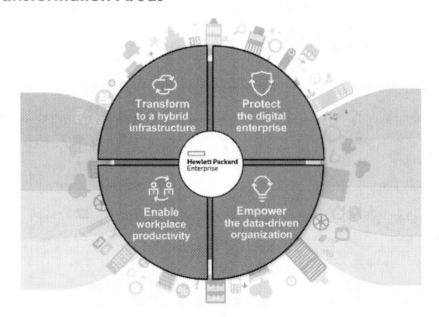

Figure 1-1 HPE transformation areas

To succeed in today's idea economy, organizations need to develop new IT capabilities, create new outcomes, proactively manage risk, be predictive, and create a hyper-connected workplace. IT is transforming quickly to keep pace with the fast-changing customer needs and competitive pressures. In order to help customers achieve superior business outcomes, HPE identified four Transformation Areas that reflect what customers consider most important as illustrated by Figure 1-1:

● **TA1–Transform to a hybrid infrastructure**—A hybrid infrastructure enables customers to get better value from the existing infrastructure and delivers new value quickly and continuously from all applications. This infrastructure should be agile, workload optimized, simple, and intuitive.

To transform to a hybrid infrastructure, customers must focus on the following activities:

  – Define their right mix—Defining a hybrid roadmap to guide the business and IT, allowing for a better balance of CapEx and OpEx

  – Power their right mix—Providing the foundation for future IT services

  – Optimize their right mix—Consuming services at the right place, at the right time, at the right cost

 Note

  HPE servers play a key role in helping customers transform to a hybrid environment. Consequently, the material in this book focuses on this Transformation Area.

● **TA2—Protect the digital enterprise**—Customers consider it a matter of when, not if, their digital walls will be attacked. The threat landscape is wider and more diverse than ever before. A complete risk management strategy involves security threats, backup and recovery, high availability, and disaster recovery.

To protect the digital enterprise, customers must focus on the following features:

  – **Built-in resilience**—Automated and integrated data protection and security controls, robust security governance, and high-availability infrastructure strengthen overall security practices.

  – **Planned ecosystem**—Strategic planning and investment in sophisticated enterprise security, latest protection topologies, and tools for compliance help protect against security threats.

  – **Adaptive and federated systems**—Integrated tools, elastic pools of protection capacity, and analytics-based optimization help balance performance.

  – **Integrated solutions**—Regular assessments of capabilities ensure people, technology, and processes are aligned to deliver better business outcomes.

- **TA3–Empower the data-driven organization**—Customers are overwhelmed with data; the solution is to obtain value from information that exists. Data-driven organizations generate real-time, actionable insights.

  To empower the data-driven organization, customers must focus on:

  - **Agility and scalability**—An investment road map enables the rapid deployment of powerful open hardware and software at a lower cost with more flexibility to scale.

  - **Actionable analytics**—Predictive insights should be constantly refined and highly relevant to multiple facets of the business.

  - **Data-driven decisions**—Powerful analytics solutions (traditional or cloud based) connect to virtually any data source quickly and easily.

  - **Information as an asset**—Information is governed in a secure end-to-end life cycle, balancing value, cost, and risk.

- **TA4–Enable workplace productivity**—Many customers are increasingly focused on enabling workplace productivity. Delivering a great digital workplace experience to employees and customers is a critical step.

  To enable workplace productivity, customers must focus on:

  - **Greater efficiency**—Software-defined infrastructure and user-based management reduce costs and improve user experience.

  - **Universal accessibility**—High-performance wireless devices, new working practices, and cross-device collaboration improve communication.

  - **Anywhere workforce**—The flexibility to work anywhere means accessing resources on any device with secure, tested, and monitored apps.

  - **Adaptable investment strategy**—This strategy involves evolving capabilities, devices, and applications in line with business needs.

Transform to a hybrid infrastructure

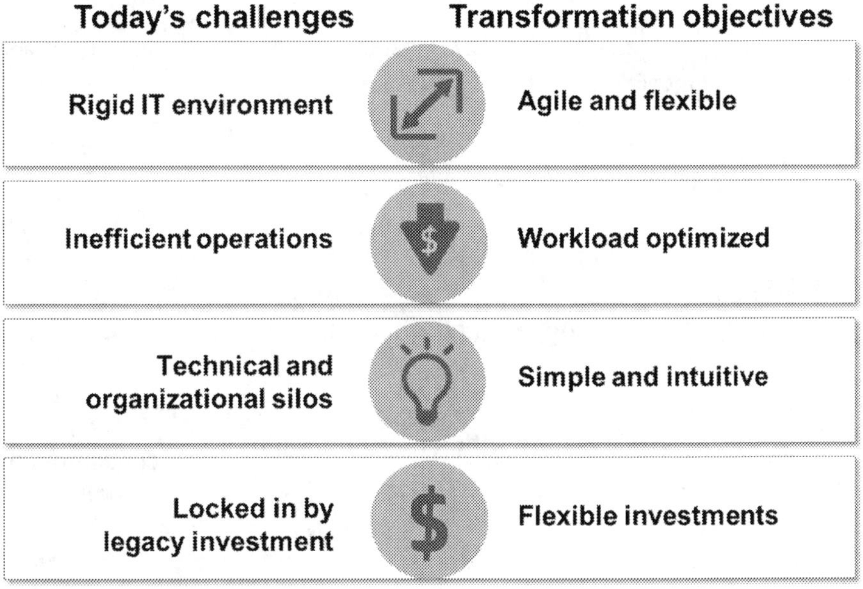

Figure 1-2 Challenges solved by transforming to a hybrid infrastructure

Getting the most out of hybrid infrastructure opportunities requires planning performance, security, control, and availability strategies. For this reason, organizations must understand where and how a hybrid infrastructure strategy can most effectively be applied to their portfolio of services. Figure 1-2 shows some of the challenges faced by customers and the hybrid transformation objectives.

Many customers struggle with:

- **Rigid IT environment**—Legacy hardware scales poorly and slows deployment of apps and workloads.

- **Inefficient operations**—Data center has high operating costs and overhead, slow IT services, and poor utilization with patchy availability and performance.

- **Technical and organizational silos**—Inefficiencies and lack of collaboration mean IT is dedicated to "keeping the lights on."

- **Being locked in by legacy investment**—Proprietary systems and depreciation schedules limit upgrade opportunities.

When transforming to a hybrid infrastructure, customers must focus on:

- **Agility and flexibility**—A converged and virtualized hybrid infrastructure scales easily and delivers continuous value to make IT a service provider.

- **Workload optimization**—Modern infrastructure offers better utilization, adjusting performance and availability dynamically.

- **Simplicity and intuitiveness**—Software-defined controls, along with automation and converged management, free up IT resources.
- **Flexible investments**—Open-standards-based systems and new IT consumption models enable continuous business innovation.

Enterprise challenges

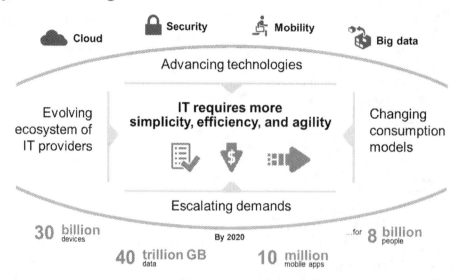

Figure 1-3 IT must transform to enable service delivery

The increasing adoption of cloud, security, mobility, and big data solutions are disrupting the IT and server industry. A 2012 report from the International Data Corporation (IDC) predicts that by 2020, there will be 30 billion devices, 40 trillion gigabytes of data, and 10 million mobile apps for 8 billion people. This forecast demonstrates how critical these trends are to an organization's bottom line. Increasing user demands, emerging IT delivery and consumption models, and advancing technologies are accelerating the design, architecture, and capabilities of servers as illustrated by Figure 1-3.

These disruptive forces require that IT be more agile, efficient, and simple. Existing IT solutions are designed based on administrator needs. Businesses are seeking new ways to deliver and consume compute, such as cloud (user-driven control) and software-defined solutions (application-driven control). Some companies are even revamping their overall IT strategy to increase speed, flexibility, and sustainability.

Many IT organizations struggle to deliver services rapidly enough to achieve their desired business outcomes. Thus, the gap between IT supply and business demand for simple, fast, cost-effective, and value-added services continues to grow. This is creating numerous customer challenges, including:

- **Inefficient operations constrain growth**—Infrastructure and operations consume too much of the overall IT budget, which leaves little for new or innovative projects.

- **Slow product and service delivery stifles competitive advantage**—Siloed infrastructure, management, and processes lead to more steps, time, and opportunity for error. There are too many nonstandard, manual tasks and an ever-expanding backlog of projects. Delays in service delivery result in lost revenue opportunities.

- **Suboptimal business performance**—IT decision makers say that slow or poor application performance is one of their top data center challenges. Weak business application performance can cause decreases in revenue and adversely impacts the ability of IT to meet changing business demands.

IT must be the business partner for value creation

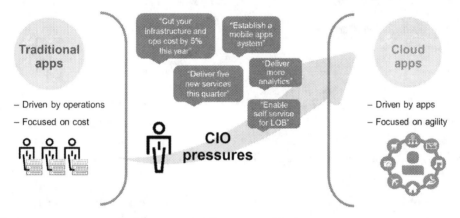

Figure 1-4 Two different sets of infrastructure

Traditional business applications are built to run the business. There are applications such as enterprise resource planning (ERP) that have been prepackaged and pretested by the software vendor and the hardware vendor; these apps typically go through release cycles maybe once or twice a year. IT has been built around these applications for the last 30 years.

In the current IT environment, applications do not support the business—they are the business. These applications for mobility and new cloud-native applications drive revenue. Many chief information officers (CIOs) and IT organizations are being challenged to deliver traditional business applications while supporting new applications such as mobile and cloud-native apps, as illustrated by Figure 1-4. Gartner calls this *bi-modal computing* in which the company maintains existing infrastructure for traditional applications while supporting a different set of infrastructure and tools for the new applications.

Supporting two different sets of infrastructure is unsustainable. The HPE vision is to merge these environments and produce one infrastructure that can deliver all the applications today while enabling the agility and fluidity of infrastructure that customers need for tomorrow.

The infrastructure must be transformed in the compute era

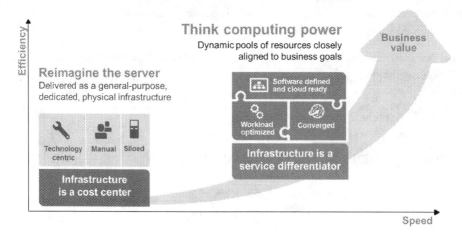

Figure 1-5 Infrastructure transformation

Traditionally, infrastructure has two high-level attributes—it is functionally defined and purpose built. Purpose-built devices have limited ability to change functions as needs change. This "status quo IT" only perpetuates silos and increases complexity with an overload of products and tools that lack interoperability. The incremental, generational approach has characterized the server industry for the last few decades.

Current operations need to adapt to the *compute era*. The compute era requires resources to transform by increasing performance to address current and future workload needs. As illustrated by Figure 1-5, this is part of a larger transformation in which IT infrastructure evolves from being a cost center to a service differentiator, ultimately adding business value to the enterprise. The HPE compute portfolio offers software-defined, cloud-ready, workload-optimized, and converged infrastructure solutions.

Key design principles for the compute era

HPE designs solutions for the compute era based on three design principles:

- **Software defined and cloud ready**—HPE offers software-defined capabilities to orchestrate and manage all IT resources optimally across their lifecycle with almost no administration overhead. The HPE cloud-ready design helps eliminate costly manual device-by-device configurations and speeds time to scale for the hybrid world. This approach enhances agility, efficiency, productivity, and accuracy.

- **Workload optimized**—For performance that translates into business value, HPE builds servers, networking, and storage in workload-optimized designs with a common modular architecture. This approach delivers superior business performance, often by tuning and optimizing every element of infrastructure for a specific workload.

- **Convergence**—The HPE approach to infrastructure provides a common modular architecture and converged management across servers, storage, and networking. It reduces IT complexity and accelerates time to value. An HPE Converged Infrastructure breaks down technology silos and brings together all IT resources into flexible pools of assets that can be shared by many applications and managed as a service. This approach helps businesses optimize resources for speed, efficiency, and performance.

Compute power is the engine of the infrastructure

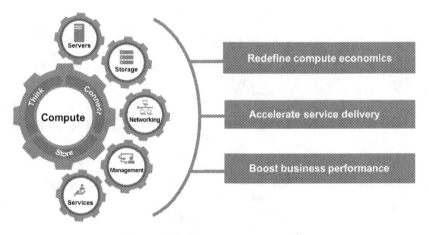

Figure 1-6 Compute power goals

HPE solution design principles help lower the cost of services, reduce time to provision services, and increase the value of services. Instead of a generational approach, HPE identifies compute resources as the processing engine for business value and results. For example, HPE ProLiant Gen9 servers are focused on optimizing the customer's business, rather than merely optimizing the server.

In order for enterprises to succeed, the platform for compute power must achieve three goals as illustrated by Figure 1-6:

- **Redefine compute economics**—With an end-to-end portfolio that delivers compute resources scaled to business requirements

- **Accelerate service delivery**—With workload flexibility and optimization, where compute resources are:

  - Delivered in flexible and dynamic blocks of capacity and capability

  - Automatically provisioned to deliver the right computing, storage, and networking capacities to address dynamic workload needs

- **Boost business performance**—With significant performance improvement across the entire portfolio and up the workload stack, providing:

  – Advanced computing across the spectrum

  – The highest performance at the lowest cost

  – Improved and automated operations

HPE compute portfolio

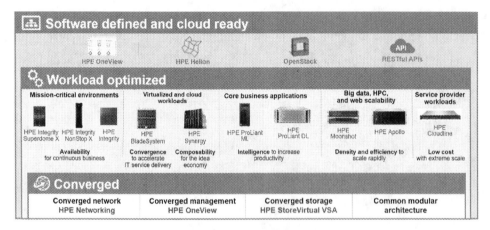

Figure 1-7 HPE compute portfolio

As shown in Figure 1-7, HPE has a broad portfolio that can address customers' needs in the compute era. The server portfolio addresses a variety of workloads:

- HPE continues to be committed to high availability and scale-up for mission-critical environments with HPE Integrity Superdome X, HPE Integrity NonStop X, and HPE Integrity servers.

- HPE BladeSystem is a leading cloud and virtualization platform, using convergence, federation, and automation to reduce costs, minimize downtime, and speed service delivery.

- HPE Synergy, the first platform built from the ground up for Composable Infrastructure, offers an experience that empowers IT to create and deliver new value instantly and continuously.

- HPE ProLiant ML and DL rack and tower servers offer efficiency, value, and positive return on investment (ROI).

- HPE Moonshot and HPE Apollo lines are purpose-built for the density and ultra-density of high-performance computing (HPC).

- HPE Cloudline addresses the compute requirements of large service providers by delivering low total cost of ownership (TCO) and extreme scale along with service and support.

Note

This book will cover HPE BladeSystem, Synergy, ProLiant ML and DL, and Apollo systems.

HPE ProLiant Gen9 servers

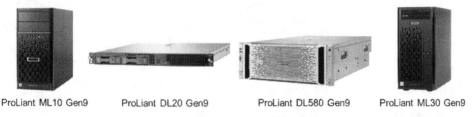

ProLiant ML10 Gen9   ProLiant DL20 Gen9   ProLiant DL580 Gen9   ProLiant ML30 Gen9

Figure 1-8 HPE ProLiant Gen9 servers

ProLiant Gen9 servers, shown in Figure 1-8, extend the HPE compute approach and meet key customer needs, including:

- **Reducing IT costs** through larger and more efficient compute resources and storage. ProLiant Gen9 servers allow customers to tailor their compute solution for efficiency by offering flexible choices for servers, storage, networking, power options, and more.

- **Delivering faster setup, deployment, and maintenance** with advancements in embedded management, as well as on-premises and on-cloud management.

- **Increasing workload performance** across the Gen9 architecture, including improved processor and memory performance, optimized storage performance, and increased networking performance with lower latency capabilities.

Advancing compute with HPE ProLiant Gen9 servers

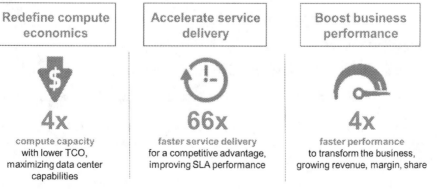

Redefine compute economics	Accelerate service delivery	Boost business performance
4x	**66x**	**4x**
compute capacity with lower TCO, maximizing data center capabilities	faster service delivery for a competitive advantage, improving SLA performance	faster performance to transform the business, growing revenue, margin, share

Figure 1-9 Advancing compute capabilities

With ProLiant Gen9 servers, HPE is advancing compute capabilities to meet growing business demands. ProLiant Gen9 servers meet compute goals in several ways, as illustrated by Figure 1-9. ProLiant servers:

- Redefine economics of compute resources by providing:

  - Four times the compute capacity with lower TCO, maximizing data center capabilities at the lowest cost of service[1]

  - Twice as much compute per watt using 12 Gb/s serial attached SCSI (SAS) solid state drives (SSDs) and HPE DDR4 SmartMemory[2]

  - Tailored compute for efficiency across multiple workloads with flexible choices in server, storage, networking, and power options

  - Lower energy and floor space consumption

  - Decreased energy costs with efficient cooling

  - Reduced power consumption without impacting IT performance using HPE Flexible Slot Power Supplies (96% efficiency)

  - Increased compute density with HPE PCIe Workload Accelerators

- Accelerate service delivery by providing:

  - Up to 66 times faster service delivery with simple automation, which saves administrative time and reduces manual errors[3]

  - Automated lifecycle management

  - Faster cloud deployment

  - Reliable, secure, embedded management

  - Rapid provisioning with software-defined HPE OneView management

- Boost business performance by providing:

  - Four times faster workload performance[4]

  - Better computing, memory, and IO performance

  - Increased networking performance and lower latency

  - Support for 1 million IOPS with 12 Gb/s HPE Smart Array fast controllers[5]

  - Quick data access with up to four times read and write workload acceleration provided by HPE SmartCache[4]

  - Up to 14% better memory performance provided by HPE DDR4 SmartMemory and ProLiant rack servers[6]

Notes

1. Based on HPE internal analysis.

2. HPE internal analysis compared the DL380 Gen9 server with the DL380 G7 server. The source for system wattage is the IDC Qualified Performance Indicator. The performance was taken from SPECint_rate_base2006 industry benchmark. Calculation: performance/watt, August 2014.

3. Based on an IDC white paper sponsored by HPE, *Achieving Organizational Transformation with Converged Infrastructure Solutions for SDDC*, published in January 2014. For this customer, the time to build and deploy infrastructure for 12 call centers was reduced from 66 days to one day.

4. HPE Smart Storage engineers calculated HPE SmartCache performance with equivalent controller in a controlled environment. Data was taken from HPE internal SmartCache webpage.

5. Internal performance lab testing used lometer and the HPE Smart Array P840 with RAID 0, 4k random reads, and Microsoft Windows 2012 R2; testing is ongoing with changes in firmware. This number is current as of 21 July 2014.

6. This was calculated based on HPE internal calculations. The performance compares DIMMs running on an HPE server compared to a third-party server with DDR4. Intel recommends running DDR4 server memory (two DIMMs per channel at 1866 MHz) with Haswell processors.

Activity: Securing growth with HPE servers

To further your understanding of how customers implement ProLiant servers in their data centers, watch this video about how HPE customer Cheetah Mobile used HPE servers to improve its processes.

 Note

This video was published prior to the Hewlett-Packard Company separation. To access the video, scan this QR code or enter the URL into your browser.

https://www.youtube.com/watch?v=R31yOV-PVsw

Be prepared to pause the video as needed to answer these questions:

1. What are the company's IT requirements and challenges?

   _____

   _____

   _____

   _____

2. How many monthly active users does Cheetah Mobile serve?

   _____

   _____

3. Which HPE server did Cheetah Mobile use to provide high-density and flexible operational capabilities? What other benefits did the business experience from this server?

   _____

   _____

   _____

   _____

4. According to Hu Kai, why is it important for customers to have one complete solution?

   _____

   _____

   _____

   _____

5. Which areas does HPE focus on to help businesses shift from traditional business models?

   _____

   _____

   _____

   _____

6. How might the same server benefit a customer in a different business?

_____

_____

_____

_____

HPE Converged Infrastructure

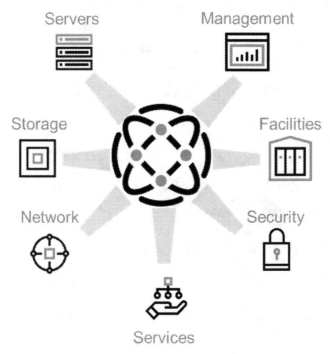

Figure 1-10 HPE Converged Infrastructure

HPE Converged Infrastructure helps customers to reduce costs and increase operational efficiency by eliminating silos and freeing available compute, storage, and networking resources. As shown in Figure 1-10, in a Converged Infrastructure the entire IT infrastructure can be converged—including servers, storage, networking, security, and facilities—and managed from a central point. HPE Converged Infrastructure solutions are backed by years of experience in focused engineering, research and development (R&D) investments, and data center experience. These solutions are making data centers simpler, more flexible, more efficient, and less expensive to operate.

A converged infrastructure:

- Helps businesses accelerate time to application and business value

- Combines server, storage, and networking into resource pools

- Unifies management tools, policies, and processes, allowing resources and applications to be managed in a holistic, integrated fashion

- Includes built-in security and tight integration to offer protection against external and internal security threats

- Merges power and cooling management capabilities, enabling systems and facilities to work together

- Makes data centers simpler, more flexible and efficient, and less expensive to operate

The path to a composable infrastructure

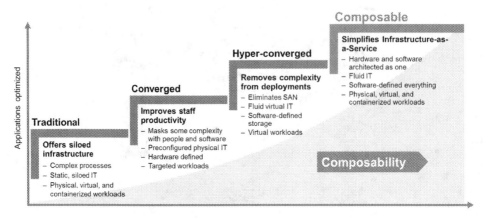

Figure 1-11 Composable infrastructure

Figure 1-11 illustrates how IT infrastructure has evolved through traditional, converged, and hyper-converged architectures. Traditional architectures use siloed approaches to address increased workloads and data, often at the expense of overprovisioning resources and complicating management. Converged architectures use preconfigured physical components to mask some complexity with people and software. Hyper-converged architectures address velocity and remove complexity from deployments. Through software-defined storage, a hyper-converged environment optimizes virtual workloads and eliminates the need for a SAN.

HPE Converged Infrastructure, software-defined management, and hyper-converged systems reduce costs and increase operational efficiency by eliminating silos and freeing available compute, storage, and networking resources. Building on a converged infrastructure, HPE has designed the Composable Infrastructure, which is a next-generation approach designed around three core principles:

- **Fluid resource pools** are combined into a single structure that boots up ready for any workload with fluid pools of compute, storage, and network fabric that can be instantly turned on and flexed. They effortlessly meet each application's changing needs by allowing for the composition and recomposition of single blocks of disaggregated compute, storage, and network infrastructure.

- **Software-defined intelligence** embeds intelligence into the infrastructure. It uses workload templates to compose, re-compose, and update in a repeatable, frictionless manner to speed deployment and eliminate unnecessary downtime. It provides a single management interface to integrate operational silos and eliminate complexity.

- **Unified application programming interface (API)** provides a single interface to discover, search, inventory, configure, provision, update, and diagnose the Composable Infrastructure. A unified API allows the infrastructure to be programmed with code so it can become infrastructure as a service.

HPE Synergy is designed to bridge traditional and new IT with the agility, speed, and continuous delivery needed for today's applications. With Synergy, a single line of code enables full infrastructure programmability and can provision the infrastructure required for an application, for breakthrough time-to-market and deployment flexibility. A Composable Infrastructure facilitates the move to a continuous services and application delivery model and enables applications to be updated as needed, rather than just once or twice a year. It enables IT to operate like a cloud provider to lines of business and the extended enterprise, consistently meet service-level agreements (SLAs), and provide the predictable performance needed to support core workloads.

HPE position and market share

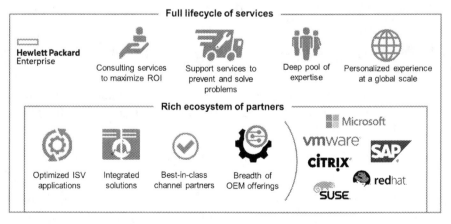

Figure 1-12 HPE position and market share

As illustrated by Figure 1-12, HPE is one of the world's largest providers of IT infrastructure, software, services, and solutions. In addition to offering the broadest server portfolio in the industry, HPE offers a full lifecycle of services, including consulting services to maximize ROI and support services to prevent and solve problems. Customers receive assistance from a deep pool of expertise and a personalized experience on a global scale.

HPE also leverages a rich ecosystem of best-in-class business partners who provide optimized independent software vendor (ISV) applications, integrated solutions, and a breadth of OEM offerings. HPE holds the number one position in servers shipped running VMware vSphere and Microsoft Hyper-V and was the first manufacturer to offer:

- Integration of management for server, storage, and networking infrastructure within management tools from VMware, Microsoft, and Red Hat

- More server and storage models certified for key partners than any other vendor

- Broad partner-certified professional designations for field technical resources

HPE server market share

IDC released a quarterly update to its Worldwide Quarterly Server Tracker, citing market share figures for the fourth quarter of 2015. Key facts include:

- **Fact 1:** HPE continues to lead in the worldwide server market by revenue. HPE is number one in worldwide revenue for the eighth consecutive quarter.

  - Dell was number two, behind HPE by 8.2 percentage points.

  - In 2015, HPE was also first in worldwide revenue, growing at a rate of 5.8% when compared with the previous full year.

- **Fact 2:** For nearly 14 consecutive years, HPE has been the number one vendor in worldwide server shipments.

  - HPE shipped a server every 13.4 seconds (24x7) and shipped nearly one in every four servers worldwide.

  - In 2015, HPE was also first in server shipments with more than 2,350,000 units shipped. HPE shipped 331,000 more servers than Dell (ranked second), 1,400,000 more than Lenovo (ranked third), and 2,000,000 more than Cisco.

- **Fact 3:** HPE leads the total server blade market with a number one position in both revenue and units.

  - HPE has led the server blade market in revenue and units for more than nine consecutive years. HPE had 13.8 percentage points more revenue share from blades than Cisco (ranked second).

- **Fact 4:** HPE ranked first in density-optimized servers.

  - HPE grew revenue in density-optimized servers by 63.4% year over year, gaining 5.0 percentage points during this time (two times the market rate).

  - In 2015, HPE was also first in worldwide revenue for density-optimized servers, growing at a rate of 89.6% and gaining 8.3 percentage points when compared to the previous full year.

  - In 2015, HPE gained more revenue from density-optimized servers than its top three competitors (Dell, IBM, and Lenovo) combined.

- **Fact 5:** HPE leads in combined density-optimized and server blades.

  - HPE is number one in revenue market share with 35.1%, which is twice as much as the closest competitor, Cisco. IDC identified this segment as a significant and growing market driven by demand in public cloud hyperscale environments, as well as private cloud and integrated systems.[1]

 Note

[1]IDC Press Release published 26 February 2014.

- **Fact 6:** HPE ProLiant has been the x86 server market share leader in both revenue and units for more than 19.5 consecutive years.[1]

  - Sustained market leadership is a hallmark of HPE ProLiant innovation. HPE had a revenue share of 29.0%, which is 8.5 points more than Dell (ranked second) and 19.9 percentage points more than Lenovo (ranked third).

- **Fact 7:** HPE leads in EPIC+RISC server blade revenue.

  - HPE Integrity blades maintained the number one position in revenue for the EPIC+RISC blade segment with 84.5% worldwide share, gaining 5.1 percentage points year over year.

- **Fact 8:** UNIX, Windows, and Linux representing 99.9% of all servers shipped worldwide. For these three major operating environments, HPE is number one worldwide in server revenue and unit shipments.

 Note

1. Includes Compaq ProLiant from Q1 FY96 through Q2 FY02 and HPE ProLiant from Q3 FY02 through Q4 FY15.

Note

These market share figures are for the fourth quarter of 2015 unless otherwise noted. This data represents worldwide results as reported by the IDC Worldwide Quarterly Server Tracker in March 2016. For more information on this report, scan this QR code or enter the URL into your browser.

https://www.idc.com/getdoc.jsp?containerId=prUS41076116

Ranking of HPE servers

In December 2015, Gartner analysts ranked HPE across several categories. One category focused on the range of x86 servers available from HPE. Gartner reported that HPE x86 servers cover several markets in various form factors and are flexible to meet different budgets. For these reasons, Gartner attributed HPE server offerings as one reason HPE is the volume leader worldwide for most of its product segments.

 Note

For more information about this Gartner analysis, scan this QR code or enter the URL into your browser. Registration at the Gartner website is necessary to view this document.

https://www.gartner.com/doc/reprints?id=1-2V82M98&ct=160104&st=sb

HPE Technology Services

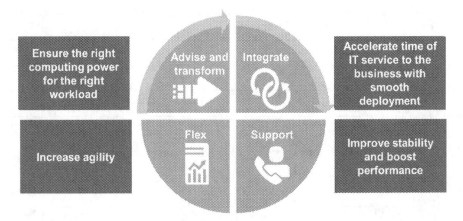

Figure 1-13 HPE Technology Services

In the new compute era, HPE offers IT consulting services and support to architect, deploy, and optimize an enterprise's technology assets. The HPE expertise in the new generation of technology enables IT organizations to remain relevant. With specialized departments in critical aspects of IT—such as hybrid infrastructure, data science, information security, and hyper-connected architectures—HPE Technology Services can enhance business outcomes and maximize returns on technology investments.

Customers have many options when determining the right way for IT to support business requirements. HPE Technology Services support customers as they transform existing IT systems. After a plan has been designed, the next step is to integrate new technology into existing processes without losing momentum or interrupting business. HPE implementation and deployment teams execute the plans and help customers get started quickly. This includes education services where needed to refresh technical skills and to help systems administrators learn new ways to do things.

After the new infrastructure is in place, it must meet changing needs and increasing demands. As the rate of change accelerates, IT staff should focus on simplifying operations and providing a stable, well-performing IT infrastructure. HPE Services connects the customer's IT systems to HPE support technology to automate and streamline the process of making changes. These services provide assistance with identifying, diagnosing, and resolving problems. The HPE support portfolio also enables customers to prevent problems, receive enhanced access to technical resources, and reduce risk to the business.

As shown in Figure 1-13, these steps typically are on a continuous cycle in order to manage an ever-changing environment. HPE Technology Services provides innovative ways to consume compute capacity and ensure that business needs are met, giving customers more choice, options, and flexibility.

Learning check

1. Which areas should customers focus on when transforming to a hybrid infrastructure?

_____

_____

_____

_____

2. Which features of HPE ProLiant servers improve workload performance?

_____

_____

_____

_____

3. A converged infrastructure unifies management tools, policies, and processes, allowing resources and applications to be managed in a holistic, integrated fashion.

☐ True

☐ False

Learning check answers

1. Which areas should customers focus on when transforming to a hybrid infrastructure?

- **Agility and flexibility—A converged and virtualized infrastructure that scales easily**

- **Workload optimization—A modern infrastructure offering better utilization**

- **Simplicity and intuitiveness—Software-defined controls, automation, and converged management**

- **Flexible investments—Open-standards-based systems and new IT consumption models**

2. Which features of HPE ProLiant servers improve workload performance?

- **Optimized storage performance**

- **Four times faster workload performance**

- **Better computing, processor, and IO performance**
- **Increased networking performance and lower latency**
- **Support for 1 million IOPS with 12 Gb/s HPE Smart Array fast controllers**
- **Quick data access with up to four times read and write workload acceleration provided by HPE SmartCache**
- **Up to 14% better memory performance provided by HPE DDR4 SmartMemory and ProLiant rack servers**

3. A converged infrastructure unifies management tools, policies, and processes, allowing resources and applications to be managed in a holistic, integrated fashion.

☐ **True**

☐ False

Summary

- HPE identified four Transformation Areas that reflect what customers consider most important: protect the digital enterprise, empower the data-driven organization, enable workplace productivity, and transform to a hybrid infrastructure. HPE servers play a key role in helping customers transform to a hybrid environment.

- The increasing adoption of cloud, security, mobility, and big data solutions are disrupting the IT and server industry. IT organizations are being challenged to deliver traditional business applications while supporting new applications such as mobile and cloud-native apps. The HPE compute portfolio offers software-defined, cloud-ready, workload-optimized, and converged infrastructure solutions.

- ProLiant Gen9 servers extend the HPE compute approach and meet key customer needs. They redefine economics of compute resources, accelerate service delivery, and boost business performance.

- Building on a converged infrastructure, HPE has designed the Composable Infrastructure. This approach is designed around fluid resource pools, software-defined intelligence, and unified API.

- HPE is one of the world's largest providers of IT infrastructure, software, services, and solutions. HPE continues to lead the worldwide server market by revenue.

- With specialized departments in critical aspects of IT—such as hybrid infrastructure, data science, information security, and hyper-connected architectures—HPE Technology Services can enhance business outcomes and maximize returns on technology investments.

2 HPE Converged Management

WHAT IS IN THIS CHAPTER FOR YOU?

After completing this chapter, you should be able to:

✓ Explain the HPE approach to converged management for the infrastructure lifecycle

✓ Name the on-system tools used to manage an HPE ProLiant system

  ✓ Unified Extensible Firmware Interface (UEFI)

  ✓ HPE representational state transfer (REST) application programming interface (API) and HPE RESTful Interface Tool

  ✓ HPE Intelligent Provisioning

  ✓ HPE integrated Lights-Out (iLO) 4

  ✓ HPE Smart Update Manager (HPE SUM)

  ✓ HPE Service Pack for ProLiant (SPP)

  ✓ Agentless management

  ✓ Smart Storage Administrator (SSA)

✓ Explain how to perform on-premises management with HPE OneView

✓ Explain how to perform remote management with on-cloud management tools:

  ✓ HPE Insight Online

  ✓ HPE Insight Remote Support

OPENING CONSIDERATIONS

Before proceeding with this section, answer the following questions to assess your existing knowledge of the topics covered in this chapter. Record your answers in the space provided here.

1. What system management tools have you used in the past? What challenges did you face? What worked well for you?

   _____

   _____

   _____

   _____

   _____

2. How familiar are you with the portfolio of HPE server management tools?

   _____

   _____

   _____

   _____

   _____

HPE Converged Management

Whether in an SMB or enterprise environment, customers need management tools targeted for their business needs. The HPE approach to infrastructure management includes a complete portfolio of server management solutions that address the challenges of ever-increasing business demands and complexity in today's data center.

HPE server management capabilities are specifically designed to manage the entire HPE server portfolio, from towers to racks to blades. With HPE converged management, IT staff can gain precise control of their infrastructure through built-in intelligence that can be easily accessed remotely. HPE Converged Infrastructure management covers the lifecycle of critical operations: configuration and provisioning for rapid deployment, system health monitoring with proactive failure notification, firmware updates, and automated simplified support management.

Server lifecycle management for ProLiant

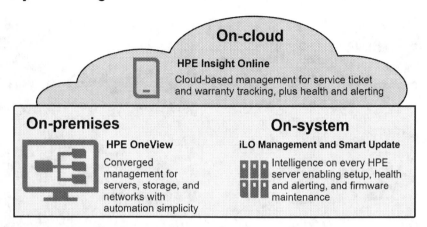

Figure 2-1 ProLiant server lifecycle management

As illustrated by Figure 2-1, HPE ProLiant Gen9 management innovations target three environments to ensure that customers have complete lifecycle management, for their current environment and in the future as their business grows.

- **On-system**—Built-in intelligence and automation for increased server admin productivity. It provides on-system management to provision, monitor, and troubleshoot servers, as well as remote and out-of-band management. The on-system management tools available for ProLiant Gen9 servers include:

  – UEFI

  – HPE integrated Lights-Out (iLO)

  – HPE RESTful Interface Tool

  – HPE SUM

  – HPE SPP

  – Agentless management

  – Smart Storage Administrator (SSA)

- **On-premises**—Converged to simplify management across servers, storage, and networks, and automated to streamline delivery of IT services and transitions to a hybrid cloud. HPE OneView for the software-defined data center provides rapid, repeatable, reliable operations at low cost.

- **On-cloud**—Automated, efficient support experience. An easy, all-in-one personalized dashboard enables you to track IT health and automated cases for servers, storage, and networks. On-cloud tools include HPE Insight Remote Support.

Software-defined data center

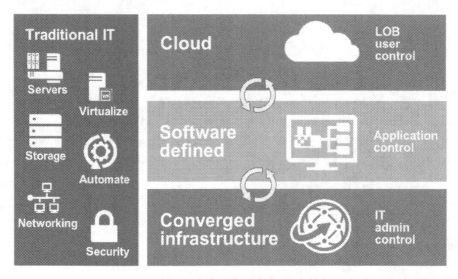

Figure 2-2 Enhance IT execution capabilities

A software-defined data center (SDDC) is a concept in which the infrastructure of an organization's data center extends the use of virtualization technology by abstracting, pooling, and automating all of its data center resources, as illustrated by Figure 2-2. Implementing an SDDC is effectively delivering an IT as a Service (ITaaS) solution.

In an SDDC, the various elements of the infrastructure are virtualized and delivered as services. Although ITaaS might represent an outcome of an SDDC, an SDDC solution is intended to benefit data center architects and IT staff more than the users or the consumers of the resources. Software abstraction in the data center infrastructure is not visible to the consumers.

An SDDC encompasses a variety of concepts and data center infrastructure components, where each component can be provisioned, operated, and managed through a programmatic user interface. An SDDC is not the same thing as a private cloud because a private cloud only has to offer a virtual machine self-service solution, accompanied by the storage and networking associated with the VMs. Within a private cloud, IT administrators could use traditional provisioning and management interfaces. Instead, the SDDC envisions a data center that could potentially support private, public, and hybrid cloud offerings.

The core architectural components that comprise a given vendor's SDDC solution might include:

- **Virtualization of compute**—A software implementation of a computer's processor, memory, and IO resources. This is commonly referred to as *hypervisor software*.

- **Software-defined networking (SDN) or network virtualization**—Could involve provisioning VLANs on a switch, Ethernet ports operating as a single or an aggregated link, ports supporting access or VLAN trunking, security settings, and so forth.

- **Software-defined storage or storage virtualization**—Could involve provisioning storage LUNs on a storage array and host bus adapter (HBA) zoning on a SAN switch.

- **Management and automation software**—Enables an administrator to provision, control, and manage all SDDC components.

Some of the commonly cited benefits of an SDDC include improved efficiencies by extending virtualization across all resources, increased agility to provision resources for business applications more quickly, improved control over application availability and security through policy-based definitions, and the flexibility to run new and existing applications in multiple platforms and clouds. In addition, an SDDC implementation could reduce a company's energy usage by enabling servers and other data center hardware to run at decreased power levels or be turned on and off as needed.

UEFI

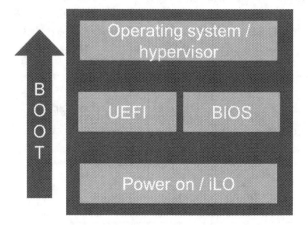

Figure 2-3 Server boot process

Unified Extensible Firmware Interface (UEFI) is an industry-standard specification that defines the model for the interface between the operating system or hypervisor and system firmware during the startup process as shown in Figure 2-3. Developed by a consortium of more than 100 technology companies including HPE and Microsoft, UEFI is processor architecture-agnostic, supporting x86, x64, ARM, and Itanium processors.

Note

UEFI is a specification that defines a software interface between a computer's operating system and firmware. UEFI has been developed as a potential replacement for the BIOS firmware interface commonly found on many servers and PCs. In practice, most UEFI firmware images provide legacy support for BIOS services. UEFI can support remote diagnostics and repair of computers, even without another operating system. The UEFI interface includes data tables that contain platform-related information along with boot and runtime service calls that are available to the operating system and its loader. These provide a standard environment for booting an operating system and running pre-boot applications.

UEFI standardizes interfaces within platform initialization firmware and within the pre-boot UEFI environment/shell. It provides a pre-boot GUI that standardizes the environment for booting an operating system and pre-boot applications (boot loaders, diagnostics, setup scripts, and so forth). UEFI also provides a pre-operating system network stack, Secure Boot, and support for expanded storage.

The goal for implementing UEFI in ProLiant Gen9 servers is to modernize platform firmware and provide an interface that is not architecture-specific. ProLiant Gen9 servers introduce UEFI as the default firmware interface, although they continue to support legacy BIOS settings. HPE recommends that the UEFI default be used with all ProLiant Gen9 servers.

UEFI is responsible for initializing the hardware of ProLiant Gen9 servers and then handing full hardware control over to the operating system/hypervisor. Every ProLiant Gen9 server is a UEFI Class 2 solution, supporting both legacy and UEFI boot modes, allowing users to switch between either mode. Each member of the ProLiant Gen9 server family defaults to UEFI.

 Note

Many option cards are only supported in UEFI mode and need UEFI compliant option ROMs. One example is the HPE Smart Array B140i controller.

UEFI benefits

UEFI benefits for the ProLiant Gen9 server family include the ability to:

- **Use drives larger than 2.2 TB**—Hard drives in UEFI use Global Unique Identifier Partition Table (GPT), which provides far greater boot drive capacities, allowing the use of high-capacity drives for storage and system booting. UEFI offers complete access to the system hardware and resources, allowing UEFI diagnostics and troubleshooting applications to be run before loading an operating system.

- **Configure UEFI with standard boot methods for enhanced flexibility**—UEFI supports pre-boot execution environment (PXE) boot for IPv6 networks allowing a unified network stack to PXE boot from any network controller while maintaining backward compatibility and continued support for IPv4 PXE. UEFI allows PXE multicast boot support for image deployment to multiple servers at the same time. Servers with an Embedded User Partition (a general-purpose disk partition on nonvolatile flash memory, which is embedded on the system board) can be configured using iLO. After the partition is formatted, it can be used for read and write access from the server operating system.

- **Enable Secure Boot to improve security measures**—UEFI protects against unauthorized operating systems and malware rootkit attacks, validating that the system only runs authenticated option ROMs, applications, and operating system boot loaders that have been digitally signed. UEFI uses a public key to verify UEFI drivers loaded from PCIe cards, drivers loaded from mass storage devices, pre-boot UEFI shell applications including firmware updates and operating system UEFI boot loaders.

- **Take advantage of the UEFI shell and HPE RESTful API for scalable configuration deployment**—UEFI includes the UEFI shell, a CLI application that allows scripting, file manipulation, obtaining system information, and running other UEFI applications. The UEFI shell is based on the UEFI Shell Specification 2.0, but is enhanced with an extended command set for additional functionality. The UEFI shell includes a programming API that can be used to create custom UEFI applications. UEFI supports the HPE RESTful API, an industry-recognized architectural style, for standardized interaction to configure servers at scale using an HTTPS Web protocol.

- **Perform industry-standard server configurations with fewer reboots**—Testing in HPE Labs found that configuring BIOS, iLO, and NICs on a ProLiant DL380 Gen9 server with UEFI BIOS required two system reboots, compared with a ProLiant DL380 Gen8 server with legacy BIOS, which required four system reboots.

Operating systems that support UEFI include:

- Microsoft Windows Server 2008 (x64 only)

- Windows Server 2008 R2 (x64 only)

- Windows 2012

- Windows 2012 R2

- Red Hat Enterprise Linux (RHEL) 6.X and later

- SUSE Linux Enterprise Server (SLES) 11.X and later

- Ubuntu Linux 10.X and later

- VMware ESX 5.0 and later

- Solaris 11.1 and later

- Oracle Linux 6.5

UEFI System Utilities

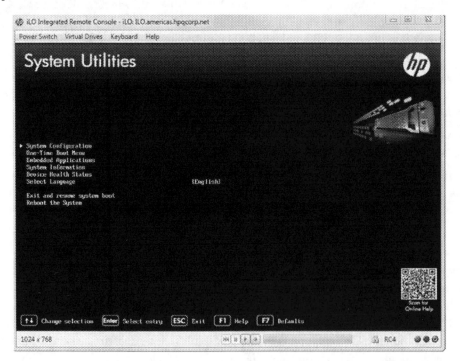

Figure 2-4 UEFI System Utilities

The System Utilities screen shown in Figure 2-4 is the main screen in the UEFI menu-driven interface. Press the up or down arrow keys to select a menu option. A selected option changes color from white to yellow. Press **Enter** to display submenus and other configuration options for your selection. The System Utilities screen displays menu options for the following configuration tasks:

- **System Configuration**—Displays options for viewing and configuring the BIOS/Platform Configuration (RBSU) menu and the iLO 4 Configuration Utility.

- **One-Time Boot Menu**—Displays options for selecting a boot override option and running a UEFI application from a file system.

- **Embedded Applications**—Displays options for viewing and configuring embedded applications, including Intelligent Provisioning and firmware updates.

- **System Information**—Displays options for viewing the server name and generation, serial number, product ID, BIOS version and date, power management controller, backup BIOS version and date, system memory, and processors.

- **Device Health Status**—Displays options for viewing the current health status of all devices in the system.

- **Select Language**—Enables you to select a language to use in the user interface. English is the default language.

- **Exit and resume system boot**—Exits System Utilities and continues the normal boot process.

- **Reboot the System**—Exits System Utilities and reboots by going through the UEFI Boot Order list and launching the first bootable option in the system. For example, you can boot an operating system, if enabled and listed as the first bootable option in the list.

To access UEFI System Utilities:

1. Reboot the server. The server starts up and the ProLiant POST screen appears.

2. Press **F9** in the ProLiant POST screen. The System Utilities screen appears.

3. Use the up and down arrows to change a selection.

4. Press **Enter** to select an entry.

5. Press **Escape** to go back to the previous screen.

 Note

> To access *HPE UEFI System Utilities and Shell Command Mobile Help for HPE ProLiant Gen9 Servers*, scan the QR code on the bottom of the System Utilities screen with a mobile device.

To exit the System Utilities screen and reboot the server, press **Esc** until the main menu is displayed and then select one of the following options:

- Exit and resume system boot—Exits and continues the normal boot process. The system continues through the boot order list and launches the first bootable option in the UEFI boot menu.

- Reboot the system—Exits and reboots the system.

 Note

> Scan this QR code or enter the URL into your browser for documentation on HPE UEFI System Utilities.

http://www.hp.com/go/ProLiantUEFI/docs

Embedded UEFI shell

Figure 2-5 Embedded UEFI shell

The system firmware in all ProLiant Gen9 servers includes an embedded UEFI shell in the ROM, as shown in Figure 2-5.

Based on the UEFI shell specification, the shell environment provides an API and CLI that allow scripting, file manipulation, and obtaining system information. The shell can also be used to run other UEFI applications. These features enhance the capabilities of the UEFI System Utilities.

Access to the UEFI shell is enabled by default. It can be accessed in one of the following ways:

- During server POST, press **F11** (Boot Menu) in the ProLiant POST screen, select **Embedded UEFI Shell**, and press **Enter**.

- Using a (virtual) serial console connection, boot to the UEFI shell and open a connection in an SSH client application using the server's IP address. When the logon prompt appears, enter the user name and password. When the hpiLO-> prompt appears, enter vsp and press **Enter**. The UEFI Shell> prompt appears.

- Using a serial port on the server.

Embedded UEFI shell commands

The UEFI shell is a mini operating system and includes many commands to assist in the management of the pre-boot environment. These commands include:

- help –b—The help command, among others, can be used with the –b option to display one page at a time. Press **Enter** to continue or **q** to exit help.

- alias—Displays, creates, or deletes UEFI Shell aliases.

- attrib—Displays or changes the attributes of files or directories.

- cd—Displays or changes the current directory.

- cls—Clears standard output and optionally changes background color.

- comp—Compares the contents of two files on a byte for byte basis.

- cp—Copies one or more files or directories to another location.

- date—Displays and sets the current date for the system.

- devices—Displays the list of devices managed by UEFI drivers.

- devtree—Displays the UEFI Driver Model compliant device tree.

- dh—Displays the device handles in the UEFI environment.

- dmem—Displays the contents of system or device memory.

- drivers—Displays the UEFI driver list.

- echo—Controls script file command echoing or displays a message.

- edit—Full screen editor for ASCII or UCS-2 files.

- eficompress—Compresses a file using UEFI compression algorithm.

- efidecompress—Decompresses a file using UEFI decompression algorithm.

- else—Identifies the code executed when "if" is false.

- endfor—Ends a "for" loop.

- endif—Ends the block of a script controlled by an "if" statement.

- exit—Exits the UEFI Shell or the current script.

- for—Starts a loop based on "for" syntax.

- fwupdate—Invokes an HPE UEFI Shell utility to update system BIOS firmware.

- getmtc—Gets the monotonic counter (MTC) from BootServices and displays it.

BIOS/Platform Configuration screen

Figure 2-6 BIOS/Platform Configuration screen

The BIOS/Platform Configuration menu shown in Figure 2-6 replaces the ROM-Based Setup Utility (RBSU) on ProLiant Gen9 servers. Use this menu to access and use both UEFI and legacy BIOS options. You can configure system BIOS settings from the BIOS/Platform Configuration screen through the various menus.

System Options

From the System Options menu, you can configure system settings such as:

- **Serial port options**—Assign COM port number and associated resources to the selected physical serial port.

- **USB options**

  – Configure how USB ports and embedded devices operate at startup (USB Enabled [default], External USB Port Disabled).

  – Configure USB Boot Support to prevent the system from booting any connected USB devices and disable booting the iLO virtual media.

- Select whether the system should attempt to boot external USB drive keys, internal USB drive keys, or the internal SD card slot first.

- Control the Virtual Install Disk, which contains server-specific drivers that an operating system can use during installation. If this option is enabled, Windows Server automatically locates required drivers and installs them.

- Control the Embedded User Partition, which is a general purpose disk partition on nonvolatile flash memory embedded on the system board. After it is enabled, the partition can be formatted using the server operating system or by using the HPE RESTful Interface tool. After the partition is formatted, it can be accessed for read and write access from the server operating system.

- Set the operating mode of USB 3.0 ports.

- **Processor options**—Configure processor options such as configuring Intel Hyperthreading, processor core enablement, and Intel programmable interrupt controller (x2APIC) support.

- **SATA Controller Options**—Configure options such as selecting the Embedded SATA configuration and configuring SATA Secure Erase.

- **Virtualization Options**—Configure virtualization options such as Virtualization Technology, Intel VT-d, and SR-IOV.

- **Boot Time Optimization**—Configure Boot Time Optimizations such as Dynamic Power Capping and Extended Memory Test.

- **Advanced Memory Protection**—Configure Advanced ECC Support (default) or Online Spare with Advanced ECC Support.

Boot Options

From the Boot Options menu, you can configure settings such as:

- **Boot Mode**—Set either **UEFI Boot Mode** (default on ProLiant Gen9 servers) or **Legacy BIOS Boot Mode**. The operating system can only boot in the mode in which it is installed.

 Note

> The following options require booting in UEFI Boot Mode: Secure Boot, IPv6 PXE Boot, booting from disks larger than 2.2 TB in AHCI SATA Mode, and HPE Dynamic Smart Array RAID.

- **UEFI Optimized Boot**—Must be set to **Disabled** for compatibility with Windows Server 2008 and Windows 2008 R2 if the system is configured for UEFI Boot Mode. If enabled, the system BIOS boots using native UEFI graphics drivers. If disabled, the system BIOS boots using INT10 legacy video support. This option must be enabled for VMWare ESXi and for Secure Boot to operate properly.

- **Boot Order Policy**—Control system behavior when attempting to boot devices per the Boot Order and no bootable device is found.

- **UEFI boot order list**—Change the order of the UEFI boot list.

- **Advanced UEFI Boot Maintenance options**—Configure advanced UEFI boot order options, such as manually adding or deleting boot options.

- **Setting the Legacy BIOS Boot Mode order**—If a server is configured in Legacy BIOS Boot Mode, the order for those settings can be changed. This setting defines how the server looks for operating system boot firmware.

Embedded UEFI Shell

- **Setting the Embedded UEFI Shell**—Enable or disable the Embedded UEFI Shell. The Embedded UEFI Shell is a pre-boot command line environment for scripting and running UEFI applications, including UEFI boot loaders.

- **Adding Embedded UEFI Shell to the boot order**—Add the Embedded UEFI Shell as an entry in the boot order list. This option is only accessible when the Embedded UEFI Shell is enabled and boot mode is set as UEFI.

- **Enabling the UEFI Shell Script Auto Start**—Enable or disable automatic execution of the default UEFI shell startup script during shell startup. When enabled, the shell looks for the startup.nsh (similar concept to autoexec.bat) file in any of the FAT16 or FAT32 file systems available.

Power Management

- **HPE Power Profile**—Select a profile based on power and performance characteristics.

- **HPE Power Regulator**—Configure only when the power profile is set to **Custom**. Select from HPE Dynamic Power Savings Mode (default), which automatically varies processor speed, and power usage based on processor utilization, which allows reduced overall power consumption with little or no impact to performance, and does not require operating system support.

- **Minimum Processor Idle Power Core C-State**—Select the processor's lowest idle power (C-State) that the operating system uses. The higher the C-State, the lower the power usage of that idle state. This option can be configured only if the HPE Power Profile is set to **Custom**.

- **Minimum Processor Idle Power Package C-State**—Configure the lowest processor idle power state. The processor automatically transitions into package C-States based on the Core C-States, in which cores on the processor have transitioned.

- **Advanced Power Management Options**—Access advanced power options to enable such features as Channel Interleaving and Collaborative Power Control. You can set the QPI link frequency to a lower speed and set the processor idle power state.

Performance Options

- **Intel Turbo Boost Technology**—Set the processor to a higher frequency than its rated speed if it has available power and is within temperature specifications. The default is Enabled.

- **Setting ACPI SLIT Technology**—Describes the relative access times between processors, memory subsystems, and IO subsystems. Operating systems that support the System Locality Information Table use this to allocate resources and workloads more efficiently. The default is Disabled.

- **Accessing Advanced Performance Tuning Options**—Access the Advanced Performance Tuning options menu where you can configure options for performance tuning.

Server Security

- **Setting the Power On Password**—Set a password for accessing the server during the boot process.

- **Setting an Administrator Password**—Set an administrator password to protect server configuration.

- **Setting the F11 One-Time Boot Menu**—Specify a boot override option for this boot only. This option does not modify normal boot order settings.

- **Disabling Intelligent Provisioning (F10 Prompt)**—Disable access to Intelligent Provisioning from the ProLiant POST screen.

- **Setting Embedded Diagnostics**—Enable or disable Embedded Diagnostics, which is available from the Boot menu. Embedded diagnostics include System Health, System Tests, Component Tests, Test Logs, and IML Log.

- **Configuring the Embedded Diagnostics Mode**—Configure Embedded Diagnostics to display in graphical Auto mode (default) or Text Console mode.

- **Protecting a System from Viruses**—Protect a system against malicious code and viruses by marking memory as non-executable unless the location contains executable code. This option requires operating system support.

 Note

Certain operating systems (including Windows 2012 and Windows 2012 R2) require this option to be enabled and override this setting.

- **Accessing Secure Boot options**—Ensure that each component launched during the boot process is digitally signed and the signature is validated against trusted certificates embedded in UEFI.

- **Accessing the Trusted Platform Module (TPM)**—Enable the Trusted Platform Module (if installed) and BIOS secure startup.

Additional options

Other options configurable from the BIOS/Platform Configuration screen are:

- **PCI Device Enable/Disable**—Enable or disable embedded and add-in devices. Disabling devices re-allocates the resources that are normally allocated to the device. By default, all devices are enabled.

- **Server Availability**—Enable the automatic server recovery (ASR) status and timeout, configure POST, and set the power button mode and power-on delay.

- **BIOS Serial Console and EMS**—View POST error messages and run RBSU remotely through a serial connection to the server COM port or iLO virtual serial port. The remote server does not require a keyboard or a mouse.

- **Server Asset Information**—Modify server information, administrator contact information, service contact information, and the system startup message.

- **Advanced Options**—Access Advanced Options to configure ROM Selection, Video Options, Embedded Video Connection, Fan and Thermal Options, and Advanced System ROM options.

Secure Boot on ProLiant Gen9 servers

Secure Boot based on UEFI is a feature supported in ProLiant Gen9 servers in which the system firmware, option card firmware, operating systems, and software collaborate to greatly enhance platform security.

Secure Boot provides software identity checking at every step of the boot process, including platform firmware, option cards, and operating system boot loader. After the operating system boot loader has run (securely), the responsibility for security is passed on to the operating system itself—it is not within the scope of UEFI to cover operating system security.

Without UEFI Secure Boot, malware developers can take advantage of several potential vulnerabilities in the pre-boot environment, including the system-embedded firmware itself, as well as the time between the initialization of the firmware and the booting of the operating system. Malware introduced at this point can provide an environment in which an operating system—no matter how secure—cannot run safely.

Secure Boot is completely implemented in the BIOS and does not require special hardware such as trusted platform module (TPM), although it can work with TPM if required.

Secure Boot ensures that each component launched during the boot process is digitally signed and that the signature is validated against a set of trusted certificates embedded in the UEFI BIOS. It validates the software identity of the following components in the boot process:

- UEFI drivers loaded from PCIe cards
- UEFI drivers loaded from mass storage devices
- Pre-boot UEFI shell applications
- Operating system UEFI boot loaders

Only firmware components and operating systems with boot loaders that have an appropriate digital signature can execute during the boot process. Only operating systems that support Secure Boot and have a UEFI boot loader signed with one of the authorized keys can boot when Secure Boot is enabled.

 Note

Scan this QR code or enter the URL into your browser for more information on UEFI system utilities and Secure Boot.

http://www.hp.com/ctg/Manual/c04398276.pdf

A physically present user can customize the certificates embedded in the UEFI BIOS by adding or removing their own certificates. This can also be performed by remotely connecting to the server using the iLO4 Remote Console.

The HPE RESTful API provides a secure, programmatic method to configure Secure Boot.

Note

All ProLiant Gen8 and later servers are fully compliant with the National Institute of Standards and Technology (NIST) 800-147B—BIOS Protection Guidelines for Servers.

Secure Boot verification

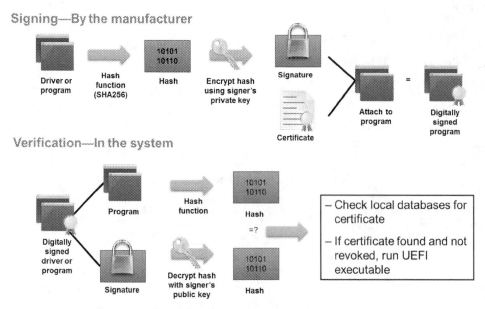

Figure 2-7 Secure Boot verification

The creator of the driver or the program is required to create a signature and certificate and embed it into the program image to produce a digitally signed program. When these programs are loaded into the system during the boot process, the system firmware checks the signed images and compares them to the certificates stored in the local databases (these are loaded into the system during the manufacturing process), as illustrated by Figure 2-7.

If the certificate is found and not revoked, the image is executed. If the certificate is not found or has been revoked, the image will not execute and the boot process comes to a halt.

Activity: Performing basic shell operations in UEFI

To gain a deeper understanding of how to perform UEFI operations, watch this YouTube video about using UEFI shell commands.

 Note

> This video was published prior to the Hewlett-Packard Company separation. To access the video, scan this QR code or enter the URL into your browser.

https://www.youtube.com/watch?v=cekq7bDZw14

Be prepared to pause the video as needed to answer these questions:

1. How do you enable or disable the embedded UEFI shell?

_____

_____

_____

_____

2. Which command should be used to retrieve a list of available commands?

_____

_____

3. In general, which types of tasks can be accomplished using commands available in the UEFI shell?

_____

_____

_____

_____

4. How can the help reset command be used?

_____

_____

_____

_____

HPE iLO

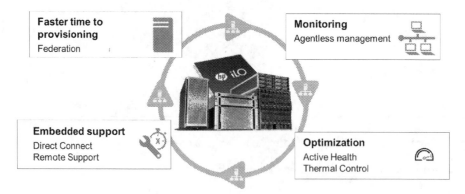

Figure 2-8 HPE iLO management technologies

The HPE Integrated Lights-Out (iLO) subsystem is a standard component of ProLiant servers that simplifies initial server setup, server health monitoring, power and thermal optimization, and remote server administration. The iLO subsystem includes an intelligent microprocessor, secure memory, and a dedicated network interface. This design makes iLO independent of the host server and its operating system.

iLO 4 enables you to monitor and manage servers and other network equipment remotely even when the server is off and regardless of whether the operating system is installed or functional. It allows access to BIOS settings and the reinstallation of the operating system.

As illustrated by Figure 2-8, iLO 4 enables you to:

- **Provision**—Inventory and deploy servers using virtual media and iLO Federation remotely with the iLO web interface, remote console, command line interface (CLI), or mobile app.

- **Monitor**—iLO provides health and performance protection with advanced power and thermal control for maximum power efficiency. Agentless Management monitors core hardware and related alerts without installation of agents or providers on the host operating system.

- **Optimize**—iLO provides an Integrated Remote Console for remote administration so you can control your server from any location through the iLO web interface, remote console, CLI, or mobile app. Integrated Remote Console capabilities include Keyboard, Virtual Media, Global Team Collaboration, and Video Record/Playback. To remotely manage groups of servers at scale, iLO Federation offers built-in rapid discovery of all iLOs, group configurations, group health status, and the ability to determine which servers have iLO licenses. With an iLO Advanced license, you can enable the full implementation of iLO Federation management for features such as Group Firmware Updates, Group Virtual Media, Group Power Control, Group Power Capping and Group License Activation.

- **Support**—iLO provides core instrumentation that operates whether the operating system is up or down. Should something go wrong, you can view the Integrated Management Log through the iLO web interface or download Active Health System logs and send them to HPE Support for faster problem identification and resolution.

HPE iLO 4 provides the core foundation and intelligence for all ProLiant Gen8 and Gen9 servers. iLO 4 is ready to run and does not require additional software installation. iLO 4 management technologies simplify server setup, enable health monitoring, provide power and thermal control, and promote remote administration. iLO 4 management technologies support the complete lifecycle of all ProLiant Gen9 servers, from initial deployment through ongoing management and service alerting. The iLO 4 capabilities that ship standard on all ProLiant Gen8/Gen9 servers include:

- **Agentless management**—Is key to the value of ProLiant Gen8 and Gen9 servers. The base hardware monitoring and alerting capability is built into the system (running on the iLO chipset) and starts working the moment that a live power cord is connected to the server. When a network cable is connected to the dedicated iLO network port, iLO 4 allows you to implement agentless management with SNMP alerts from iLO 4, regardless of the state of the host server. When enabled, SNMP alerts are sent directly by iLO 4, regardless of the host operating system or whether a host operating system is installed.

- **Active Health System**—Offers 24 x 7 mission control for ProLiant servers. To help mitigate the risk of costly unplanned downtime, Active Health and Insight Online automatically analyze the health of ProLiant Gen8 and Gen9 servers across 1600 data points, enabling clients to resolve unplanned downtime issues much faster than ever. iLO also monitors firmware versions and the status of fans, memory, the network, processors, power supplies, and internal storage.

- **Embedded remote support**—Monitors servers and proactively sends alerts to notify of potential problems. HPE Insight Remote Support works with iLO 4, HPE Active Health System, and HPE Agentless Management for simple configuration of remote support for ProLiant Gen8 and Gen9 servers. Embedded remote support software is available on ProLiant Gen8 and Gen9 servers with iLO 4, regardless of the operating system software and without installing operating system agents on the server. You can access a high-performance and secure Integrated Remote Console to the server from anywhere in the world. Use the shared .NET Integrated Remote Console to collaborate with up to four server administrators.

- **Thermal control**—iLO monitors temperatures in the server and sends corrective signals to the fans to maintain proper server cooling. The Adaptive ProLiant Management Layer (APML) feature was designed for ILO 4 on ProLiant Gen9 servers. It improves thermal management and reduces the time spent creating/maintaining thermal solutions. This abstraction method provides a user-transparent method to update system health and fan data without flashing the system ROM. APML permits online updates with no system re-boot for fan/thermal data. APML uses one unified platform definition per server and rule-based adaptations for configuration differences to remove arbitrary limits on the number of sensors and fans. APML enables more flexible fan and power supply redundancy rules, and includes inter-integrated circuit (i2C) bus topology information that was previously stored in system management BIOS records.

In addition, iLO 4 enables you to:

- Remotely mount high-performance virtual media devices to the server.

- Use Virtual Power and Virtual Media from the GUI, the command line interface (CLI), or the iLO scripting toolkit for many tasks, including the automation of deployment and provisioning.

- Securely and remotely control the power state of the managed server.

- Monitor the power consumption and server power settings.

- Register a ProLiant Gen8 or Gen9 server for HPE Insight Remote Support.

- Configure Kerberos authentication, which adds the HPE Zero Sign In button to the login screen. Clicking the **Zero Sign In** button logs the user in to iLO without requiring the user to enter a user name and password.

Connecting to iLO 4

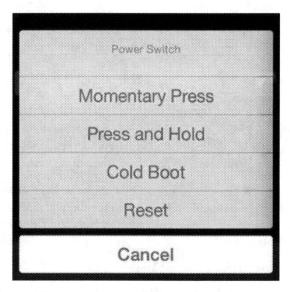

Figure 2-9 iLO mobile app for iPhone

iLO can be accessed through several user interfaces:

- **Web-based interface**—The iLO web interface groups similar tasks for easy navigation and work-flow. The interface is organized in a navigational tree view located on the left side of the page. To access the iLO 4 web interface, use local user accounts or domain user accounts. Local user accounts are stored inside iLO 4 memory when the default user administrator is enabled.

- **Secure Shell (SSH)**—With the SSH interface, you can use the most important iLO 4 features from a text-based console.

- **iLO scripting and command line**—You can use the iLO scripting tools to configure multiple iLO systems, to incorporate a standard configuration into the deployment process, and to control servers and subsystems.

- **HPE RESTful API**—iLO 4 2.00 and later includes the HPE RESTful API, which is a management interface that server management tools can use to perform configuration, inventory, and monitoring of an ProLiant server via iLO. A REST client such as the HPE RESTful Interface Tool sends HTTPS operations to the iLO web server to GET and PATCH JSON-formatted data, and to configure supported iLO and server settings, such as the UEFI BIOS settings. iLO 4 2.30 is Redfish 1.0-conformant and remains backward compatible with the existing HPE RESTful API.

- **iLO mobile app**—The HPE iLO mobile app (Figure 2-9) provides access to the remote console of a ProLiant server from an iOS or an Android device. The mobile app interacts directly with the iLO processor on ProLiant servers, providing total control of the server at all times as long as the server is plugged in. For example, you can access the server when it is in a healthy state, or when it is powered off with a blank hard drive. IT administrators can troubleshoot problems and perform software deployments from almost anywhere.

Note

HPE iLO mobile is the name of the application in the HPE app store; iLO Console is the name of the icon.

Using the Integrated Remote Console

The iLO Integrated Remote Console is a graphical remote console that turns a supported browser into a virtual desktop, allowing full control over the display, keyboard, and mouse of the host server. Using the Remote Console also provides access to the remote file system and network drives.

With Integrated Remote Console access, you can observe POST boot messages as the remote host server restarts, and initiate ROM-based setup routines to configure the remote host server hardware. When you are installing operating systems remotely, the Integrated Remote Console (if licensed) enables you to view and control the host server monitor throughout the installation process.

iLO provides the following Integrated Remote Console access options:

- **.NET IRC**—Provides access to the system KVM, allowing control of Virtual Power and Virtual Media from a single console through a supported browser on a Windows client. In addition to the standard features, the .NET IRC supports Console Capture, Shared Console, Virtual Folder, and Scripted Media.

- **Java IRC**—Provides access to the system KVM, allowing control of Virtual Power and Virtual Media from a Java-based console. In addition to the standard features, the Java IRC includes the iLO disk image tool and Scripted Media.

- **Stand-alone IRC (HPLOCONS)**—Provides full iLO Integrated Remote Console functionality directly from a Windows desktop, without going through the iLO web interface. HPLOCONS has the same functionality and requirements as the .NET IRC application that is launched from the iLO web interface.

 Note

To download HPLOCONS, scan this QR code or enter the URL into your browser.

http://www.hp.com/support/ilo4

- **iLO Mobile Application for iOS and Android devices**—Provides Integrated Remote Console access from a supported mobile phone or a tablet.

Note

For more information, scan this QR code or enter the URL into your browser.

http://www8.hp.com/us/en/products/servers/ilo/mobile.html

iLO 4 Configuration Utility

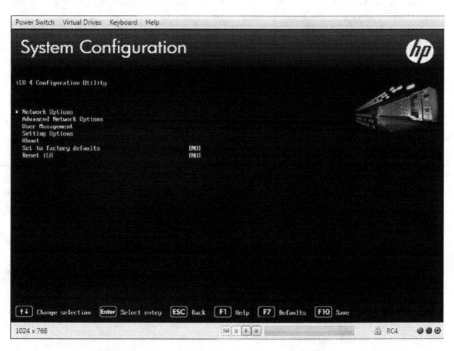

Figure 2-10 iLO Configuration Utility

HPE recommends using the iLO 4 Configuration Utility to set up iLO for the first time and to configure iLO network parameters for environments that do not use DHCP, DNS, or WINS.

As shown in Figure 2-10, the iLO 4 Configuration Utility is embedded in the system ROM of ProLiant servers that support UEFI. This UEFI menu option enables you to configure iLO 4 settings. You can access the iLO 4 Configuration Utility from the physical system console or by using an iLO Integrated Remote Console session. Options include:

- **Network Options**—Configure basic iLO network options such as IP address, subnet mask, gateway IP address, DNS name, DHCP Enable, among others.

- **Advanced Network Options**—Configure advanced iLO network options such as Gateway from DHCP, DHCP Routes, DNS from DHCP, DNS Servers, WINS from DHCP, Domain Name, among others.

- **User Management**—Add, edit, and remove iLO user accounts.

- **Setting Options**—Configure iLO access settings including iLO 4 Functionality, iLO 4 Configuration Utility, Require Login for iLO 4 Configuration, Show iLO 4 IP Address during POST, and Local Users.

- **About**—View firmware date, firmware version, iLO CPLD version, and serial number, among other information.

- **Set to Factory Defaults**—Reset iLO to the factory default settings.

 Caution

This operation clears all user and license data.

- **Reset iLO**—You can reset iLO if it is slow to respond. Resetting iLO does not affect the running operating system or make any configuration changes, but it ends all active connections to iLO.

One-Time Boot Menu screen

Figure 2-11 One-Time Boot Menu

Through the iLO 4 Remote Console, you can select a UEFI boot option for a one-time boot override, as shown in Figure 2-11. This option does not modify predefined boot order settings. If you use a USB key or virtual media through the iLO 4 Remote Console, you must refresh this menu so the devices appear. To do so, exit by pressing **Esc** and then re-enter the One-Time Boot Menu selection from System Utilities menu. This causes the One-Time Boot Menu to refresh the content.

Options include:

- **Generic USB Boot**—This option provides a placeholder for any USB device that is bootable in UEFI. You can set the boot priority of this option and retain this priority for use with USB devices you might install in the future. Setting this priority does not affect priorities set for individual USB devices in the UEFI Boot Order list. Newly added USB devices appear at the bottom of the list by default, and you can move those entries in the list and boot from them as well.

Note

This option is available in UEFI Mode only. The system attempts to boot all UEFI bootable USB devices in the order you specify in the Generic USB Boot entry, even if installed individual USB devices are configured lower in the boot order.

- **Run a UEFI application from a file system**—Select this option to a run a UEFI application from a file system by browsing all FAT file systems that are available in the system. It is also possible to select an X64 UEFI application (a file with a .EFI extension) to execute. It can be an operating system boot loader or any other UEFI application.

- **Legacy BIOS One-Time Boot Menu**—Choose a specific override option for this boot only. This option does not modify boot order mode settings. The server needs to be rebooted for this change to take effect.

Embedded Applications screen

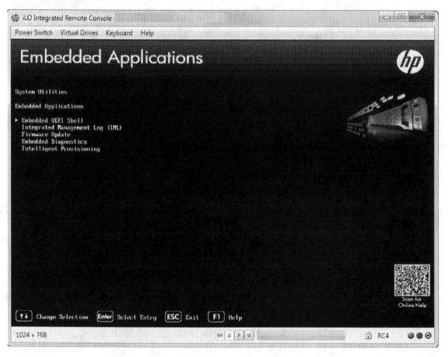

Figure 2-12 Embedded Applications

As shown in Figure 2-12, options available from the Embedded Applications screen include:

- **Embedded UEFI Shell**—Use this option to access the Embedded UEFI Shell screen.

- **Integrated Management Log (IML)**—The IML provides a record of historical events that have occurred on the server. Entries in the IML can help with diagnosing issues or identify potential issues. The IML timestamps each event with one-minute granularity.

- **Firmware Update**—Use this option to update firmware components in the system. Other components can be updated, such as Smart Array and NIC using a binary ROM (or NIC or Smart Array) update file obtained from HPE.

- **Active Health System Log**—Use this option to download an Active Health System Log. By default, the system downloads logs from the previous seven days.

- **Embedded Diagnostics**—Use this option to access the ProLiant Hardware Diagnostics menu. Use the Embedded Diagnostics to access health summary status, run system tests and component tests, and view test logs.

- **Intelligent Provisioning**—Intelligent Provisioning is an essential single-server deployment tool embedded in ProLiant Gen9 servers that simplifies ProLiant server setup, providing a reliable and consistent way to deploy ProLiant server configurations. This option lets you select the Intelligent Provisioning host override option for this boot only, and does not modify the normal boot order or boot mode settings.

System Information screen

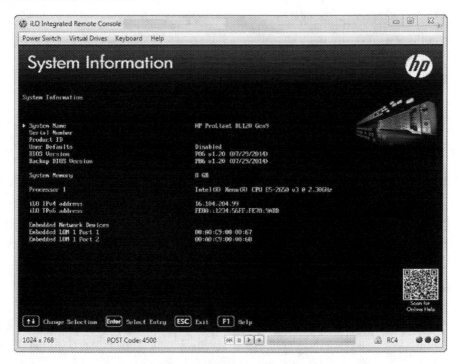

Figure 2-13 System Information

The System Information menu, as shown in Figure 2-13, displays server details and can be used to check that the firmware version was updated after applying an update. Information displayed includes:

- System name and generation
- Serial number
- Product ID
- User defaults
- BIOS version
- Backup BIOS version and date
- System memory (GB)
- Processor 1–4
- iLO IP address
- Embedded network devices
- PCI Device Information
- Firmware Information

 Note

You can also view firmware information by using the HPE RESTful Interface Tool.

Pre-boot Health Summary

```
                    HP ProLiant Pre-boot Health Summary
ProLiant DL380p Gen9
  Serial Number: 0123456789abcdef          Product ID: nfciscooler
  iLO IP:          16.101.6.138 [fe80::9eb6:54ff:fe8e:4330]
  iLO Hostname:   iloluisnfc
  iLO Firmware:   2.00 pass 23+ Mar 28 2014
  System ROM:     P89 03/11/2014           Backup:  03/11/2014
  CPLD:           0x01                     Embedded Smart Array: 0.01

Critical Integrated Management Log Events
  [C] 02/01/2014 13:17  Option ROM POST Error: 1785-Slot 0 Drive Array Not
      Configured
  [C]  CLOCK NOT SET    Option ROM POST Error: 1785-Slot 0 Drive Array Not
      Configured
  [C]  CLOCK NOT SET    Option ROM POST Error: 1785-Slot 0 Drive Array Not
      Configured
  [C]  CLOCK NOT SET    Option ROM POST Error: 1785-Slot 0 Drive Array Not
      Configured
  [C]  CLOCK NOT SET    Option ROM POST Error: 1785-Slot 0 Drive Array Not
      Configured
  [C]  CLOCK NOT SET    Option ROM POST Error: 1785-Slot 0 Drive Array Not
      Configured
  [C]  CLOCK NOT SET    Option ROM POST Error: 1785-Slot 0 Drive Array Not
      Configured
```

Figure 2-14 Pre-boot Health Summary

If a ProLiant Gen9 server does not start up, you can use iLO 4 to display diagnostic information on an external monitor. This feature is supported on servers that support external video and have a UID button or an SUV connector. When the server is off and power is available, iLO runs on auxiliary power and can take control of the server video adapter to show the Pre-boot Health Summary, as shown in Figure 2-14.

To view the Pre-boot Health Summary:

1. Verify that the server is off and power is available.

2. Do one of the following:

   • Press the UID button on the server.

⚠️ **Caution**

To use this feature, press and release the UID button. Holding it down at any time for more than five seconds initiates a graceful iLO reboot or a hardware iLO reboot. Data loss or NVRAM corruption might occur during a hardware iLO reboot.

- Log on to the iLO web interface. Change the UID state to **UID ON** by clicking the UID icon at the bottom right corner of any iLO web interface window.

- Plug in a Serial USB VGA (SUV) connector.

The ProLiant Pre-boot Health Summary screen is displayed on the server monitor and remains on until the server is powered on, the UID state is changed to **UID OFF**, an SUV connector is removed, or an iLO reboot completes. The following information is listed:

- Server model number

- Server serial number

- Product ID

- iLO IP address (IPv4 and IPv6)

 Note

The IP address information is displayed only if **Show iLO IP during POST** is set to **Enabled** on the Administration → Access Settings page in iLO.

- iLO hostname

- iLO firmware version

- ProLiant System ROM version

- ProLiant System ROM backup version

- iLO complex programmable logic device (CPLD) version

- System CPLD version

- Embedded Smart Array version number

 Note

This value is displayed only if server POST has successfully completed since the last auxiliary power cycle.

- Critical events

 Note

The most recent critical events from the IML are displayed with the most recent event displayed first.

Using iLO to update firmware

Firmware updates enhance server and iLO functionality with new features, improvements, and security updates. You can update firmware by using the following methods:

- **Online firmware update**—When you use an online method to update firmware, you can perform the update without shutting down the server operating system. Online firmware updates can be performed in-band or out-of-band. HPE recommends using online mode when possible.

  - **In-band**—Firmware is sent to iLO from the server host operating system. The HPE ProLiant Channel Interface Driver is required for in-band firmware updates. During a host-based firmware update, iLO does not verify login credentials or user privileges because the host-based utilities require a root login (Linux and VMware) or Administrator login (Windows).

  - **Out-of-band**—Firmware is sent to iLO over a network connection. Users with the Configure iLO Settings privilege can update firmware by using an out-of-band method. If the system maintenance switch is set to disable iLO security, any user can update firmware with an out-of-band method.

 Note

Out-of-band management increases the stability and security of system management for firmware updates and Agentless Management.

- **Offline firmware update**—When you use an offline method to update the firmware, you must reboot the server by using an offline utility. You can use the following offline firmware update methods:

  - **SPP**—Use SPP to install firmware.

  - **Scripting Toolkit**—Use the Scripting Toolkit to configure several settings within the server and update firmware. This method is useful for deploying to multiple servers.

In previous versions, iLO 4 has been referenced as *out-of-band management. In-band management,* by contrast, runs on software that must be installed on the system. It works only after the operating system has been booted. In-band is a cheaper solution, but it does not allow access to BIOS settings or the reinstallation of the operating system, and it cannot be used to fix problems in the boot process.

iLO Federation

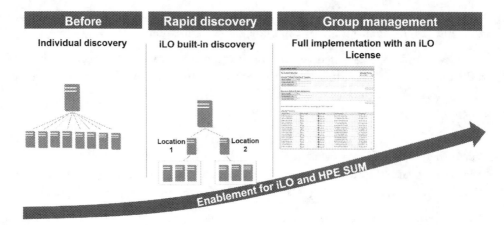

Figure 2-15 iLO Federation

Today's enterprise IT administrators face management problems directly related to scale-out environments that continue to be managed with existing tools not designed for these environments. These problems include how to communicate with thousands of servers to discover and manage these systems in a timely manner, using server data that is current and relevant. Traditional management environments are based on outdated hierarchical models and present a single point of failure.

Monitoring and managing server status in large data center environments with traditional methods, such as the use of host files or ping sweeps (using direct interrogation) as illustrated on the left-hand side of Figure 2-15, are time-consuming. Server status information can be inaccurate by the time it is reported. Ping sweep approaches to iLO discovery and software updates take between one and two minutes per server. This means that in large server farms, essential management tasks can take days. In addition, direct interrogation is used to discover additional devices, which might not be on the same subnet. These conditions allow many existing solutions to cross network boundaries. This IT infrastructure discovery solution does not scale well.

Administrators have typically managed large infrastructures by using scripts and DHCP. Current approaches also use trust systems that typically employ back doors or impose the burden of a public key infrastructure (PKI) to configure secure communication. These approaches are limited in their ability to scale and deployment complexity.

iLO Federation eliminates the need for adjusting scripts during server migration and data center re-architecture efforts. It also removes reliance on tools for external communication. iLO Federation standardizes several fields within the protocol so that a ping sweep approach is still possible, and adds extra information to support direct interrogation of responders.

In the past, iLO operated on a one-to-one approach, meaning that administrators could only look at one iLO at a time. iLO Federation is a fully distributed method for performing discovery of multiple systems, self-organizing those systems into groups, establishing trust, and securely communicating between systems.

iLO Federation uses the industry-standard multicast approach and provides multicast methods, allowing other systems to discover iLOs as illustrated in the center of Figure 2-15. iLO uses a peer-to-peer management system, in which the iLOs communicate with each other and share the workload of managing all the systems. The closest iLO neighbor is identified as a peer. The local iLO identifies its peers through multicast discovery.

iLO Federation technologies provide reliability and interoperability, and include the following capabilities:

- **On-system intelligence**—Robust scalability, self-healing, with no single points of failure.

- **Real-time self-discovery**—With multicast discovery of any bare metal server, iLOs can be discovered after the server receives auxiliary power.

 Important

iLO Federation discovery is a standard feature that allows for queries of data and viewing of iLO information without a license. However, iLO Federation management requires an iLO Advanced or iLO Scale-Out license in order to modify data and define security groups. Licensing provides additional iLO functionality, such as graphical remote console, multi-user collaboration, and video record/playback along with many more advanced features. There are multiple levels of licensing depending on business needs.

- **Group membership**—iLOs can be configured with Federation settings and configured to be a member of a group. iLOs that are members of the same group will discover each other on the network, and begin reporting data/distributing commands. As illustrated on the right-hand side of Figure 2-15, full implementation with an iLO license unlocks additional federation functions such as group firmware update, group license activation, group virtual media, group power control, and group power capping.

- **Security**—iLO Federation uses shared key encryption to implement trust requirements and ensure high levels of security.

 Note

Any user can view information on iLO Federation pages, but some features require a license. Scan this QR code or enter the URL into your browser for more information on iLO licensing.

http://www8.hp.com/us/en/products/servers/ilo/licenses.html

Using iLO 4 Federation

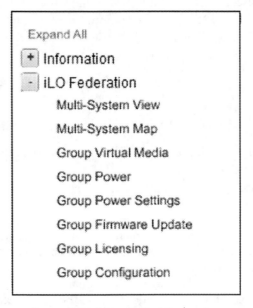

Figure 2-16 Using iLO 4 Federation

iLOs can be grouped and then activities can be directed to the group of iLOs. When iLO systems are in the same iLO Federation management group, the iLO web interface on one system can be used to manage all of the iLO systems in the group.

As shown in Figure 2-16, iLO Federation provides the following functionality in the GUI:

- **Multi-System View**—Get a summary of the status of multiple systems at one time.

- **Multi-System Map**—Display information about the iLO systems in a selected group.

- **Group Virtual Media**—Provide an ISO image to the systems in the group.

- **Group Power**—Power the systems up and down and display power status.

- **Group Power Settings**—Configure Automatic Group Power Capping for multiple servers.

- **Group Firmware Update**—Update the firmware of multiple servers.

- **Group Licensing**—Display the license status for members of a selected iLO Federation group and enter an optional key to activate iLO licensed features.

- **Group Configuration**—Add, delete, and modify federation group membership.

Important

iLO systems in the same iLO Federation group must use the same version of iLO 4 firmware.

Group Health Status

Overview

Health

⊘ OK	100% 4 Systems

Model

ProLiant BL465c Gen9	25% 1 System
ProLiant XL4500 Gen9	75% 3 Systems

Critical and Degraded Systems

There are no systems with critical or degraded status

Figure 2-17 Group Health Status

With iLO Federation, using the Multi-System View enables you to drill into various displays. Clicking a server name filters by that server. Clicking the iLO hostname or IP address launches the iLO web interface.

Group Health Status provides an overview that shows system summary information, as shown in Figure 2-17. To view the status for a configured group of servers:

1. Navigate to the **iLO Federation → Multi-System View** page.

2. From the Selected Group menu, select a group.

The page displays the following information for the servers in the selected group:

- **Health information**—The number of servers in each listed health status (OK, Degraded, Critical). The percentage of the total number of servers that is in the listed health status is also displayed. The health status value can be clicked to select a subset of systems matching that health status.

- **Model information**—The list of servers, grouped by ProLiant model number. The percentage of the total number of servers for each model number is also displayed. The model number can be clicked to select a subset of systems matching that model.

- **Critical and Degraded Systems**—The list of servers with a Critical or Degraded status.

Group Health Status—Critical and Degraded Systems

Critical and Degraded Systems

Server Name	System Health	Server Power	UID Indicator	System ROM	iLO Hostname	IP Address
WIN-7cc23c5c8c4	Degraded	ON	UID OFF	c1.00	iLOdf8882adf48c229	
WIN-7cc22b5c387	Degraded	ON	UID OFF	c1.00	iLOb64a969001baf643	
WIN-7cc2af53f6b	Critical	ON	UID OFF	P70 05/10/2012	iLO57e38f2b50885c14	
WIN-7cc3ab0621a	Critical	ON	UID OFF	P71 01/19/2012	iLO9c7a29fb55be1d42	
WIN-7cc0e452c2c	Critical	ON	UID ON	c1.01	iLOe23a91d58b706e95	
WIN-7cc0ea87ad5	Degraded	ON	UID OFF	P70 05/08/2012	iLOee11Bc7a05d74ea	

Figure 2-18 Critical and Degraded Systems

As shown in Figure 2-18, the Critical and Degraded Systems view displays additional details for any systems that are not OK, including:

- System Name, System Health, and System ROM version

- Server Power and UID Indicator

- iLO Hostname and IP Address

From the Critical and Degraded Systems view, you can determine the severity of the issue:

- A status of Degraded means that a system component has changed to a less robust state (single fan failure, single points of failure, and so forth).

- A status of Critical indicates that a condition has occurred that might cause a system to fail (overheating, all fans fail, and so forth).

Group Virtual Media

The Group Virtual Media feature enables you to connect scripted media that can be accessed by the servers in an iLO Federation group. Scripted media refers to connecting images hosted on a web server by using a URL. iLO will accept URLs in HTTP or HTTPS format. FTP is not supported.

When there is a requirement for mass deployment of an operating system, the traditional one-to-one media installation consumes a tremendous amount of time and manual effort. Using Group Virtual Media, a single operating system image can be deployed over thousands of servers with ease.

When using Group Virtual Media, note the following:

- Scripted media can be connected to the iLO systems in an iLO Federation Management group. Scripted media supports only 1.44 MB floppy images (.img) and CD/DVD-ROM images (.iso). The image must be located on a web server on the same network as iLO.

- To use the Group Virtual Media feature with an iLO Federation group, ensure that each member of the group has the Virtual Media privilege.

- Only one of each type of media can be connected to a group at the same time.

- You can view, connect, eject, or boot from scripted media.

 Note

Before using the iLO Virtual Media feature, review the operating system considerations in the HPE iLO 4 User Guide. Scan this QR code or enter the URL into your browser to access the user guide.

http://h10032.www1.hp.com/ctg/Manual/c03334051

Group Virtual Media screens

1.44 MB floppy image (.img)

Connect Virtual Floppy to 40 Systems

Media Inserted	None
Scripted Media URL	
Boot on Next Reset	☐

CD/DVD-ROM image (.iso)

Connect CD/DVD-ROM to 40 Systems

Media Inserted	None
Scripted Media URL	
Boot on Next Reset	☐

Figure 2-19 Group Virtual Media

As shown in Figure 2-19, the Group Virtual Media screens provide fields for entering a URL representing a floppy or a CD/DVD-ROM image. iLO also supports mounting a .img file on a USB key; it does not need to be a floppy .img image.

Group Power Control

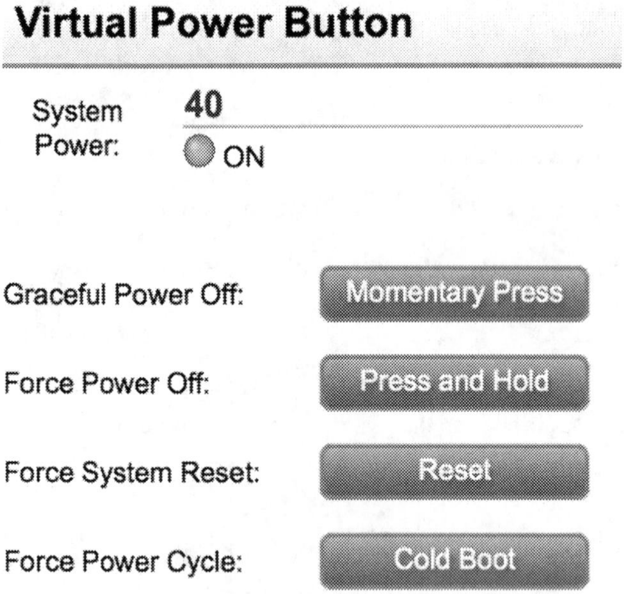

Virtual Power Button

System Power: **40**
   ⬤ ON

Graceful Power Off: Momentary Press

Force Power Off: Press and Hold

Force System Reset: Reset

Force Power Cycle: Cold Boot

Figure 2-20 Group Power Control

iLO Federation provides a way to control the power on all systems in a group or in multiple groups. It is possible to control power to individual systems as well.

The Group Power feature enables you to manage the power of multiple servers from a system running the iLO web interface. As shown in Figure 2-20, you can:

- Power off, reset, or power cycle a group of servers that are in the On or Reset state
- Power on a group of servers that are Off

Group Power Control—Affected Systems

Affected Systems

Server Name	Server Power	UID Indicator	iLO Hostname	IP Address
WIN-90e146cfdf6	◯ OFF	⬤ UID ON	iLO1f7aa0574a37c0c4	
WIN-90e1d081883	◯ OFF	⬤ UID OFF	iLOb660ca079daa48fe	
WIN-90e26e6a6c3	◯ OFF	⬤ UID ON	iLO6e13993340d31282	
WIN-90e274fdf16	◯ OFF	⬤ UID ON	iLO234a69621050ec40	
WIN-90e2e8b1be7	◯ OFF	⬤ UID OFF	iLOdd591e110566d74f	
WIN-90e6abe6c0e	◯ OFF	⬤ UID OFF	iLO4ea85c2d524b6527	
WIN-90e9daeedc7	◯ OFF	⬤ UID OFF	iLOfb5a13a73cecdcd5	
WIN-90e9d2d144b	◯ OFF	⬤ UID ON	iLO5200052b42f86980	
WIN-90ec05580c6	◯ OFF	⬤ UID ON	iLOb02a22e30999891a	

Figure 2-21 Group Power Control—Affected Systems

As shown in Figure 2-21, the Group Power Control within iLO Federation provides a view of the systems that will be affected by pushing the Virtual Power Button. This provides a bail-out mechanism before the actions are taken.

Group Power Capping

With iLO Federation, you can set dynamic power caps for grouped servers. Group Power Capping enables you to conserve energy costs by controlling power to idling systems.

A separate power cap can be set for every group. With Group Power Capping, the power caps that are set for a group operate concurrently with the power caps that can be set on the Power Settings page for an individual server.

To configure power capping settings for an iLO Federation Management group, ensure that each member of the group has granted the Configure iLO Settings privilege to the group. When a group power cap is set, the grouped systems share power in order to stay below the power cap. More power is allocated to busy servers and less power is allocated to servers that are idle. When a power cap is set, the average power reading of the grouped servers must be at or below the power cap value.

Group Power Capping Settings

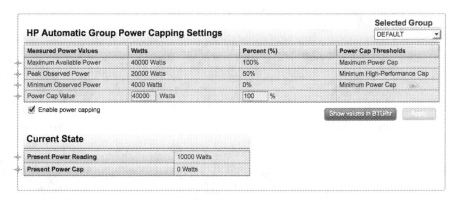

Figure 2-22 Group Power Capping Settings

The Automatic Group Power Capping Settings section, as shown in Figure 2-22, enables you to view measured power values, set a power cap, and disable power capping.

The Measured Power Values section lists the following:

- **Maximum Available Power**—Is the total power supply capacity for all servers in a group.

- **Peak Observed Power**—Is the maximum observed power for the servers in a group.

- **Minimum Observed Power**—Is the minimum observed power for the servers in a group.

- **Power Cap Value**—Is the value of the power cap that has been set for the servers in a group.

During POST, the ROM runs two power tests that determine the peak and minimum observed power values.

The Power Capping Settings section allows you to configure the power capping settings. The Current State section shows the current power consumption.

- **Present Power Reading**—Is the current power reading for all servers in a group.

- **Present Power Cap**—Is the configured power cap for all servers in a group. This value is 0 if the power cap is not configured.

Group Firmware Update

This feature adds value to the task of upgrading the firmware on multiple systems. Rather than having to spend several days upgrading individual systems, you can use the Group Firmware Update feature to update the firmware of multiple servers from a system running the iLO web interface.

Firmware types supported for update are:

- iLO

- ProLiant System ROM (BIOS)

- System Programmable Logic Device (CPLD)

- Power Management Controller

- SL chassis firmware

 Note

The firmware images (raw .bin or .flash files) must be hosted on a web server on the same network as the iLO, similar to virtual media, and entered as a URL on the Group Firmware Update page.

Group Firmware Update views

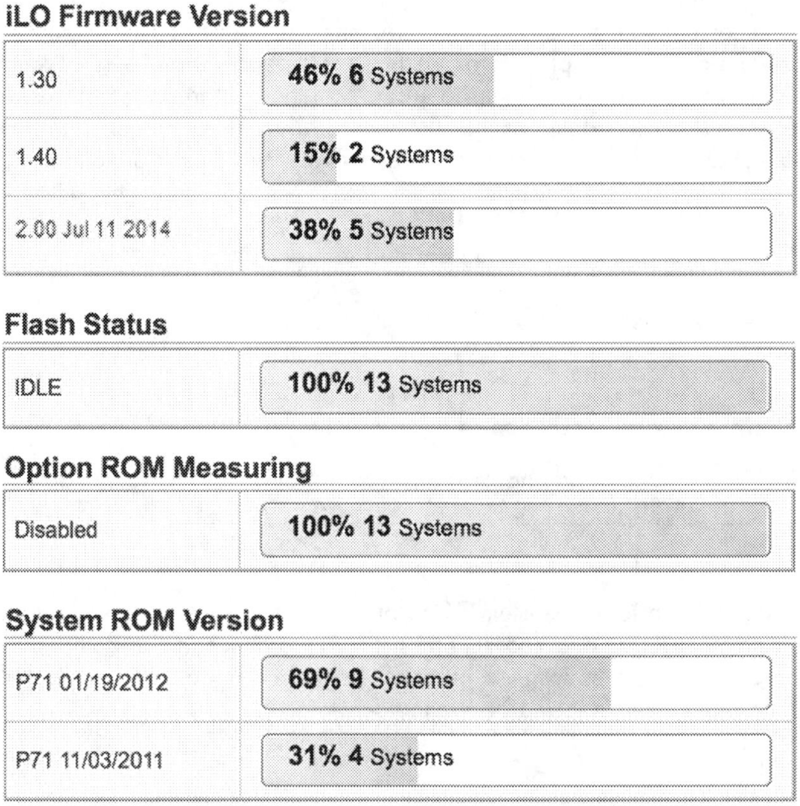

iLO Firmware Version

1.30	**46% 6** Systems
1.40	**15% 2** Systems
2.00 Jul 11 2014	**38% 5** Systems

Flash Status

IDLE	**100% 13** Systems

Option ROM Measuring

Disabled	**100% 13** Systems

System ROM Version

P71 01/19/2012	**69% 9** Systems
P71 11/03/2011	**31% 4** Systems

Figure 2-23 Group Firmware Update

When the firmware is being updated, the iLO Federation Group Firmware Update screen, as shown in Figure 2-23, reports on the progress in real-time.

The Group Firmware Update feature enables you to:

- View the number of severs with each firmware version. The percentage of the total number of servers with the listed firmware version is also displayed.

- View the flash status for the grouped servers. The percentage of the total number of servers with the listed flash status is also displayed.

- View the Trusted Platform Module (TPM) status for the grouped servers. The percentage of total servers with the listed TPM status is also displayed.

Note

A TPM is a computer chip that securely stores artifacts used to authenticate the platform. These artifacts can include passwords, certificates, or encryption keys. TPM can also be used to store platform measurements to ensure that the platform remains trustworthy. On a supported system, iLO decodes the TPM record and passes the configuration status to iLO. The iLO Overview page displays the following TPM status information:

- Not Supported—A TPM is not supported.

- Not Present—A TPM is not installed.

- Present—This indicates one of the following statuses:

   - A TPM is installed but is disabled.

   - A TPM is installed and enabled.

   - A TPM is installed and enabled, and option ROM measuring is enabled.

 Note

If you attempt to perform a system ROM or an option ROM update on a server with Option ROM Measuring enabled, iLO prompts you to cancel the update, verify that you have a recovery key, and suspend BitLocker before the update. Failure to follow these instructions might result in losing access to your data.

Video demonstration—using management processors for on-system management

Watch this 21-minute demonstration that shows how to how to use the Onboard Administrator (OA) to view data about a BladeSystem enclosure, server blades, and power and thermal conditions.

The video also shows the iLO functionality of a ProLiant server that enables you to manage a system individually, and walks through the available UEFI menu options for system management.

 Note

To view this demonstration, scan this QR code or enter the URL into your browser.

https://youtu.be/dmH5nVlh8Lw

HPE RESTful API

The HPE RESTful API is a management interface that server management tools can use to configure, inventory, and monitor a ProLiant Gen9 server using iLO 4 2.00 or greater. It is an architectural style consisting of a coordinated set of architectural constraints applied to components, connectors, and data elements within a distributed hypermedia system.

The open, industry-standard HPE RESTful API provides a programmable interface and lightweight data model specification that is simple, remote, secure, and extensible. REST has become a popular communication protocol as it enables IT staff to quickly and securely customize server configuration and provisioning, and at the same time provide a common interface for integration to HPE Helion and cloud ecosystems such as OpenStack.

REST is a web service that uses basic Create, Read, Update, Delete, and Patch (CRUD) operations performed on resources using HTTP Post, Get, Put, Delete, and Patch operations. REST is a set of conventions describing a way to create, read, update, or delete information on a server using simple HTTP calls. It is an alternative to more complex programming mechanisms such as SOAP, CORBA, and RPC. Simply put, a REST call is an HTTP request to a server.

A REST client sends HTTPS operations to the iLO web server to GET and PATCH JSON-formatted data, and to configure supported iLO and server settings, such as UEFI BIOS settings.

The REST architecture generally runs over HTTP, although other transports can be used. What the HPE OneView user interface allows you to do graphically, the RESTful API enables you to do programmatically. For example, you can use a scripting language such as Microsoft PowerShell to perform tasks by using RESTful API calls that you might otherwise complete through the web-based UI.

HPE provides two options for RESTful API programming:

- Use the HPE RESTful API for direct programming with total tool-less access to do scripting, or write integration tools with open programming options

- Use the HPE RESTful Interface Tool for simple scripting with command shells

Using the HPE RESTful API

Figure 2-24 Using the RESTful API

Today, many available tools for server management that use scripting bring with them limitations around automation, orchestration, and management. Because scripting interfaces are not common across HPE management tools, HPE is using the HPE RESTful API as a standardized scripting solution to address key challenges around:

- **Unsecure remote capabilities**—Remote scripting is often not secure, triggering the need for another mechanism to transport scripts to target nodes.

- **Learning and deployment**—This can be time-consuming because a single command utility may not work across server components with existing scripting tools. The learning curve increases because administrators are required to learn different types of interfaces across the data center.

- **Scripting efficiency**—Using different tools creates complexity. Running the server through PXE for updates also delays scripting. Running scripts on a large number of servers is not readily scalable.

Designed for ProLiant Gen9 servers, the HPE RESTful API directly addresses scripting challenges in a way that is:

- **Simple**—Easier access to information eliminating multiple tools to run scripts and provision server

- **Remote and secure**—Capabilities leveraging industry-proven HTTPS protocol

- **Extensible**—Ability to script and expose new functionality with few or no firmware upgrade dependencies

Figure 2-24 shows the Postman REST client plug-in for Google Chrome being used to communicate with the API.

HPE RESTful API for iLO

The HPE RESTful API for iLO is designed using the Hypermedia as the Engine of Application State (HATEOAS) REST architecture. This architecture allows the client to interact with iLO through a simple fixed URL—rest/v1. This has the advantage of the client not needing to know a set of fixed URLs. When creating a script to automate tasks using the HPE RESTful API for iLO, you only need to hardcode this simple URL and design the script to discover the RESTful API URLs that are needed to complete a task.

 Note

Scan this QR code or enter the URL into your browser for more information on HATEOAS.

http://en.wikipedia.org/wiki/HATEOAS

The HPE RESTful API for iLO is the main management API for iLO-based HPE servers. Using this API, it is possible to take full inventory of the server, control power and reset, configure BIOS and iLO settings, and fetch event logs, in addition to performing many other functions.

This API follows the trend in moving to a common pattern for new software interfaces. Many web services in a variety of industries use RESTful APIs because they are easy to implement and easy to consume, and they offer scalability advantages over previous technologies. HPE OneView, OpenStack, and many other server management APIs are now RESTful APIs. Most HPE management software offerings, as well as the entire software defined data center (SDDC) architecture, are built upon RESTful APIs.

The HPE RESTful API for iLO has the additional advantage of consistency across all present and projected server architectures. The same data model works for traditional rack-mount servers and blades, as well as newer types of systems such as Moonshot. The HPE RESTful API for iLO provides this advantage because the data model is designed to self-describe the service's capabilities to the client and has room for flexibility designed in from the start.

HPE RESTful Interface Tool

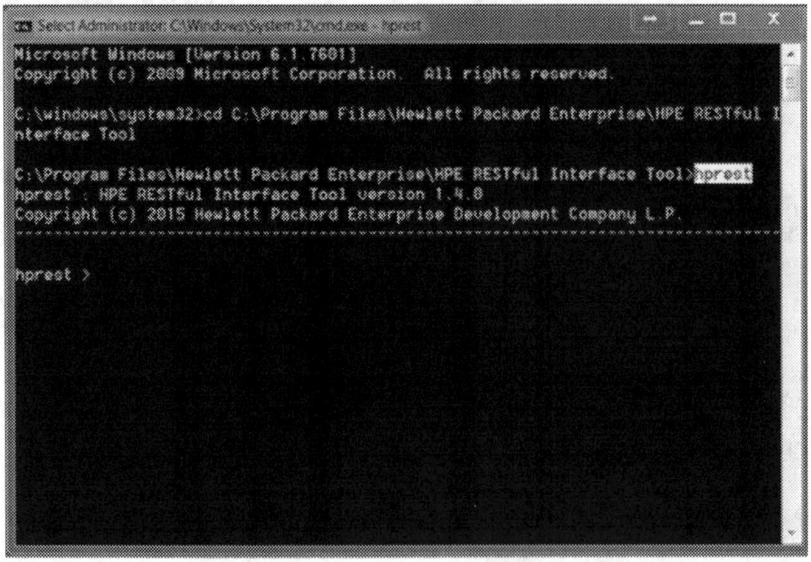

Figure 2-25 HPE RESTful Interface Tool

The HPE RESTful Interface Tool simplifies server configuration by using industry-recognized RESTful APIs, enabling you to script provisioning on ProLiant Gen8 (running iLO 4) and ProLiant Gen9 servers. As shown in Figure 2-25, the RESTful Interface Tool offers a single command-line interface to display and set parameters for iLO 4 and BIOS/UEFI (including secure boot).

The RESTful Interface Tool is the key to enabling software-defined computing for the new idea economy. Benefits include:

- **Easy customization**—A single command line to simplify customizing workflows and scripts by standardizing a set of commands that interacts with all server components

- **Reduced travel costs**—Capability to remotely manage servers

- **Reduced deployment complexity**—Enablement of any of the three modes—interactive, scriptable, or file-based—to program and execute scripts easily

- **Simplified scripting**—Self-descriptive tool to reduce the learning curve

Other HPE scripting tools

Figure 2-26 Scripting Toolkit for Windows and Linux

Two additional scripting tools are available:

- **Scripting Tool for Windows PowerShell**—A powerful set of utilities that you can use to perform various configuration tasks on ProLiant servers, the Scripting Tool for Windows PowerShell is designed for customers familiar with Windows PowerShell. It is the Microsoft task automation framework, consisting of a command-line shell and associated scripting language built on Microsoft .NET Framework. It follows the standard PowerShell syntax and scripting model, making it easy for customers to incorporate these functions into their administrative scripts.

  The HPE Scripting Tool for Windows PowerShell uses lightweight commands (cmdlets) that better enable integration with the current IT ecosystem, allowing retrieval of firmware versions from multiple enclosures and servers. PowerShell Onboard Administrator cmdlets enable retrieval of firmware versions from multiple BladeSystem enclosures and servers, and pipe information to cmdlets that update enclosures, blade, and enclosure options.

  Ideal for enterprises looking for fast and effective HPE hardware configuration using the standard PowerShell architecture and scripting model, the Scripting Tool for Windows PowerShell is available for free download.

- **Scripting Toolkit (STK) for Windows and Linux**—As shown in Figure 2-26, a server deployment product that allows customers to automate the configuration and installation for high-volume ProLiant server and BladeSystem infrastructure deployments. It includes command line utilities for configuring and deploying servers in a customized, predictable, and unattended manner. STK automates firmware, drivers, and server maintenance. It uses industry-recognized APIs, enabling you to script provisioning across generations of servers. It enables customers to duplicate the configuration of a source server on target servers with minimal user interaction.

HPE Intelligent Provisioning

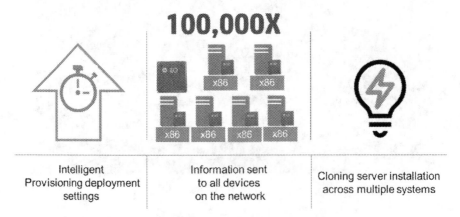

| Intelligent Provisioning deployment settings | Information sent to all devices on the network | Cloning server installation across multiple systems |

Figure 2-27 HPE Intelligent Provisioning

Intelligent Provisioning enables out-of-the box single-server deployment and configuration without the need for media containing firmware, drivers, and tools. Intelligent Provisioning includes these tools on the system so the server is immediately ready for provisioning. As illustrated by Figure 2-27, it addresses the complexity of server maintenance and offers improved, embedded server configuration and operating system deployment.

 Note

Intelligent Provisioning provides in-built access to firmware, drivers, and tools. It does not include operating system media.

It eliminates much of the complexity required to deploy a bare-metal server and allows a system to be deployed faster than with conventional methods. New features for ProLiant Gen9 servers include a refreshed GUI, and the ability to access 1 TB of HPE StoreVirtual virtual storage appliance (VSA) storage through Intelligent Provisioning at no additional cost.

Intelligent Provisioning is a single-server deployment tool embedded in ProLiant Gen8 and Gen9 servers that replaces the SmartStart CDs and Smart Update Firmware DVD used with previous generations of ProLiant servers. It simplifies ProLiant server setup by providing a reliable and consistent way to deploy ProLiant servers.

Intelligent Provisioning uses iLO Federation for discovery and reporting activities. When iLO Federation management is configured on the network, you can store Intelligent Provisioning server profiles on the network. If a profile is stored by an iLO Federation management group member on the network on one server, the same profile can be installed on any other server in the group by selecting it from the drop-down menu on the Deployment Settings page. Advantages of Intelligent Provisioning include:

- Eliminates many of the steps it takes to deploy a bare-metal server

- Allows a system to be deployed and online much faster

- Clones server installation setup and files transfer from one server to another simultaneously across numerous servers at once instead of updating them independently of one another

- Uses features within iLO Federation for rapid discovery of new devices

Accessing Intelligent Provisioning

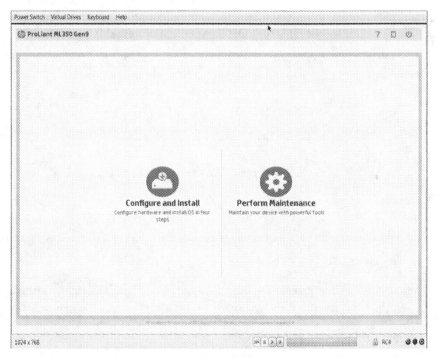

Figure 2-28 Accessing Intelligent Provisioning

To access Intelligent Provisioning, power on or reboot the server and press **F10** when prompted during the server POST. When you access Intelligent Provisioning, one of the following actions takes place:

- If this is the first time you are using Intelligent Provisioning, on-screen prompts provide guidance through initial configuration and registration tasks.

- If Intelligent Provisioning was previously accessed and the initial configuration and registration tasks are complete, the Intelligent Provisioning home page is displayed as shown in Figure 2-28. In the home screen, select one of the following menus to use Intelligent Provisioning:

- Configure and Install menu

- Perform Maintenance menu

To exit Intelligent Provisioning, reboot the server by clicking the power icon at the top right of the page.

Intelligent Provisioning setup—Step 1

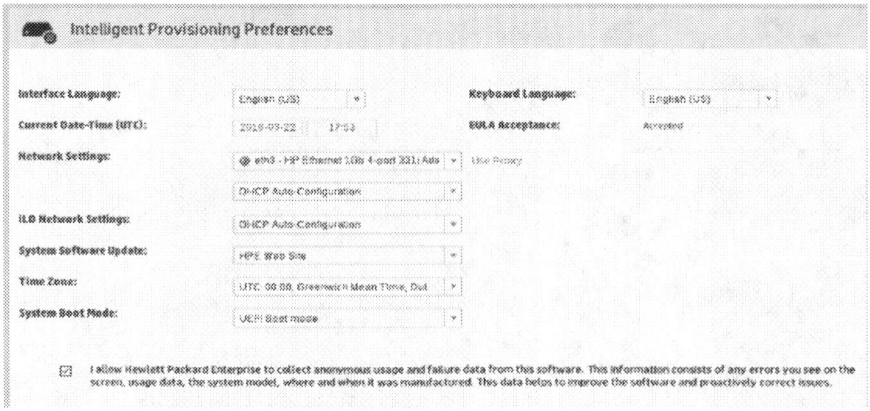

Figure 2-29 Intelligent Provisioning setup step 1

The Set Preferences screen, as shown in Figure 2-29, appears automatically the first time Intelligent Provisioning runs on a server. To set up the software, you must perform the following steps:

1. Choose the interface language and keyboard language.

2. Confirm that the system date and time are accurate. To change the date or time, click the displayed date or time and use the displayed calendar or clock to select the new values.

3. Read and accept the end user license agreement (EULA).

4. Enter network settings. Select the active NIC from the list, and then choose from one of the following IP addressing schemes:

   - **DHCP Auto-Configuration**—HPE recommends selecting DHCP to have IP addresses assigned automatically to servers.

   - **IPv4 Static**—Selecting IPv4 adds four new fields: the static IPv4 address, network mask, gateway address, and DNS address.

   - **IPv6 Static**—Selecting IPv6 adds two fields: the static IP address and the gateway address.

5. Specify whether a proxy is being used. If there is a proxy on the network, it might need to be configured for use with features that communicate across the network. If **Use Proxy** is chosen, enter a proxy address and port.

6. Enter the iLO network settings. Select one of the following iLO network IP addressing schemes:

   - **DHCP Auto-Configuration**—HPE recommends selecting DHCP to have IP addresses assigned automatically to servers.

   - **IPv4 Static**—Selecting IPv4 adds four new fields: the static IPv4 address, network mask, gateway address, and DNS address.

   - **Off**—Selecting Off makes this server unavailable through iLO.

7. Select a delivery option for System Software Updates for Intelligent Provisioning.

   - **HPE website**—HPE recommends selecting this option to be prompted when updates are available and download all software updates for the server from hp.com.

   - **HTTP/FTP**—When prompted, enter an address in the URL field.

   - **Disable**—Select this to disable automatic updates if it is planned to update system software manually.

8. Select the correct Time Zone.

9. Select the desired System Boot Mode.

 Note

Changes to the system boot mode are implemented during operating system installation or in the next POST.

Click the **Continue** right arrow to proceed automatically to Step 2: Activating Intelligent Provisioning.

Intelligent Provisioning setup—Step 2

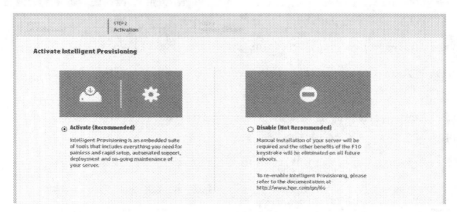

Figure 2-30 Intelligent Provisioning setup Step 2

To activate Intelligent Provisioning and make it available during server POST:

1. Select **Activate (Recommended)** as shown in Figure 2-30.

2. Click the **Continue** right arrow to proceed automatically to Step 3: Registering for HPE Remote Support.

To disable Intelligent Provisioning:

1. Select **Disable (Not Recommended)**.

2. Click the **Continue** right arrow. The server reboots. During POST, F10 is in red text on the screen, indicating that the F10 key is disabled and Intelligent Provisioning is no longer accessible.

To re-enable Intelligent Provisioning:

1. Reboot the server and, when prompted, press **F9** to access the UEFI System Utilities.

2. From the System Utilities screen, select **System Configuration** → **BIOS/Platform Configuration (RBSU)** → **Server Security** → **Intelligent Provisioning (F10 Prompt)** and press **Enter**.

3. Select **Enabled**.

Intelligent Provisioning setup—Step 3

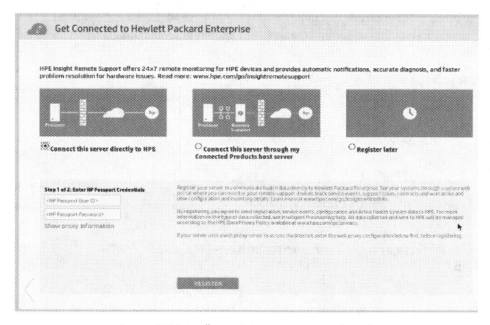

Figure 2-31 Intelligent Provisioning setup Step 3

HPE Insight Remote Support provides automatic submission of hardware events to HPE support to prevent downtime and enable faster issue resolution. The screen shown in Figure 2-31 allows you to configure the connection to HPE Insight Remote Support:

- **Insight Online direct connect**—Register a server to communicate directly with Insight Online without the need to set up an Insight Remote Support centralized hosting device in the local environment. Insight Online will be the primary interface for remote support information. Insight Online is an HPE Support Center feature that enables IT staff to view remotely monitored devices anywhere, anytime. It provides a personalized dashboard for simplified tracking of IT operations and support information, including a mobile dashboard for monitoring on the go.

- **Insight Remote Support central connect**—Register a server to communicate with HPE through an Insight Remote Support centralized hosting device in the local environment. All configuration and service event information is routed through the hosting device. This information can be viewed using the local Insight Remote Support Console or the web-based view in Insight Online (if it is enabled in Insight Remote Support).

Intelligent Provisioning—Installing an operating system

To use Intelligent Provisioning to configure the hardware and install an operating system on a ProLiant server, follow the on-screen prompts in the Configure and Install menu to complete the tasks in the following four screens:

1. Hardware Settings

2. OS Selection

3. OS Information

4. Review

Each Configure and Install screen provides a guided method of configuring the server, installing an operating system, and updating the system software.

Intelligent Provisioning—Hardware Settings

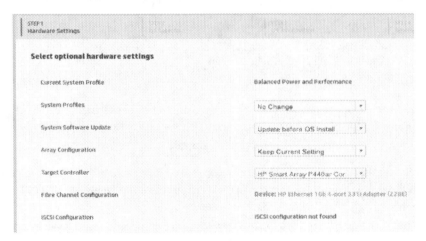

Figure 2-32 Intelligent Provisioning – Hardware Settings

In the first configuration screen shown in Figure 2-32, enter global settings to control power use, software updates, and array configuration.

1. Select the power management system profile to use. These profiles set a basic policy for performance versus power usage without having to configure individual settings through the UEFI System Utilities menus. The current (or a recommended) profile is displayed, but to change the settings, the options include:

   • No Change

   • Balanced Power and Performance

- Minimum Power Usage

- Maximum Performance

The suggested default varies. If Intelligent Provisioning detects existing settings on the server that match one of these profiles, that profile is displayed in the System Profiles field. If Intelligent Provisioning detects settings that do not match one of the profiles, **No Change** is displayed in this field.

2. Select whether to perform a software update before the operating system is installed.

3. Enter array configuration specifications for the server's storage subsystem. Options include:

   - **Keep Current Setting**—Uses existing settings to maintain any previously constructed arrays. Use this option when reprovisioning a server. This option is displayed only when valid logical drives are present on the server. For new server installations, this option is not displayed.

   - **Recommended Settings**—The HPE Smart Storage Administrator polls any drives that are present and builds an appropriate array for those drives. For example, if two drives are connected to the Smart Array card, the setup defaults to RAID 1. HPE recommends selecting this option when initially provisioning a server.

 Caution

Selecting this option resets all disks (and arrays, if any are present). Because no arrays or disk data are present during a first-time setup, this does not affect the server. However, if this option is chosen when reprovisioning a server, data and any disk array settings can be lost.

   - **Customize**—Opens HPE Smart Storage Administrator (after clicking the **Continue** right arrow) and allows you to choose array settings.

4. Select the desired target controller from the drop-down menu if more than one is available.

5. Confirm Fibre Channel and iSCSI configuration settings. If discovered, the Fibre Channel and iSCSI information is displayed at the bottom of the screen. Intelligent Provisioning supports installation to iSCSI targets and to shared storage devices. Before an installation is started, the devices must be set up outside of Intelligent Provisioning, using options that appear during POST, or through their setup applications. In addition, the boot controller order needs to be set correctly in the UEFI System Utilities before installation.

6. Confirm SD Card Configuration settings. If a supported SD card is installed, the device details appear.

7. Click the **Continue** right arrow.

Intelligent Provisioning—Operating system selection

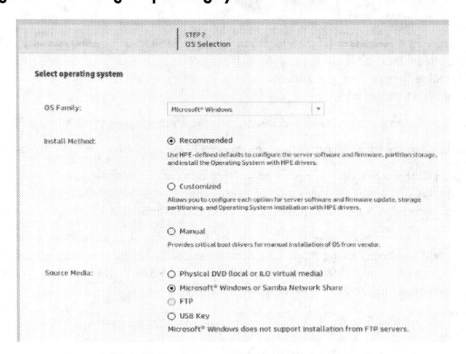

Figure 2-33 Intelligent Provisioning—Operating system selection

1. On the screen shown in Figure 2-33, specify the operating system family to install and the installation method. Available supported operating system families include:

    - Microsoft Windows

    - Red Hat Enterprise Linux

    - SUSE Linux Enterprise Server

    - VMware ESXi/vSphere Custom Image

    If the operating system you are looking for is not listed, it might not be supported for use with the controller model, or you might need to update ProLiant firmware.

Important

Download an HPE custom ISO before the installation from the HPE website before installing VMware. Scan this QR code or enter the URL into your browser to download an HPE custom ISO. Ethernet port 0 must be active.

http://www8.hp.com/us/en/products/servers/
solutions.html?compURI=1499005#tab=TAB4

2. Select an installation method. Depending on the operating system family selected, installation choices vary, including the suggested default.

 Note

If updates are available, a message appears on the screen providing guidance on the proper steps for incorporating them into Intelligent Provisioning.

Options may include:

*   **Recommended**—Uses HPE defaults to configure the server software and firmware, partition storage, and install the operating system with HPE drivers. HPE recommends selecting this option for first-time server setup.

*   **Customized**—Enables individual configuration of the options for server software and firmware updates, storage partitioning, and operating system installation with HPE drivers. Select this option if there are specific parameters that differ from the recommended settings, such as for storage partitioning.

*   **Manual**—Installs the operating system from a custom operating system CD/DVD. After selecting this option, insert the provided CD/DVD so that the server can reboot from the operating system CD/DVD. The Virtual Install Disk (VID) is disabled by default. If you enable VID, a USB mass-storage device appears with the name of the VID during the operating system installation process. The VID contains a limited set of storage and networking drivers, so any required SAS/iSCSI/FCoE adapter can be loaded in the event that the operating system disc does not have the appropriate drivers. Because the VID and the custom CD/DVD might not contain all of the needed drivers, you might need to create a driver CD/DVD to ensure that all required drivers are installed and that the operating system can install successfully.

3. Select the source media from which the operating system is being installed. Media types include:

- **Physical DVD or iLO virtual media (default)**—A standard bootable operating system DVD/CD-ROM media, and virtual media through iLO

- **Microsoft Windows or Samba Network share**—The network share that contains the operating system installation files

- **FTP**—The FTP server that contains operating system installation files

- **USB Key**—The USB flash drive that contains the operating system installation files

Note

Only FAT-formatted USB drives are supported. For operating system image files that cannot copy to the USB unless it is NTFS-formatted, use a different source media, such as a DVD, a network share, or an FTP server.

4. If installing from a CD/DVD disk or a USB drive, insert the media.

5. Click the **Continue** right arrow to go to the next screen, which varies, depending on the media type.

Intelligent Provisioning—Operating system information

Figure 2-34 Intelligent Provisioning—Operating system information

Depending on the operating system that is being installed, you might be prompted to perform the following steps shown in Figure 2-34:

1. Select the operating system and the keyboard language.

2. Enter the product key (not displayed for all operating systems). This is the Product ID number. If a product key is not entered and one is required, the operating system installation pauses indefinitely, prompting for the key to be entered. The installation resumes after the product key is entered.

3. Enter the computer name (optional) and an administrator password (optional).

4. Click the **Continue** right arrow to proceed

When installing an operating system, you can deploy VSA software, which enables IT staff to create fully featured shared storage on a virtualized server. HPE StoreVirtual VSA is a virtual machine that supports hypervisor environments. VSA provides shared storage for both VMware ESX/ESXi and Microsoft Hyper-V hypervisor environments.

Intelligent Provisioning—Review

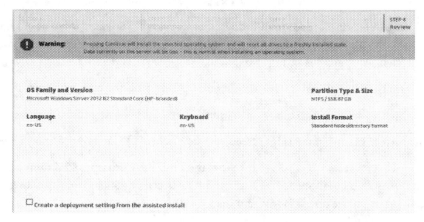

Figure 2-35 Intelligent Provisioning—Review

As shown in Figure 2-35, the Review screen displays hardware and operating system settings. Continuing past this screen installs the operating system and configures the server.

 Caution

Continuing past this screen resets the drives to a newly installed state and installs the selected operating system. Any existing information on the server is erased. This does not affect a first-time setup, because there is no data present on the server.

---

1. Review the information on the screen and confirm its accuracy.

2. If these same settings need to be used at a later time to install and configure a different server, select **Create a deployment setting from the assisted install**.

Note

> Because deployment settings profiles support only the settings provided in the Recommended installation methods, if this is a customized installation, settings that are unique to the customized installation method are not captured.

3. Click the **Continue** right arrow to begin the automated installation and configuration process. Depending on the deployment settings, a variety of screens are displayed, providing progress information about the installation.

During the installation and configuration process, consider the following:

- A EULA might be displayed for Windows installations. Also, messages about an untested Windows version and hpkeyclick messages might be displayed while the drivers are installed. This is expected behavior. No action is required.

- The Firmware Update screen might be displayed, depending on the following two system settings:

  - In the Step 1: Set Preferences screen, **System Software Update** must have been enabled.

  - In the Step 1: Hardware Settings screen, **Update before OS Install** must have been selected.

If the Firmware Update screen is displayed, follow the on-screen prompts to obtain and install the latest firmware on server components. When the updates are complete, the Installing OS page is displayed, ready to begin the operating system installation.

If attempting to deploy an operating system on a server with no installed drives, the server reboots and, after POST, a page is displayed indicating that the settings are being applied. The deployment does not proceed, but messages are written to the Integrated Management Log (IML).

Perform Maintenance

Figure 2-36 Perform Maintenance

The Intelligent Provisioning Perform Maintenance screen, as shown in Figure 2-36, provides access to numerous maintenance-related tasks:

- **Active Health System download**—Download Active Health System telemetry data from the server onto a USB key in the form of an Active Health System log file. After you download the Active Health System log, the log file can be sent to HPE when support cases are opened to assist with troubleshooting. HPE support uses the log file for problem resolution.

- **Firmware Update**—ProLiant Gen9 servers and their installed hardware options are pre-loaded with the latest firmware, but updated firmware might be available. Use the Firmware Update utility to find and apply the latest firmware for ProLiant server and installed options.

- **Intelligent Provisioning Preferences**—Change basic preferences, including the interface and keyboard languages, network and share setting, system date and time, and software update settings. In addition, the EULA is accessible from this screen.

- **Deployment Settings**—Create a server configuration package that can be deployed to one or more ProLiant Gen9 servers and server blades using a USB key and iLO scripting. Using deployment settings is an alternative to using the HPE Scripting Toolkit.

- **HPE Smart Storage Administrator**—These utilities provide high-availability configuration, management, and diagnostic capabilities for all Smart Array products.

- **Insight Diagnostics**—Captures system configuration information and provides detailed diagnostic testing capabilities. Insight Diagnostics provides a comprehensive suite of offline system and component tests, providing in-depth testing of critical hardware components for devices such as processors, memory, and hard drives. During offline testing, the user-installed operating system is not running.

- **Quick Configs**—Set a power management policy through Intelligent Provisioning without having to configure individual settings through the UEFI System Utilities.

- **iLO configuration**—View and change iLO settings through Intelligent Provisioning, instead of through the iLO web interface.

- **HPE Insight Remote Support**—Insight Remote Support provides automatic submission of hardware events to HPE to prevent downtime and enable faster issue resolution. Use this screen to register or unregister for Insight RS.

- **Erase Utility**—Clear hard drives and the Active Health System logs, and reset the RBSU settings in the UEFI System Utilities.

- **License Management**—Activate the iLO Advanced License Pack and the HPE SmartCache License Pack.

HPE Smart Storage Administrator

Figure 2-37 HPE Smart Storage Administrator

As shown in Figure 2-37, the Smart Storage Administrator (SSA) is a web-based application that helps you configure, manage, diagnose, and monitor Smart Array controllers and HBAs. Starting with Intelligent Provisioning 1.50, SSA is the main tool for configuring arrays on Smart Array controllers. It replaces the HPE Array Configuration Utility (ACU) with an updated design and functionality. You should only make minimal changes to existing ACU scripts such as calling the appropriate binary or executable to maintain compatibility.

Additional features of SSA include:

- GUI, CLI, and scripting interfaces

- English, French, German, Italian, Japanese, Simplified Chinese, and Spanish languages

- The ability to run on any machine that uses a supported browser

All formats provide support for standard configuration tasks. SSA also supports advanced configuration tasks, but some of its advanced tasks are available in only one format. The diagnostic features in SSA are also available in the stand-alone software HPE Smart Storage Administrator Diagnostics Utility CLI.

Additional SSA features and functions include:

- **Support for HPE Secure Encryption**—Is a data encryption solution for ProLiant Gen8 and Gen9 servers that protects data at rest on any bulk storage attached to a Smart Array controller.

- **SSD Over Provisioning Optimization**—Optimizes solid state drives (SSDs) by deallocating all used blocks before data is written to the drive. The optimization process is performed when the first logical drive in an array is created and when a failed drive is replaced with a physical drive.

- **Rapid Rebuild Priority**—Determines the urgency with which a controller treats an internal command to rebuild a failed logical drive. SSA offers four settings: low, medium, medium high, and high.

- **Auto RAID 0**—Creates a single RAID 0 volume on each physical drive specified, enabling the user to select multiple drives and configure as RAID 0 simultaneously.

SSA home page

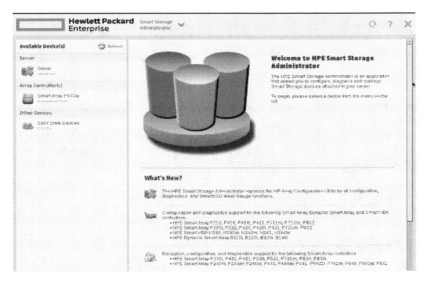

Figure 2-38 SSA home page

The Smart Storage Administrator quick navigation menu is in the top left-hand corner of the screen as shown in Figure 2-38. Clicking the down arrow displays the available devices, and clicking one of the available devices displays additional information and options for the device. Return to a server home screen, or choose **Configuration** or **Diagnostics** for a device listed.

Available devices are listed on the left-hand side of the screen. Clicking a server or an array controller displays the available actions, alerts, and summary for that device. Point to the status alerts to see details on an alert. The What's New? section summarizes the changes since the HPE Array Configuration Utility became HPE Smart Storage Administrator, and since the previous versions of HPE SSA.

The Refresh button is near the top right of the screen. After adding or removing devices, click **Refresh** to update the list of available devices. The Help button is near the top right of the screen.

SSA Controller actions screen

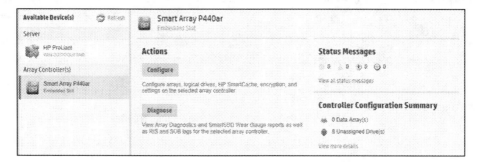

Figure 2-39 SSA Controller actions screen

Selecting a controller from the left-hand pane displays the actions page for that controller as shown in Figure 2-39. Available actions include:

- **Configure**—Modify Controller Settings, Advanced Controller Settings, Modify Spare Activation Mode, Clear Configuration, among others

- **Diagnose**—Array diagnostic report and SmartSSD Wear Gauge Report

SSA Configure screen

Figure 2-40 SSA Configure screen

To access the SSA Configure screen shown in Figure 2-40, either click a device under Configuration in the quick navigation menu, or select an available device from the Home screen, and then click **Configure** under the available options.

The Configure screen displays the GUI elements from the Welcome screen and lists available actions, status messages, more detailed information, and a controller configuration summary for a selected controller. When a controller is selected, the following elements appear:

- **Selected Controller, Controller Devices, and Tools**—This panel, at the left, displays systems, controllers, arrays, physical drives, logical drives, and a cache and license manager.

- **Actions**—This panel, in the middle, provides the following information:

  - Tasks that are available for the selected device based on its current status and configuration.

  - Options and information pertinent to the task, after a task is selected.

- **Status Messages**—This panel provides:

  - Status icons with the number of individual alerts for each category.

  - A view all status messages link that displays device-specific alerts.

- **Controller Configuration Summary**—This panel provides a summary of the following elements:

  - Data arrays

  - Data drives and logical drives

  - Unassigned drives

SSA Diagnostics screen

Figure 2-41 SSA Diagnostics screen

When you select either the Array Diagnostic Report or the SmartSSD Wear Gauge Report on the SSA Diagnostics page, the available actions on the Actions panel include viewing the report or saving the report, as shown in Figure 2-41.

 Note

> The SSA Diagnostics feature replaces the Array Diagnostic Utility supported by SmartStart 8.20 and earlier.

SSA generates the following reports and logs:

- **Array diagnostic report**—This report contains information about all devices, such as array controllers, storage enclosures, drive cages, as well as logical, physical, and tape drives. For supported SSDs, this report also contains SmartSSD Wear Gauge information.

- **SmartSSD Wear Gauge report**—This report contains information about the current usage level and remaining expected lifetime of SSDs attached to the system.

- **Serial output logs**—This log details the serial output for the selected controller.

For each controller, or for all of them, the following tasks can be selected:

- View Diagnostic Report
- Save Diagnostic Report
- View SmartSSD Wear Gauge Report
- Save SmartSSD Wear Gauge Report

For the view tasks, SSA generates and displays the report or the log. For the save tasks, SSA generates a report without the graphical display.

For either task, the report can be saved. In online and offline environments, SSA saves the diagnostic report to a compressed folder, which contains an XML report, a plain text report, and a viewer file so the report can be displayed and navigated using a web browser.

Each SSA Diagnostics report contains a consolidated view of any error or warning conditions encountered. It also provides the following detailed information for every storage device:

- Device status
- Configuration flags
- Firmware version numbers
- Physical drive error logs

SSA Diagnostics never collects information about the data content of logical drives. The diagnostic report does not collect or include the following information:

- File system types, contents, or status

- Partition types, sizes, or layout

- Software RAID information

- Operating system device names or mount points

SSA Array Details screen

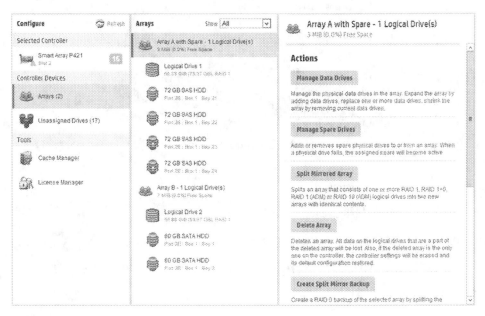

Figure 2-42 SSA Array Details screen

The Array Details page, as shown in Figure 2-42, displays logical drives and their member physical drives. Actions include:

- Manage data drives

- Manage spare drives

- Split mirrored array

- Delete array

- Create split mirror backup

Creating a logical drive

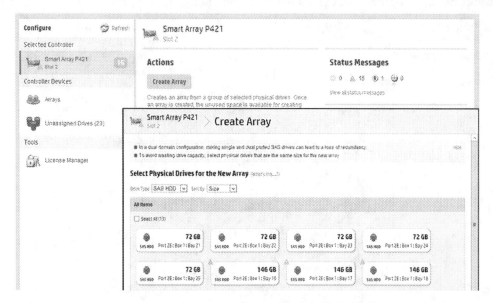

Figure 2-43 Creating a logical drive

To create a new logical drive (array):

1. Select a controller and click **Create Array.**

2. Select the physical drives for the new logical drive as shown in Figure 2-43.

3. Make selections for RAID Level, Strip Size/Full Stripe Size, Sectors/Track, and Size.

4. Click **Create Logical Drive.**

HPE Smart Update solution

In developing the Smart Update solution, HPE addressed the primary issues IT departments encounter with system management, including the need for:

- Consistent, integrated, and fully supported update sets (service packs) for system firmware and software

- Simple and powerful system update technology that can update systems while they are online and does not require management agents installed on target systems

- Scalable system maintenance updates that can reach thousands of target systems through integration with system management platforms such as HPE OneView and others

Smart Update is a re-engineering of the system maintenance process for HPE servers and infrastructure that solves these and other challenges and provides an extensible platform for system maintenance. There are three distinct elements to the Smart Update solution:

- **Smart Components**—Each firmware or driver update is a self-contained executable that takes care of updating the existing firmware or driver with a newer release, and double checks that it is indeed executing against the right hardware. The Smart Components contain the intelligence to perform the update when the operating system (Windows, Linux, or VMware) is up and running.

- **SPP and firmware bundles**—These are collections of Smart Components and HPE SUM. Each downloadable set is heavily tested for coherency and interdependencies in the HPE labs. These convenient bundles are released two to four times per year as new products are introduced that are tested as a set and available for download from the Web. The results of the interdependency testing are coded into HPE SUM.

- **HPE Smart Update Manager**—This is the installer that guides the user through the necessary steps to install a new set of updates. It is easy to use with a browser-based GUI, as well as command line and scripting capabilities. A deployment screen provides details on components that need updates, including estimated deployment time. HPE SUM can perform updates immediately or on a schedule and can reboot immediately after update, drive the operating system to delay the reboot by up to an hour, or wait for a reboot that is part of a regularly scheduled maintenance window.

HPE Smart Update Manager

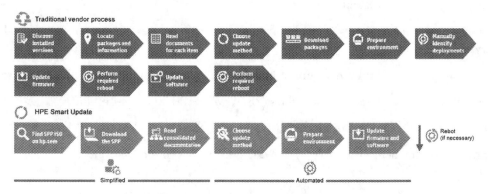

Figure 2-44 HPE Smart Update

HPE SUM keeps firmware, drivers, and agents up to date, while taking into account all interdependencies, with a minimum of impact on system uptime and productivity. HPE SUM allows you to provision multiple systems based on user-defined templates, with only required updates being deployed, ultimately reducing deployment time. Figure 2-44 illustrates the difference between the traditional vendor update process and HPE Smart Update.

HPE SUM provides a single interface for firmware driver and software updates across the HPE server portfolio. It is preloaded with all relevant interdependency information from extensive HPE testing before each SPP release. It does not require an agent for remote installations, because it copies a replica of itself to each of the target servers only for the duration of the installation.

HPE SUM has an integrated hardware and software discovery engine that finds the installed hardware and current versions of firmware and software on target servers and identifies associated targets that should be updated together, to avoid interdependency issues. HPE SUM installs updates in the correct order and ensures that all dependencies are met before deploying an update, including updates for Onboard Administrator and Virtual Connect. It prevents an installation if there are version-based dependencies that it cannot resolve.

HPE SUM supports online updates of all ProLiant firmware, drivers, agents, and tools (for Windows and Linux targets; some firmware can be updated online in VMware ESXi 5.0, vSphere 5.1 and later). Reboots can be "forced always," "as needed," "delayed" (operating system-controlled, up to 60 minutes) or "not" (assuming a manual reboot at a later time).

HPE SUM features

HPE SUM increases the ease of server management through these features:

- Easy discovery of all supported devices

- Integrated acquisition of the latest updates

- Scalability to 50 nodes

HPE SUM deployment capabilities include:

- The ability to deploy firmware and software from a Windows workstation/server to Linux servers

- The ability to deploy firmware from a Windows or Linux workstation/server to VMware servers

- Local offline firmware updates using the HPE SUM or SPP bootable-ISO image

- Remote offline deployment using the HPE Scripting Toolkit or iLO Virtual Media

- Enhanced deployment experience by viewing deployment logs during the deployment process (also known as *live logs*)

- Enhanced HDD firmware deployment for selective HDD update (limited to firmware packages that support this)

Note

Scan this QR code or enter the URL into your browser for more information and documentation on HPE SUM.

http://www.hp.com/go/hpsum

HPE SUM and iLO Federation

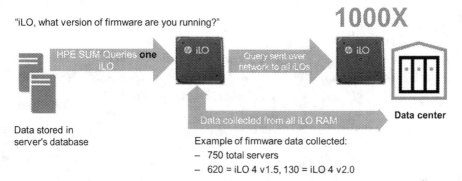

Figure 2-45 HPE SUM and iLO Federation

When you log on to HPE SUM, it automatically searches for iLO Federation groups on connected networks as shown in Figure 2-45. HPE SUM searches each group and displays the nodes that respond to the HPE SUM search. The Edit scalable update group screen is used to enter the IP address and user credentials for one node in the iLO Federation group that HPE SUM uses as the interface for inventory and deployment to the nodes in the group.

When you select a group, HPE SUM displays information about the group, including a description, server types, number of servers, and installed firmware versions. HPE SUM displays the PMC, CPLD, System ROM, and iLO firmware.

The HPE SUM iLO Federation feature relies on proper configuration of iLO Federation groups before launching HPE SUM. Having multiple iLO Federation groups with the same name or fragmented iLO Federation groups can result in HPE SUM only working with a portion of the expected systems.

Other systems ask iLOs for data on a periodic basis, and store that data in a database. They might request server data such as temperatures, profiles, or firmware versions from iLO once an hour and store that in the server's database or HPE OneView appliance. This is less scalable than the iLO Federation approach, which includes virtual real-time updates.

iLO Federation management provides scalability enhancements when used with HPE SUM:

- Automatically discover iLO Federation groups on the management network.

- Update the iLO and ROM firmware online on ProLiant servers in the iLO Federation Group through the iLO

- Update all applicable firmware on ProLiant servers in the iLO Federation Group through the iLO using offline firmware deployment

- Ability to apply updates to all members of an iLO Federation group (Advanced iLO license required)

HPE Service Pack for ProLiant

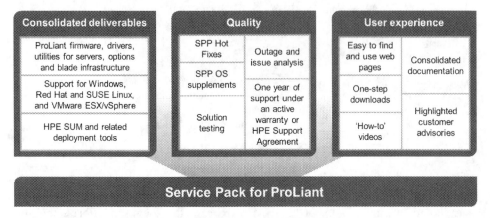

Figure 2-46 HPE Service Pack for ProLiant

SPP is a comprehensive package that includes firmware, drivers, and tools across ProLiant servers and infrastructure including many generations of ProLiant BL/DL/ML/SL series servers as illustrated by Figure 2-46.

SPP provides this consolidated set of solution-tested ProLiant system software available as a single download. You spend less time on maintenance with resulting confidence in the update's stability.

SPP is available for customers to download and install on products that are under an active warranty or an HPE Support Agreement. For latest component updates, check the server product-specific webpage. SPP will, in general, carry the current generation of an operating system plus one previous generation.

SPP is pretested for component dependencies, customizable for the environment, and supported for one year.

In between each full SPP release, you might need to apply hot fixes as necessary to address specific issues. Hot fixes are supported as part of the SPP because each hot fix component is tested individually against the latest SPP and all other SPPs released within the last 12 months.

Maintenance Supplement Bundles (MSBs) are used together with an associated SPP and are released in April and October. This includes any supplements or hot fixes that have been issued since the associated SPP was issued. Applying an MSB extends the support period of the SPP. SPP supplements and other (non-hot fix) components might also be needed to support new operating system releases or new functionality. These types of releases can be combined with a full SPP release to create a custom solution.

On-premises management with HPE OneView

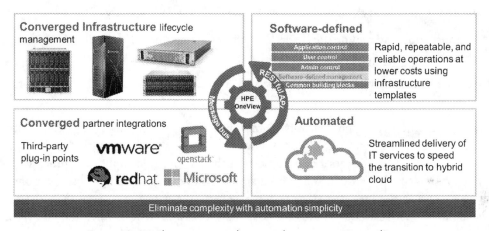

Figure 2-47 Eliminate complexity with automation simplicity

HPE OneView is a software-defined management platform that addresses the challenges of manual operation, human error, and limited extensibility in virtualized BladeSystem and rack server environments as illustrated by Figure 2-47. HPE OneView features a unique automation hub that consists of a state change message bus (SCMB) and the RESTful API. This means all of the information that HPE OneView collects, all of the changes that HPE OneView knows about, and all of the templates and control functions that HPE OneView delivers can be accessed programmatically.

By capturing processes, configurations, and best practices in software, HPE OneView creates a modern collaborative management approach that automates the deployment and management of infrastructure—repeatedly, reliably, and at scale. With HPE OneView, basic tasks take seconds and stay simple at any scale, radically accelerating all data center processes.

The software-based approach to lifecycle management in HPE OneView automates operations to reduce the cost and time to deliver IT services. The HPE RESTful API allows you to create customized workflows and scripts, as well as configuration profiles for push-button builds that instantly deliver resources without mistakes or variation.

HPE OneView offers a single integrated platform that provides one view of the server, network, and storage environment for a simple, integrated user experience. It is available as a virtual appliance running in a VMware ESXi or Microsoft Hyper-V virtual machine.

HPE OneView provides converged management for the software-defined data center through the following features:

- **Server profile templates**—Centralized configuration and change management
- **Server profile mobility**—Move server profiles across different generations and server types
- **Online firmware and driver updates**—Updates that align with IT process
- **Storage snapshots and clones**—For rapid data protection and recovery
- **SAN monitoring and diagnostics as well as Fibre Channel over Ethernet (FCoE) support**—More connectivity options for better problem diagnosis
- **Virtual Connect management**—Enhanced VLAN scale, partially stacked domains, and QoS, dual hop FCoE

The integrated SCMB messaging platform enables dynamic and responsive integrations with solutions such as HPE Insight Control for VMware vCenter and OpenStack for HPE Helion CloudSystem. The SCMB provides a communication channel between application processes and the underlying systems that it manages. It uses asynchronous messaging to notify subscribers of changes to both logical and physical resources. For example, you can program applications to receive notifications when new server hardware is added to the managed environment or when the health status of physical resources changes—without having to poll the appliance continuously for status using the RESTful API.

HPE OneView integrates with many commonly used tools, applications, and products, including service desk, reporting, monitoring, and configuration management database (CMDB) tools; Helion CloudSystem software; and Microsoft, VMware, and Red Hat Enterprise Linux hypervisor solutions.

 Note

A CMDB is a repository that acts as a data warehouse for an IT organization. It contains information describing IT assets such as software and hardware products and personnel.

The automation hub supports two types of converged management:

- **Infrastructure lifecycle management**—Support for HPE ConvergedSystem, HPE 3PAR storage, ProLiant DL servers, and BladeSystem products

- **Partner integrations**—Out-of-the-box plug-ins for VMware, Microsoft, Red Hat, and OpenStack management products

The software-defined capabilities provide support for policy-driven infrastructure templates that drive automated activities. The templates capture best practices and allow for rapid, repeatable, and reliable automated operations that reduce operational costs. This means that IT organizations can easily incorporate HPE OneView functions into workflows that extend beyond infrastructure management.

HPE OneView key functions

HPE OneView streamlines the delivery of IT services to help support the business more efficiently and speed the transition to hybrid cloud. HPE OneView key functions include:

- Server provisioning

    - Easily enforce configuration consistency using server profile templates to monitor, flag, and update multiple servers.

    - Quickly view server profile information with access to additional details from storage and networking.

- Storage provisioning and automation

    - Enable a hypervisor-like user experience through storage volume snapshots and clones.

    - Monitor the health of SAN connections with administrator alerts to data path failures and reports on unauthorized access.

    - Automate zoning SAN fabrics and attaching storage volumes to server profiles.

- Automated change management

    - Automate system software updates at scale using template-driven firmware and device driver management.

    - Quickly see when systems are not in compliance with the master template.

    - Flexibly migrate and recover workloads using profile mobility across compute platforms and generations.

    - Install updates efficiently and reliably with HPE Smart Update tools.

- Virtual Connect networking

  - Consolidate management, number of adapters, and interconnects and reduce cables and utilization of upstream switch ports with Virtual Connect dual-hop FCoE support with 10/40 Gb.

  - Prioritize designated networking traffic flows and guarantee performance levels by using Virtual Connect Quality of Service (QoS) priority queuing.

  - Eliminate the need to stack interconnects within an enclosure and provide air-gap isolation between Ethernet fabrics by using partially stacked domains to allow better utilization of uplinks with Flex-10 and FlexFabric modules.

  - Ease the transition to advanced HPE OneView capabilities with less effort and human error by migrating HPE Virtual Connect Manager (VCM) domains to HPE OneView.

  - Provide rich type-length-value elements with support for ProLiant Gen9 profiles and Link Layer Discovery Protocol (LLDP) enhancements.

- Network monitoring

  - Monitor Cisco Nexus 5xxx/6xxx TOR switches with support for the Cisco fabric extender (FEX) for BladeSystem module (also known as the *B22HP interconnect module*) from HPE OneView (with resources, alerts, statistics, monitoring, topology, and physical connectivity views).

The HPE OneView approach to infrastructure lifecycle management

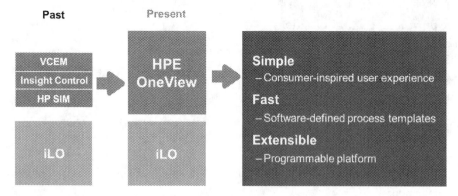

Figure 2-48 The HPE OneView approach to infrastructure lifecycle management

HPE OneView provides converged management capabilities that reduce infrastructure complexity through automation. This modern management architecture is designed to replace legacy management tools such as VCEM, Insight Control, and HP SIM as shown in Figure 2-48.

HPE OneView provides a resource-oriented solution that focuses on the entire hardware lifecycle from initial configuration to ongoing monitoring and maintenance, and accelerates IT operations for managing servers, storage, and network resources. It is purpose-built to support scenarios such as:

- Deploying bare-metal servers

- Deploying hypervisor clusters from bare metal

- Performing ongoing hardware maintenance

- Responding to alerts and outages

HPE OneView uses templates that automate the configuration and propagation of server, storage, and networking profiles. Its underlying infrastructure components—including networking, power management, and servers—support virtualization, cloud computing, big data, and mixed computing environments. This converged, software-defined, and automated platform reduces operating expenses (OPEX), improves agility, and frees up resources for new business initiatives.

This approach provides the following benefits:

- **Software-defined approaches to systems management** create reliable repeatability to help prevent unplanned outages caused by human error or device failure.

  - Profiles and groups capture best practices and policies to help increase productivity and enable compliance and consistency.

  - Manage the infrastructure programmatically by using APIs built on industry standards such as REST. These APIs are easily accessible from any programming language, and software development kits are provided for interfaces, PowerShell, and Python scripts.

- **Automation** can streamline the delivery of IT services and speed transition to Infrastructure-as-a-Service (IaaS) and hybrid cloud delivery.

  - As an intelligent hub, HPE OneView provides closed-loop automation with consistent APIs, data modeling, and state change message bus (SCMB). HPE OneView also supports lights-out automation.

- **Convergence** reduces the number of tools required to learn, manage, deploy, and integrate the infrastructure.

  - A single, open platform supports multiple generations of ProLiant DL servers, HPE 3PAR StoreServ storage, HPE ConvergedSystem and HPE BladeSystem c7000 solutions, as well as HPE and Cisco top-of-rack (TOR) switches.

  - Microsoft System Center integration adds a single integrated System Center Virtual Machine Manager Fabric Management/Storage Add-in with support for partial domains, FCoE dual hop, server profile templates, and logical enclosures.

  - VMware vCenter Server integration adds server profile-based deployment and automated HPE StoreOnce deployment for secure backup and recovery. You can automate control of compute, storage, and networking resources using VMware vCenter or Microsoft System Center without detailed knowledge of every device. Tasks, processes, and projects are accomplished faster and with more consistency than the older patchwork approaches to management.

– Red Hat Enterprise Virtualization gives administrators insight into and control over their HPE infrastructure while supporting their Red Hat virtualized environment from a single screen.

• **Lifecycle management** automates day-to-day responsibilities by simplifying time-consuming tasks, which leads to increased productivity and reduced operational costs. This capability includes:

– Server profile templates

– Firmware and driver updates

– Logical tracking and linking of resources

Converged infrastructure acceleration

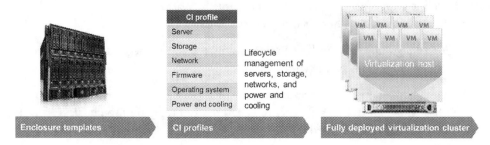

Figure 2-49 Reduce cycle times to provision and manage infrastructure

HPE OneView spans both logical and physical resources. Logical resources are items such as networks, server profiles, and connections. Physical resources are tangible items such as server hardware, interconnects, and enclosures. This advanced architecture, as illustrated by Figure 2-49, connects the resources with a common, domain-specific representation of the servers, networks, and storage. It also models the associations and interdependencies of the resources. This enables each area to contribute to "one view" of the converged infrastructure.

HPE OneView aids converged infrastructure acceleration by helping administrators and systems experts reduce the cycle times involved in provisioning and managing the infrastructure.

A simple progression of IT management tasks includes the following steps:

1. Step 1

• Define a container object known as a *logical interconnect group* to represent a collection of Virtual Connect module types and the associated LAN and SAN connectivity the interconnects will provide.

- Define a container object known as an *enclosure group*, which includes a logical interconnect group.

- Import an HPE BladeSystem c7000 enclosure and assign the enclosure group to it.

- Apply a firmware baseline to the enclosure components (HPE Onboard Administrator and Virtual Connect).

2. Step 2

- Create one or more server profile templates for virtualized hosts based on your best practices.

- Preprovision the server profile to particular server blades in the enclosure, or assign on-demand as needed.

3. Step 3

- Fully deploy a virtualized cluster using VMware vSphere stateless auto deploy.

Designed for simplicity

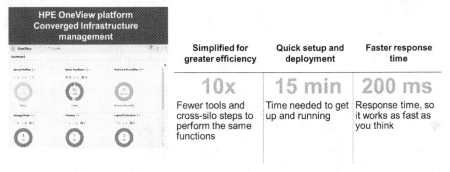

Figure 2-50 HPE OneView converged infrastructure management

The software-based approach to lifecycle management in HPE OneView includes a fully program-mable interface to create customized workflows, scripts, and configuration profiles for push-button builds that instantly deliver resources without mistakes or variations. This results in greater efficiency, rapid deployment and setup, and speedy response times as shown in Figure 2-50. Workflow templates capture best practices and policies to increase productivity and enable compliance and consistency. Built-in intelligence automates common BladeSystem management tasks usually performed by users, and connections to HPE SUM automate firmware and system software maintenance.

The HPE OneView GUI reduces the number of tools needed for everyday management of server, storage, and network resources. One tool can perform numerous tasks within a single operating model that provides a consistent way of viewing resources and results.

HPE OneView increases productivity for the IT team by facilitating collaboration, removing friction, and accelerating time to value. Administrators responsible for different areas (including servers, storage, networks, and virtualization) are authorized to use the same tool to manage resources. This allows them to collaborate and make authorized decisions without stepping on each other's responsibilities. For example, virtualization administrators can automate control of HPE compute, storage, and networking resources without requiring detailed knowledge of each device.

The HPE OneView appliance provides several software-defined resources, such as groups and server profiles, to enable an administrator to capture the best practices of an organization's experts across a variety of disciplines. An administrator can reduce cross-silo miscommunication by defining server profiles, networking objects, and other resources.

HPE OneView has also been optimized for productivity with a response time target of 200 milliseconds for all user interface interactions. Tasks that take longer than 200 milliseconds to process are labeled *asynchronous* tasks, meaning they can finish independently in the background. This allows the administrator to move on to other activities.

Consumer-inspired user experience

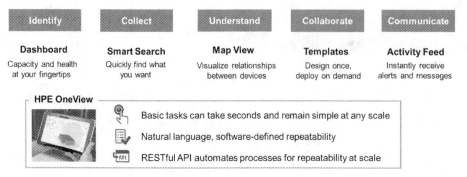

Figure 2-51 Common tasks can be simplified to accelerate IT processes

Figure 2-51 shows five key data center tasks: identify, collect, understand, collaborate, and communicate. HPE OneView key features correspond to those tasks:

- **Dashboard**—Provides a variety of capacity and health status information that is well organized and easily accessible. The dashboard offers a high-level overview of the status of the converged infrastructure components you are authorized to view. Clicking an object's status gives you an at-a-glance perspective of the event. The dashboard can display a health summary of the following:

  - Server profiles and server hardware

  - Enclosures

- Logical interconnects

- Storage pools and volumes

- Appliance alerts

The status of each resource is indicated by an icon: OK, Warning, or Critical.

- **Smart Search**—Enables the administrator to quickly locate configured objects and device information. For instance, you can locate or search for devices based on physical media access control (MAC) addresses and World Wide Names.

- **Map View**—Allows the administrator to visualize the relationship between devices and the related objects representing them. A "follow the red" status methodology bridges the logical objects to the physical systems, which is especially useful for troubleshooting and during support calls.

- **Templates**—Offers a way to design boilerplates for the underlying network, server, and storage objects that manage systems in a converged infrastructure. In general, templates are used to define best practices. Templates focus on many-to-one relationships, such as the association between an enclosure group and its constituent logical interconnect groups. This part of the architecture helps support the need for documentation and compliance through consistency.

- **Activity Feed**—Allows the administrator to quickly receive alerts and other messages as conditions arise. Activity Feed functions similarly to Twitter in terms of collaboration and communication. An administrator can add notations to events and assign them to an appropriate user.

All of these HPE OneView elements are designed to change the management approach from how devices are managed to how teams work together to complete tasks.

Single integrated platform

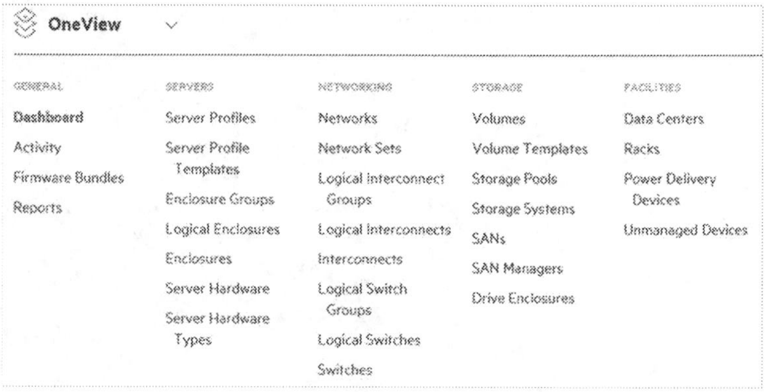

Figure 2-52 OneView unified interface

The key benefit of HPE OneView is that it is one tool that uses one dataset to present one view to the administrator, combining complex and interdependent data center management capabilities in a unified interface, as shown in Figure 2-52. HPE OneView provides core enterprise management capabilities, including:

- Availability features

- Security features

- Graphical and programmatic interfaces

- Integration with other HPE management software

HPE OneView simplifies common data center tasks:

- **Provisioning the data center**—The OneView appliance provides several software-defined resources such as enclosure groups, logical interconnects, network sets, and server profiles. This enables you to capture best practices for implementation across networking, storage, hardware configuration, and operating system build and configuration. Role-based access control and various configuration elements in the form of groups, sets, and server profiles allow system administrators to provision and manage several hundred servers without having to involve the networking and storage systems experts in every server deployment.

- **Managing and maintaining firmware and configuration changes**—The appliance provides simple firmware management across the data center. When you add a resource to the appliance, to ensure compatibility and seamless operation, the appliance automatically updates the resource firmware to the minimum version required to be managed by the appliance.

- **Monitoring the data center and responding to issues**—With HPE OneView, you can use the same interface that you use to provision resources for data center monitoring. There are no additional tools or interfaces to learn. When you add resources to the OneView appliance, they are automatically configured for monitoring, and the appliance is automatically registered to receive SNMP traps. You can monitor resources immediately without performing additional configuration or discovery steps.

Open integration with existing tools and processes

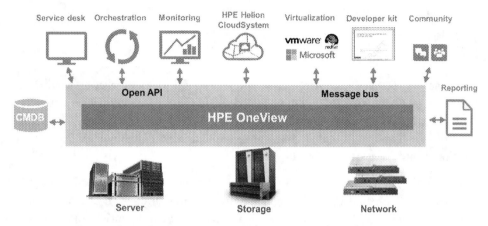

Figure 2-53 Open Integration with existing tools and processes

With the HPE OneView converged infrastructure management solution at its foundation, IT can become more responsive and can operate more efficiently and predictably. HPE OneView provides an open development platform in the RESTful API, which is designed to unlock the full power of the management architecture. The consistent APIs, data model, and SCMB in HPE OneView are architected to provide a converged management approach.

Traditional environments require predefined, serialized workflows and different tools for different tasks. HPE OneView is a scalable, resource-oriented solution focused on the lifecycle of logical and physical resources. As illustrated by Figure 2-53, HPE OneView features for managing a converged infrastructure include:

- Converged management architecture for servers, storage, and networks (designed to avoid enterprise silos)

- Software-defined control (profiles, templates, groups, and sets)

- Automated server and storage provisioning

- Flexible zoning with policy-based, automatically created aliases generated for server and storage system ports and groups

- Environmental (power and thermal) management

- System health monitoring

- Virtual Connect management and migration tool

- Pervasive Smart Search and Map View

- Remote management with HPE iLO Advanced

- Firmware updates and configuration change management

- Support for HPE Helion partner tools such as Chef, Puppet, Docker, and Ansible

System health monitoring

Efficient data views and effective control enable you to respond to issues when managing the health of ProLiant servers. When managed resources are added to the appliance, they are automatically set up for monitoring, including the automatic registration of SNMP traps and scheduling of health data collection. ProLiant Gen8 and Gen9 servers are monitored immediately without requiring you to invoke additional configuration or discovery steps.

All monitoring and management of data center devices is agentless and out-of-band for increased security and reliability. Operating system software is not required, open SNMP ports on the host operating system are not required (for Gen8 and Gen9), and zero downtime updates can be performed for these embedded agents. ProLiant Gen8 and Gen9 servers support agentless monitoring by iLO. HPE OneView uses SNMP in read-only mode to the iLO only, not to the host operating system. ProLiant G6 and G7 servers require host operating system SNMP agents.

 Note

Read-only mode means SNMP uses gets and traps, but not sets.

HPE OneView provides proactive alert notifications by email (instead of using SNMP trap forwarding) and automated alert forwarding. You can view, filter, and search your alerts using Smart Search. Alerts can be assigned to specific users and annotated with notes from administrators. Notifications or traps can be automatically forwarded to enterprise monitoring consoles or centralized SNMP trap collectors.

The customized dashboard capability allows you to select and display important inventory, health, or configuration information and to define custom queries for new dashboard displays. The single user interface provides additional summary views of firmware revisions and of the hardware inventory for servers, storage, and networks. Other data and inventory elements are visible through the user interface and RESTful API, and can be found using Smart Search.

Server provisioning

HPE OneView provides the right to use HPE Insight Control server provisioning, a complete provisioning solution for ProLiant servers. This virtual appliance solution can be used to install and configure ProLiant servers using resources such as OS build plans and scripts to run deployment jobs. Server provisioning features allow you to:

- Install Microsoft Windows, Linux, and VMware ESXi on ProLiant servers

- Deploy to target ProLiant Gen8 and Gen9 servers without using PXE (leveraging HPE Intelligent Provisioning), and with additional support for ProLiant G7 and G6 servers

- Deploy operating systems to virtual machines

- Create and run customized build plans to perform additional configuration tasks either before or after operating system deployment

- Run deployment jobs on multiple servers simultaneously

- Customize ProLiant deployments through an easy-to-use, browser-based interface

- Update drivers, utilities, and firmware on ProLiant servers using SPPs

- Configure ProLiant system hardware, iLOs, BIOS, Smart Array controllers, and Fibre Channel HBAs

- Use RESTful API calls to perform all of the functions available from the user interface

 Note

Server provisioning is only available for customers running HPE OneView Advanced.

Server profile templates

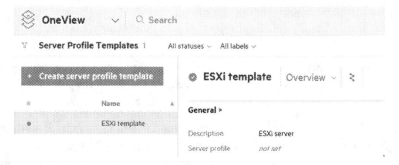

Figure 2-54 Server profile templates

Server profile templates, as shown in Figure 2-54, provide automated change management across multiple systems. You make the update once at the template level and automate updates to server profile configurations or apply new system software baselines. Server profile templates are at the center of software-defined policies and solutions. HPE OneView provides a quick glance of server connections, SAN storage, direct-attached storage, and BIOS or UEFI compliance with the profile.

With HPE OneView, administrators can quickly drill down into dashboard panels to identify issues or troubleshoot problems. By clicking the connection panel, administrators can easily move from the high-level status down to the connections summary. You can expand individual connections to reveal complete connection configuration details. Conversely, you can collapse these details to hide them.

Server profile mobility

The HPE OneView server profile mobility feature defines configurations once, in minutes, and then provisions or updates the configuration many times—consistently and reliably with no repetitive tasks—across compute, storage, and networking resource. This way, profile mobility is not limited to migrations across the same server hardware type and enclosure groups. HPE OneView provides profile mobility across:

- Different adapters
- Different server generations
- Different server blade models

The HPE OneView appliance monitors both the server profile and server profile template. It compares both elements and ensures the server profile matches the configuration of its parent server profile template.

HPE OneView storage management

HPE OneView allows you to:

- Automate HPE 3PAR StoreServ volume creation and SAN zoning
- Attach the storage volumes to server profiles

For example, a 32-server cluster can be automatically created in hours rather than manually configured in days. After deployment, storage and servers are monitored in HPE OneView, and the storage topology is visible in Map View.

Automated, policy-driven provisioning of storage resources is fully integrated with server profiles. You can use SAN managers to bring their managed SANs under HPE OneView management, and you can automatically configure SAN zoning through server profile volume attachments to mitigate configuration errors. Storage integration with server profiles saves you time and makes you more productive.

Switched fabric, direct-attach (Flat SAN), and virtual SAN topologies are supported to provide dynamic connectivity between HPE OneView managed servers and HPE 3PAR StoreServ storage systems. HPE OneView discovers the SAN paths and provides connectivity services for the following infrastructures:

- HPE 3PAR StoreServ storage systems connected directly to an enclosure using Fibre Channel
- HPE 3PAR StoreServ storage systems connected to an HPE B-series Fibre Channel SAN configuration
  - SANs managed through the HPE B-series SAN Network Advisor software

- HPE 3PAR StoreServ storage systems connected to a Brocade Fibre Channel SAN configuration
  - SANs managed through Brocade Network Advisor software
- HPE FlexFabric 5900 AF/CP switches, Cisco MDS series switches, and Brocade switches

Storage snapshots and clones

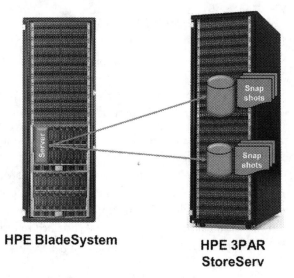

HPE BladeSystem   **HPE 3PAR StoreServ**

Figure 2-55 Storage snapshots with HPE 3PAR StoreServ

With HPE OneView, advanced automation enables an IT generalist to define and provision storage volumes, automatically zone the SAN as part of the provisioning process, and attach the volumes to server profiles.

HPE OneView storage automation makes businesses more responsive, secure, and efficient. HPE 3PAR StoreServ storage is fully integrated with HPE OneView server profiles for automated, policy-driven rollout of enterprise-class storage resources. After the storage has been rolled out, you can select an HPE 3PAR StoreServ volume in OneView and create a snapshot from that volume as illustrated by Figure 2-55.

RESTful APIs provide a simpler, stateless, and scalable approach, so users and developers can easily integrate, automate, and customize on their own.

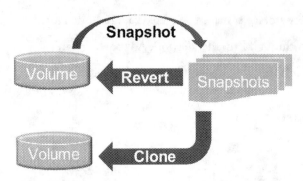

Figure 2-56 Cloning and reverting snapshots

Snapshots in OneView allow copy and provisioning access to nonstorage professionals such as database administrators, software developers, and test engineers working with systems.

As shown in Figure 2-56, users can create volume snapshots, create a volume from a snapshot by cloning the snapshot, and revert a volume from a snapshot using OneView. This means that users can restore their own copies of test data safely and quickly without relying on a storage administrator. They can easily replace and restore copies of their volumes by copying, promoting, and attaching their volumes to server profiles. This enables users to update specific snapshots with more recent snapshots, resulting in faster turnaround times for developers who need refreshed snapshots. This also alleviates the workload for storage administrators.

SAN health and diagnostics

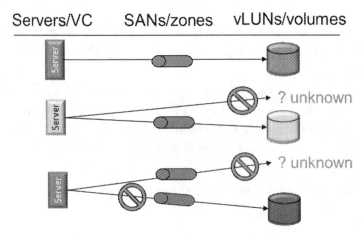

Figure 2-57 SAN health and diagnostics

Using HPE 3PAR StoreServ storage within HPE OneView is as simple as selecting a storage template and a server profile. HPE OneView automation carves out the storage volume, zones the Fibre Channel SAN, and attaches the storage to the server profile.

After they are rolled out, the SAN resources are immediately exposed in the topology map. This includes multihop Fibre Channel and FCoE architectures. In HPE OneView, proactive alerts are provided when the expected and actual connectivity and states differ or when SAN health issues are immediately visible in the topology map. HPE OneView 2.0 provides SAN configuration reports, which include guidance for SAN efficiency and help in resolving potential SAN issues before there is a business impact.

In Figure 2-57, the top row represents a healthy status through all zones from the server to the volume. The middle row represents a warning, but the path is still functional. The bottom row represents a critical issue in the data path.

SAN connectivity and synchronization with the appliance

The HPE OneView appliance monitors the health status of storage systems and issues alerts when there is a change in status. The appliance also monitors the connectivity status of storage systems. If the appliance loses connectivity with a storage system, an alert is displayed until connectivity is restored. The appliance attempts to resolve connectivity issues and clear the alert. If it cannot, you must use the Storage Systems screen to refresh the storage system manually and synchronize it with the appliance.

The appliance also monitors storage systems to ensure that they are synchronized with changes to hardware and configuration settings. However, changes to storage systems made outside the appliance (such as changing credentials) might cause the storage system to lose synchronization with the appliance, in which case you must manually refresh the storage system.

HPE OneView network management with HPE Virtual Connect

Virtual Connect interconnects continue to play an integral role in the success of HPE Converged Infrastructure solutions. Key Virtual Connect features of HPE OneView include:

- **Virtual Connect dual-hop FCoE parity support**—This feature allows FCoE traffic out of the enclosure to an external bridge device, which will handle the conversion of FCoE to native Fibre Channel traffic. It also provides benefits including cable consolidation, reduction in utilization of the upstream switch ports, and consolidation in management and number of adapters and interconnects required. This feature also supports up to 32 FCoE networks (32 virtual LANs [VLANs]) and 40 Gb FCoE uplinks of the HPE Virtual Connect FlexFabric-20/40 F8 Module.

- **Virtual Connect QoS priority queuing**—QoS is used to provide different priorities for designated networking traffic flows and guarantee a certain level of performance through resource reservation. The QoS feature allows administrators to configure traffic queues for different priority network traffic, categorize and prioritize ingress traffic, and adjust dot1p priority settings on egress traffic. Administrators can use these settings to ensure that important traffic receives the highest priority handling and less important traffic is handled at a lower priority.

 Note

QoS is a set of service requirements that the network must meet in order to ensure an adequate service level for data transmission. Network traffic is categorized and then classified. After being classified, traffic is given a priority and scheduled for transmission. The goal of QoS is a guaranteed delivery system for network traffic.

- **Partially stacked Virtual Connect domains**—This feature provides air-gap separation between Ethernet networks and enhanced active/active configurations with up to 1000 networks for the active/active pair of connections. It also removes the one-to-one relationship between the physical enclosure and the logical interconnect and eliminates the need to stack all interconnects within the enclosure.

- **Enhanced migration from VCM domains**—Migrating from VCM to HPE OneView 1.1 or earlier was a manual process that required a service outage. In addition to a migration wizard to improve this process, HPE OneView 2.0 and greater contains an embedded VCM feature that automates Virtual Connect domain migration with a single push of the button and greatly reduces downtime.

- **Fibre Channel support in Virtual Connect**—This support includes the HPE Virtual Connect 8 Gb 24-Port Fibre Channel Module and HPE Virtual Connect 8 Gb 20-Port Fibre Channel Module. HPE OneView also supports the next-generation HPE Virtual Connect FlexFabric-20/40 F8 Module.

- **Active/active configuration support for Virtual Connect**—This feature allows full use of all uplink ports in an uplink set, reduces the oversubscription rates for server-to-network-core traffic for more predictable traffic patterns, and provides faster link failure detection and failover times. Optimized for north/south traffic patterns, the active/active configuration support can be combined with the SmartLink feature to allow NIC teaming drivers to transmit on both adapter ports and maintain redundancy.

Additional advanced capabilities include support for:

- Untagged traffic
- VLAN tunneling
- Configurable Link Aggregation Control Protocol (LACP) timers
- Minimum and maximum bandwidth settings on connections
- Visibility to MAC address tables

- FlexNIC traffic statistics and performance monitoring

- Enhanced detection, protection, and reporting of pause flood and network loops

HPE OneView also provides network capabilities for BladeSystem solutions, which do not use Virtual Connect for networking. The Networking section of the HPE OneView main menu has a New Switches resource to assist in these efforts. The following switches can be monitored:

- Cisco Nexus 5548 switch

- Cisco Nexus 5596 switch

- Cisco Nexus 6001 switch

Quality of Service for network traffic

For end-to-end Quality of Service (QoS), all hops along the way must be configured with similar QoS policies of classification and traffic management. Traffic prioritization is determined by two things in an end-to-end QoS policy:

- At the interconnect, the packets are transmitted based on the associated queue bandwidth. The more the bandwidth, the higher the priority for the associated traffic at the queue.

- Egress dot1p remarking helps achieve priority at the next hops in the network. If the queue egress traffic is remarked to a dot1p value and that value is mapped to a queue in the next hops with greater bandwidth, then these packets in the end-to-end network are treated with higher priority.

QoS configuration is defined in the logical interconnect group and applied to the logical interconnect. QoS statistics are collected by the interconnect modules. A QoS configuration is applied only on Virtual Connect Ethernet and Virtual Connect FlexFabric interconnects that support QoS. On all other interconnects, the QoS settings are ignored.

Consistency state of a logical interconnect with QoS configurations

The UI displays only the currently active QoS configuration that is applied on the interconnects. In addition, two inactive QoS configurations are stored for Custom (with FCoE) and Custom (without FCoE) configuration types. These are the last known QoS configurations for the corresponding configuration types, applied previously on the associated logical interconnect and logical interconnect group.

When the consistency of a logical interconnect to its associated logical interconnect group is checked, the compliance of inactive QoS configurations are also checked (inactive QoS configurations are not visible in the UI). Even if active QoS configurations are exactly the same between a logical interconnect and associated logical interconnect group, a logical interconnect's consistency status can be shown as inconsistent (due to inconsistencies in inactive QoS configurations stored internally).

Perform an update from the group to bring the logical interconnect group and logical interconnect into a consistent state.

Network profiles, logical interconnect groups, and network sets

HPE OneView simplifies the deployment of various Virtual Connect environments:

- Single enclosure
- Racks of enclosures in a single data center
- Across multiple data centers

The software-defined approach of HPE OneView extends Virtual Connect features by using profiles, logical interconnect groups, and network sets to simplify management and capture best practices:

- **Profiles** enable servers that are licensed using HPE OneView Advanced to configure the Virtual Connect capabilities and support dynamic network changes. Virtual Connect administrators can change pre-existing connection networks and connection bandwidth without powering down server blades.

- **Logical interconnect groups** are created for configuring the Virtual Connect module with its uplinks and associated networks, enabling efficient application to multiple Virtual Connect environments.

- **Network sets** use a single name for several Ethernet links. They are used to easily update multiple networks in various profiles from a single location, rather than updating each network separately. Network sets are useful in virtual environments where each profile connection needs to access multiple networks.

 Note

A software-defined infrastructure—profiles, groups, and sets—requires the purchase of an HPE OneView Advanced license. Differences between the Standard and Advanced licenses include:

- HPE OneView Standard can monitor ProLiant DL and BL G6, G7, Gen8, and Gen9 servers.

- HPE OneView Advanced actively manages BladeSystem c7000 (G7, Gen8, and Gen9) and ProLiant DL Gen8 and Gen9 servers. ProLiant DL580 Gen9, DL560 Gen9, and BL660c Gen9 servers are also supported. HPE OneView Advanced includes a right to use HPE Insight Control server provisioning, which has deployment capabilities for multi-server operating system and firmware provisioning for ProLiant and BladeSystem servers.

Cisco Fabric Extender for HPE BladeSystem modules support

The Cisco FEX for HPE BladeSystem module is modeled as part of the logical interconnects and logical interconnect groups. HPE OneView monitors the power state of the FEX module and displays inventory and field-replaceable unit (FRU) data for FEX modules and Cisco Nexus 5000 series switches. HPE OneView also supports monitoring networks provisioned to FEX ports with FEX enabled or disabled.

Map View in HPE OneView shows the relationships between FEX and the parent TOR Cisco Nexus 5000 or 6000 series switches. Other capabilities, such as the bulk creation of Ethernet networks using the user interface and the RESTful API, can provide additional productivity for users.

Automated change management with driver and firmware updates

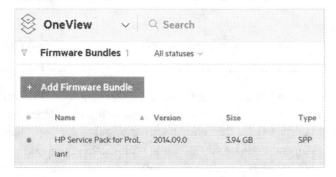

Figure 2-58 Driver and firmware updates

The firmware repository in HPE OneView allows you to manage multiple versions of SPP. An SPP is a comprehensive collection of firmware and system software components (including drivers, agents, utilities, firmware packages for ProLiant servers, controllers, storage, blades, enclosures, and other options). SPP collections are all tested together as a single solution stack.

Firmware bundles enable you to update firmware on servers blades and their infrastructure (enclosures and interconnects). An on-appliance firmware repository enables you to upload SPP firmware bundles, as shown in Figure 2-58, and deploy them across the environment according to best practices.

HPE OneView deploys SPP to provide automatic firmware updates for a variety of uses:

- Identify firmware compatibility issues.
- Set a firmware baseline on devices to establish a desired firmware state.
- Add devices while performing minimum required firmware checks and highlighting out-of-compliance devices for updates with the selected firmware baseline.

- Update firmware for an entire enclosure, or individually for components.

- Apply firmware baselines to servers as a part of the server profile while allowing specific servers to differ from an enclosure's baseline.

HPE OneView leverages and extends the Smart Update portfolio of SPP and HPE SUM, which can systematically update ProLiant servers and blade infrastructures with one click. HPE OneView extends these capabilities with software-defined approaches and with firmware baselines for efficient, reliable, nondisruptive, and simple firmware management across the data center.

Firmware updates in HPE OneView are driven by server profiles using HPE SUM for efficiency at scale. Firmware update operations do not impact the production LAN in any way because they are performed entirely through the management LAN. These same processes can be used to simplify configuration change management across your data center.

 Note

Driver and firmware updates are only available for customers running HPE OneView Advanced.

BIOS settings and firmware and driver updates can be made within an HPE OneView template and then propagated out to the server profiles created from that template. Templates provide a monitor-and-flag model. Profiles created from the template are monitored for compliance with the desired configuration. When inconsistencies are detected, the profile is flagged as no longer compliant with the template. When a new update is made at the template level, all profiles parented to that template are flagged as not compliant. From there, the administrator can bring individual or multiple nodes into compliance with the template.

Items that can be updated from a template include:

- Firmware baseline

- BIOS settings

- Local RAID settings

- Boot order

- Network and shared storage configurations

The profile can be brought into compliance from the GUI or through scripting by using Windows PowerShell, Python, or the RESTful API.

Note

> Some updates such as firmware changes require a server reboot. This reboot is not required at the time that the firmware is updated and can occur at a later time. For example, it can be scheduled during a standard maintenance reboot.

Profile compliance with the template is evaluated every time the profile or template is modified and a notification is generated automatically when a compliance issue is detected. The IT administrator has full control over remediation and can choose to update the profile from the template, resolve the inconsistency by editing the server profile directly, or dismiss the compliance warning.

HPE OneView integrations

Typically, shifting from one management tool to another, each with a partial view of available data, is both time consuming and complex. HPE OneView includes integrations that reduce the time needed to make important administrative changes. These integrations provide additional support for partner management platforms.

In addition to providing server administrators with the tools to drive day-to-day operations, HPE server management offers out-of-the-box integration with leading enterprise management solutions from HPE, VMware, Microsoft, and Red Hat, with more than 65 points of integration that help eliminate complexity and enable a single console experience.

The partner integrations available with each HPE OneView license include:

- HPE OneView for VMware vCenter, VMware vCenter Operations Manager, and VMware vRealize Log Insight

- HPE OneView for Microsoft System Center

- HPE OneView for Red Hat Enterprise Virtualization

 Note

> These integration products are not included with HPE OneView and must be purchased separately.

HPE OneView for VMware vCenter, Operations Manager, and Log Insight

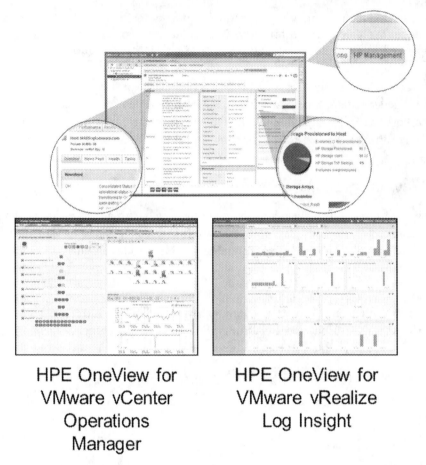

HPE OneView for
VMware vCenter
Operations
Manager

HPE OneView for
VMware vRealize
Log Insight

Figure 2-59 HPE OneView for VMware vCenter, Operations Manager, and Log Insight

HPE OneView for VMware vCenter seamlessly integrates the manageability features of ProLiant servers, BladeSystem Virtual Connect, and storage with VMware solutions as illustrated by Figure 2-59. It reduces the time needed to make changes, increase capacity, or manage planned and unplanned downtime.

The following VMware integrations are available as part of each HPE OneView license:

- **HPE OneView for VMware vRealize Operations** provides health, utilization, and performance metrics in the context of the HPE hardware hierarchy so administrators can monitor critical changes in VMware vRealize Operations Manager. Its dashboards facilitate the identification of root causes of problems and impacted resources across the Converged Infrastructure.

- **HPE OneView for VMware vCenter Log Insight** allows deep troubleshooting of your environment by analyzing unstructured data contained in iLO and Onboard Administrator (OA) logs. Information is displayed in the dashboards of VMware vCenter and vRealize Log Insight, allowing counts of critical events to be quickly identified and investigated.

- **HPE OneView for VMware Operations Manager** reveals critical trend changes. It includes dashboards that facilitate the identification of root causes of problems and impacted resources across the data center.

 Note

HPE OneView Advanced licensing allows you to use all of these VMware integrations. These integrations can be downloaded online. For more information, scan the QR code or enter the URL into your browser.

www.hpe.com/go/ovvcenter

 Note

You can use the HPE OneView for VMware vRealize Operations integration with an existing VMware vRealize Operations Manager Standard version by incorporating the limited usage VMware vRealize Operations Manager Advanced entitlement, which is provided with the purchase of HPE OneView Advanced. If you are using an earlier 5.2.x version of VMware vCenter Operations Manager, then you will need a specific license key. This license key can be obtained by registering the Partner Activation Code (PAC) from your HPE OneView license at the VMware portal.

Video demonstration—using HPE OneView for vCenter plugin for on-premises management

Watch this eight-minute demonstration that shows the management capabilities of the HPE OneView for vCenter plug-in. It demonstrates how to create a vCenter cluster by using the VMware vSphere Web Client, and how to use the HPE OneView for vCenter plug-in to grow the cluster by deploying an additional ESXi host in the environment.

Note

To view this demonstration, scan this QR code or enter the URL into your browser.

https://youtu.be/ngMKQRABQ04

HPE OneView for Microsoft System Center

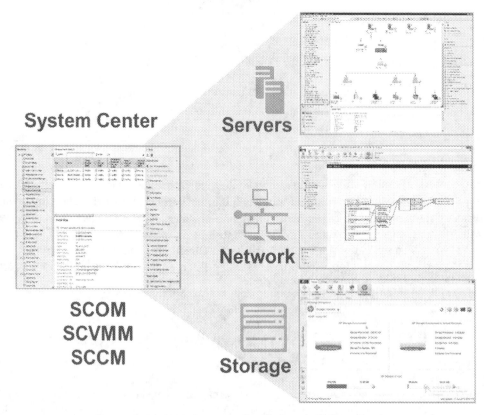

Figure 2-60 HPE OneView for Microsoft System Center

HPE OneView integrates with Microsoft System Center Server, as shown in Figure 2-60, to deliver powerful HPE hardware management capabilities directly from System Center consoles for comprehensive system health and alerting, driver and firmware updates, operating system deployment, detailed inventory, and HPE fabric visualization.

System Center consoles

The Microsoft System Center is a collection of extensions that expose HPE management features within the context of the System Center consoles:

- **System Center Virtual Machine Manager (SCVMM)**—SCVMM provides end-to-end HPE fabric visualization for virtualized environments using Virtual Connect to view from the virtual machine to the edge of the network configuration information. This support component also facilitates consistency and improves uptime with simplified driver and firmware updates through a rotating, automated workflow using the HPE ProLiant Updates Catalog.

  HPE OneView for Microsoft System Center also integrates HPE storage into SCVMM with server profile-based deployment and cluster expansion, cluster views, and additional Fibre Channel and FCoE information. Integration of HPE servers, Virtual Connect, and HPE 3PAR StoreServ storage into the SCVMM user interface enables fabric visualization and enhanced bare-metal deployment using HPE OneView server profiles. It also provides enhanced provisioning using HPE OneView server profiles to deploy Hyper-V hosts consistently and reliably, including configuration of Windows networking, Virtual Connect, and shared SAN storage. HPE OneView integration with Microsoft System Center also identifies mismatched cluster node configurations in the SCVMM, including both Windows networking and Virtual Connect.

- **System Center Operations Manager (SCOM)**—The HPE OneView for Microsoft System Center integration now adds HPE storage integration for SCOM with storage health information using HPE OneView as the data source. SCOM enables you to proactively monitor and manage hardware health and intelligently respond to hardware events on servers running Windows and Linux, as well as BladeSystem enclosures and Virtual Connect. This can be done for ProLiant Gen8 and Gen9 servers without the need for loading operating system-based SNMP agents or Web-Based Enterprise Management (WBEM) providers. This plug-in can also be used to monitor the health of servers that do not have an operating system loaded, as well as ProLiant Gen8 and Gen9 servers running any operating system that has a supported Agentless Monitoring Service (such as VMware ESXi).

- **System Center Configuration Manager (SCCM)**—SCCM provides quick and reliable Windows deployment to bare-metal HPE servers, including predeployment hardware and BIOS configuration along with postoperating system HPE driver and agent installation. You can ensure consistency and maximize uptime with simplified Windows driver and firmware updates using the HPE ProLiant Updates Catalog. In addition, the plug-in provides detailed component level inventory of every managed HPE Windows server using the HPE ProLiant Inventory Tool.

Note

HPE OneView Advanced licensing allows you to download and use HPE OneView for Microsoft System Center. For more information, scan this QR code or enter the URL into your browser.

http://www8.hp.com/uk/en/products/server-software/product-detail. html?oid=5390822

HPE OneView for Red Hat Enterprise Virtualization

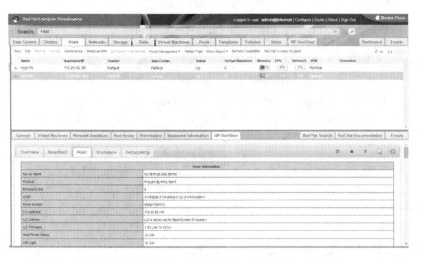

Figure 2-61 HPE OneView for Red Hat Enterprise Virtualization

The HPE OneView integration for Red Hat Enterprise Virtualization (RHEV), shown in Figure 2-61, is a user interface plug-in that seamlessly integrates the manageability features of ProLiant, BladeSystem, and Virtual Connect within the RHEV management console. This integration gives administrators control over their HPE infrastructure and their Red Hat virtualized environment from a single screen.

HPE OneView for RHEV provides the following capabilities from within the RHEV management console:

- Displays host and infrastructure inventory and health information on a single screen, reducing IT administration time and effort

- Displays network relationships from the RHEV-defined network to the physical switch

- Provides single sign-on capability to trusted HPE infrastructure tools, including Onboard Administrator, iLO, and Virtual Connect

- Offers a single screen display of host firmware including:

  - System ROM

  - iLO

  - Smart Array controllers and power management controller

  - Onboard Administrator

  - Virtual Connect module

- Logs events and delivers notifications in a dedicated newsfeed

---

Note

To download HPE OneView for RHEV, scan this QR code or enter the URL into your browser.

www.hpe.com/go/ovrhev

On-cloud management with HPE Insight Online

Figure 2-62 HPE Insight Online and Support Center

HPE Insight Online provides one-stop, secure access to the information needed to support the devices in IT environments with standard warranty and contract services. It is a new addition to the HPE Support Center portal for IT staff who deploy, manage, and support systems, in addition to HPE authorized resellers who support IT Infrastructure.

Through the HPE Support Center, shown in Figure 2-62, Insight Online can automatically display devices remotely monitored by HPE. It enables you to easily track service events and support cases, view device configurations and proactively monitor HPE contracts and warranties. This allows IT support staff and HPE authorized services partners to be more efficient in supporting HPE environments. In addition, they can do all this from anywhere and at any time. It is ideal for channel partners providing proactive remote support.

The embedded management capabilities built in to ProLiant Gen8 and Gen9 servers have been designed to seamlessly integrate with Insight Online and Insight Remote Support 7.0 and later.

Insight Online features include:

- Remote management to monitor, manage, and support IT infrastructure from outside the firewall

- Personalized dashboard for health monitoring, service alerts and support status

- Proactive contracts and warranty management

- Auto-device discovery links device and support information for one-stop access

- Seamless integration into HPE Support Center and Insight Remote Support 7.0

Learning check

1. What is the HPE RESTful API?

   _____

   _____

   _____

2. What are components of the HPE Smart Update Solution? (Select three.)

   A. Smart Components

   B. Service Pack for ProLiant

   C. Secure Encryption

   D. Smart Update Manager

   E. SmartCache

3. A customer is running Microsoft Windows Server 2012 on a ProLiant Gen9 server and needs to access the UEFI System Utilities. What must the customer do?

   A. Press **Control+B** and select **HPE UEFI System Utilities**.

   B. Reboot the server and press **F9** when the POST screen appears.

   C. Select **HPE UEFI System Utilities** from the System Management Homepage.

   D. Reboot from the SmartStart media and select **UEFI System Utilities**.

   E. Boot to the UEFI shell, open an iLO connection, enter the user name and password, and press **Enter**.

4. The SSA Diagnostics screen provides access to which functions? (Select two.)

   A. Storage RAID Report

   B. DDR4 ECC Report

   C. Power Consumption Report

   D. SmartSSD Wear Gauge report

   E. Array Diagnostic Report

5. What does HPE OneView use as the communication channel between application processes and the underlying systems that it supports for management?

   A. State change message bus (SCMB)

   B. Configuration management database (CMDB)

   C. SNMP traps

   D. Open Virtualization Appliance (OVA)

Learning check answers

1. What is the HPE RESTful API?

   - **REST is an industry-recognized architectural style for server standardized interaction to configure at scale using an HTTPS Web protocol.**

   - **The HPE RESTful API is a management interface that server management tools can use to configure, inventory, and monitor a ProLiant Gen9 server using iLO 4.**

   - **It is an architectural style consisting of a coordinated set of architectural constraints applied to components, connectors, and data elements within a distributed hypermedia system.**

2. What are components of the HPE Smart Update Solution? (Select three.)

   A. Smart Components

   B. Service Pack for ProLiant

   C. Secure Encryption

   D. Smart Update Manager

   E. SmartCache

3. A customer is running Microsoft Windows Server 2012 on a ProLiant Gen9 server and needs to access the UEFI System Utilities.

   What must the customer do?

   A. Press **Control B** and select **HPE UEFI System Utilities**.

   B. **Reboot the server and press F9 when the POST screen appears.**

    C. Select **HPE UEFI System Utilities** from the Server Management Homepage.

    D. Reboot from the SmartStart media and select **HPE UEFI System Utilities**.

4. The SSA Diagnostics screen provides access to which functions? (Select two.)

    A. Storage RAID Report

    B. DDR4 ECC Report

    C. Power Consumption Report

    D. SmartSSD Wear Gauge report

    E. Array Diagnostic Report

5. What does HPE OneView use as the communication channel between application processes and the underlying systems that it supports for management?

    A. State change message bus (SCMB)

    B. Configuration management database (CMDB)

    C. SNMP traps

    D. Open Virtualization Appliance (OVA)

Summary

- The ProLiant Gen9 server management portfolio provides agile management for accelerating IT service delivery.

- HPE server management capabilities are specifically designed to manage the entire HPE server portfolio, from towers to racks to blades. HPE Converged Infrastructure management covers the lifecycle of critical operations: configuration and provisioning for rapid deployment, system health monitoring with proactive failure notification, firmware updates, and automated simplified support management.

- ProLiant Gen9 management innovations target three environments:

  - On-system

  - UEFI

    - HPE REST API and HPE RESTful Interface Tool

    - HPE Intelligent Provisioning

- HPE iLO 4
- HPE SUM
- HPE SPP
- Agentless management
- SSA
- On-premises—HPE OneView
- On-cloud
  - HPE Insight Online
  - HPE Insight Remote Support

3 Server Technologies

WHAT IS IN THIS CHAPTER FOR YOU?

After completing this chapter, you should be able to:

✓ Provide a high-level overview of technologies in the Hewlett Packard Enterprise (HPE) ProLiant Gen9 server portfolio

✓ Describe the features of HPE servers in the following areas:

- Processors

- Memory

- Storage

- Networking

OPENING CONSIDERATIONS

Before proceeding with this section, answer the following questions to assess your existing knowledge of the topics covered in this chapter. Record your answers in the space provided here.

1. How familiar are you with HPE ProLiant Gen9 server technologies and innovations?

_____

_____

_____

_____

_____

2. What insights can you offer into third-party server technologies? How do these compare to server technologies from HPE?

_____

_____

_____

_____

_____

_____

3. What experience have you had with power and cooling technologies for ProLiant Gen9 servers?

_____

_____

_____

_____

_____

_____

_____

ProLiant Gen9 portfolio

ProLiant Gen9 servers are an integral part of the HPE portfolio, offering convergence, cloud-ready design, and workload optimization. Specifically, these servers deliver better software-defined and cloud-ready capabilities with HPE OneView enhancements, Unified Extensible Firmware Interface (UEFI) and the Representational State Transfer (REST) application programming interface (API), and integration with Microsoft and VMware software tools.

To meet growing business demands, HPE ProLiant Gen9 rack servers redefine compute economics by delivering more compute and storage capacity, along with less compute energy and floor space consumption. ProLiant rack servers provide faster compute, memory, and IO performance, coupled with increased storage and networking performance—including lower latency.

The ProLiant DL Gen9 rack server portfolio includes these models:

- DL580 Gen9

- DL560 Gen9

- DL380 Gen9

- DL360 Gen9

- DL180 Gen9

- DL160 Gen9

- DL120 Gen9

- DL80 Gen9

- DL60 Gen9

- DL20 Gen9

The ProLiant ML Gen9 tower server portfolio includes these models:

- ProLiant ML10 Gen9

- ProLiant ML30 Gen9

- ProLiant ML110 Gen9

- ProLiant ML150 Gen9

- ProLiant ML350 Gen9

ProLiant Gen9 features

ProLiant Gen9 servers advance convergence with storage virtualization enhancements such as HPE StoreVirtual VSA, HPE Smart Storage, and HPE SmartMemory solutions. They deliver improved workload-optimization capabilities with HPE PCIe Workload Accelerators.

The ProLiant Gen9 solution stack includes a common modular architecture, form factors, management software, services, storage, and networking. These solutions can help customers shift from a server-centered past to a workload-optimized future.

Customers choose ProLiant servers for a variety of reasons. HPE has the broadest server portfolio in the market. Each ProLiant server—from the entry-level server to the most scalable server—is engineered to provide meaningful, cutting-edge benefits to customers for their increasingly complex environments. ProLiant servers provide the right compute resources, for the right workload, at the right economics.

The ProLiant Gen9 server portfolio offers:

- Compute

  - Intel Xeon processors

- Networking

  - Tunnel offload

  - Remote direct memory access (RDMA) over Converged Ethernet

  - Virtual extensible LAN (VXLAN)/Network Virtualization using Generic Routing Encapsulation (NVGRE)

- 1 Gb/10 Gb/40 Gb Ethernet
- Fourteen Data Rate (FDR) InfiniBand
- 20 Gb FlexFabric adapters on BL servers
- Embedded LAN on motherboard (LOM) on DL servers

- Security
  - UEFI Secure Boot
  - HPE Secure Encryption

- Memory
  - HPE DDR4 SmartMemory registered DIMM (RDIMM)/load-reduced DIMM (LRDIMM)

- Storage
  - 12 Gb/s Smart Array controllers
  - 12 Gb/s Smart host bus adapters (HBAs) on blades
  - 12 Gb/s SAS Expander Card
  - 12 Gb/s SAS hard disk drives (HDDs)/solid state drives (SSDs)
  - PCIe Workload Accelerators
  - HPE SmartCache with SSD
  - HPE StoreVirtual VSA
  - HPE Smart Storage Battery

- Embedded and converged management
  - UEFI
  - HPE RESTful Interface Tool
  - HPE Smart Update Manager (SUM) 7.1.0 and HPE integrated Lights-Out (iLO) powered by iLO Federation
  - Location Discovery Services
  - HPE OneView
  - HPE Insight Control server provisioning (ICsp)
  - HPE Insight Online
  - Remote access mobile apps

- Flexibility
    - Universal Media Bay
    - Embedded/Flexible LOM
- Services
    - HPE Technology Services
    - HPE Care Pack Services
    - HPE Proactive Care Services

Processor support in ProLiant Gen9 servers

Designed to support up to 22 cores and memory speeds of up to 2400 MT/s, the Xeon E5-2600 v4 processor offers up to 21% performance gains for ProLiant Gen9 servers when compared with v3 processors. The Xeon E5-2600 v4 processors include virtualization enhancements, added security, and improved orchestration capabilities to help customers better manage shared platform resources. With these new processors, HPE has achieved several new #1 server benchmark performance positions in multiple categories, including results with Big Data TPC Express Benchmark Big Bench (TPCx-BB), SPECjbb2015-Composite, SPECjbb2015-Distributed, and TPC-H.

 Note

For more information on HPE benchmark results, scan this QR code or enter the URL into your browser.

http://www.hpe.com/servers/benchmarks

Processors supported in ProLiant Gen9 servers (depending on model) are:

- The Xeon processor E3 family is available with up to 4 cores and is designed to deliver the best combination of performance, built-in capabilities, and cost-effectiveness in file and print, messaging, collaboration workloads, and cloud environments. This processor family is available in the ProLiant DL20 Gen9 server.

 Note

Intel Pentium and Core i3 processors are also available in the ProLiant DL20 Gen9 server.

- The Xeon processor E5 family is available with up to 22 cores and is available in ProLiant DL60, DL80, DL120, DL160, DL180, DL360, DL380, and DL560 Gen9 servers. It is designed to deliver agile services for cloud and traditional applications and workloads with versatility across diverse workloads, including:

  - Cloud deployments that require scalability, agility, and orchestration capabilities across compute, network, and storage

  - Improved bandwidth and reduced latency for the most demanding high-performance computing (HPC) workloads and applications

  - Cloud-based network architectures supporting high throughput, low latency, and agile delivery of network services such as network functions virtualization and software-defined networking

  - Intelligent and complex storage systems requiring high performance, increased memory, and greater IO bandwidth

- The Xeon processor E7 family is available with up to 24 cores and is available in the ProLiant DL580 Gen9 server. It is designed for the scalable performance demands of complex, data-demanding workloads such as in-memory databases and real-time business analytics. These solutions accelerate performance across the data center to deliver real-time, business-critical services for the largest workloads. Target workloads include:

  - In-memory analytics applications

  - Traditional databases, enterprise resource planning (ERP), data warehousing, and online transaction processing (OLTP) applications

Intel Xeon Phi coprocessors

Figure 3-1 Intel Xeon Phi coprocessor

Intel Xeon Phi coprocessors enable seamless integration of accelerator computing with ProLiant servers for high-performance computing and large data center, scale-up deployments. The Xeon Phi 5110P coprocessor (Figure 3-1) has a peak performance of 1.011 teraflops (double-precision) with 8 GB of ECC-capable memory. The Xeon Phi 7120P coprocessor has a nominal peak performance of 1.208 teraflops (double-precision) with 16 GB of ECC-capable memory, and the capacity to exceed this nominal performance through use of a turbo clock mode.

You can use Xeon processors and Phi coprocessors together to optimize performance for almost any workload. Phi coprocessors complement the performance and energy-efficiency of the Xeon processor E5 family to make dramatic performance gains of up to 1.2 teraflops per coprocessor possible. Each coprocessor features many more and smaller cores, many more threads, and wider vector units.

Certain highly parallel applications can benefit by using Xeon Phi coprocessors. To take full advantage of Phi coprocessors, an application must scale well to more than one hundred threads and either make extensive use of vectors or efficiently use more local memory bandwidth than is available on a Xeon processor. The high degree of parallelism compensates for the lower speed of each individual core to deliver higher aggregate performance for highly parallel code.

The HPE Insight Cluster Management Utility has incorporated the Xeon Phi environmental sensors into its monitoring features so that cluster-wide accelerator data can be presented in real time, can be stored for historical analysis, and can be easily used to set up management alerts.

Memory features of ProLiant servers

IT trends such as server virtualization, cloud computing, and high-performance computing are placing significant demands on server memory speed, capacity, and availability. An IT system's reliability, performance, and overall power consumption drive companies toward business outcomes. Choosing the right memory is crucial to ensuring high reliability and delivering a faster return on IT investment.

Many businesses need a faster tier of technology to help them deal with current real-world issues such as Big Data, analytics and search workloads, medical sciences such as human genome mapping, and financial data analysis. Traditional data storage technologies are being augmented by new innovations in the hierarchy of data storage.

DDR4 SmartMemory

SmartMemory verifies whether DIMMs have passed the HPE qualification and testing processes and determines if the memory has been optimized to run on ProLiant Gen9 servers. Key technology enhancements offered by DDR4 include:

- **Increased bandwidth**—DDR4 SmartMemory provides up to 2133 MT/s bandwidth for up to a 33% increase in throughput over DDR3 memory.

- **Better data rate**—The DDR4 specification defines eventual data rates of up to 3200 MT/s, more than 70% faster than the 1866 MT/s of DDR3 memory speed.

- **1.2 volt operation**—All DDR4 memory operates at 1.2 volts, compared with 1.35 or 1.5 volts for DDR3 memory. This delivers significant system power savings, particularly in larger memory configurations.

- **16 banks of memory per rank**—Internally, the DRAMs used in DIMMs are organized into arrays of cells defined by banks, rows, and columns. DDR4 memory has 16 banks of memory in a DRAM chip compared with the eight banks in DDR3. This allows an increased number of memory requests that can be queued by the memory controller. It is one of the contributors to the lower latency of DDR4 memory.

- **Encoded Rank Selection**—DDR4 eliminates the work-around known as *rank multiplication* that DDR3 employed to enable four ranks of memory on LRDIMMs using the traditional chip select lines. When there are eight or fewer total ranks installed on a memory channel, DDR4 uses the direct chip select mode to address the correct rank. When more than eight ranks are installed, DDR4 uses a 4-bit encoded chip select value for rank selection. This encoded value is interpreted by the registers on the DIMMs to determine the correct rank to enable for the memory operation. This new encoded chip select scheme allows DDR4 memory to address up to 24 memory ranks on a memory channel.

- **Retry on error**—DDR4 memory and new memory controllers will retry a memory request whenever a memory error or address parity error occurs. This reduces the number of system halts that may have occurred due to transient errors in previous generations of memory subsystems.

 Note

DDR4 and DDR3 memory are not interchangeable.

HPE Advanced Memory Error Detection

As memory capacities increase, increases in memory errors are unavoidable. Fortunately, most memory errors are both transient and correctable. Current memory subsystems can correct up to a 4-bit memory error in the 64 bits of data that are transferred in each memory cycle.

Instead of simply counting each correctable memory error, HPE Advanced Memory Error Detection analyzes all correctable errors to determine which ones are likely to lead to uncorrectable errors in the future. This approach is able to better monitor the memory subsystem and increase the effectiveness of the Pre-Failure Alert notification. All ProLiant Gen9 servers feature Advanced Memory Error Detection.

Comparing RDIMMs and LRDIMMs

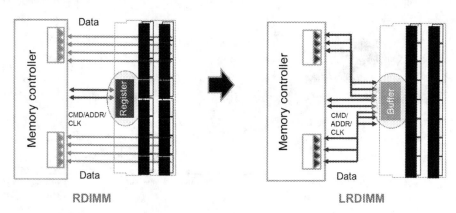

Figure 3-2 RDIMM and LRDIMM compared

As illustrated by Figure 3-2, RDIMMs improve signal integrity by having a register on the DIMM to buffer the address and command signals between the DRAMs and the memory controller. This allows each memory channel to support up to three dual-rank DIMMs, increasing the amount of memory that a server can support. With RDIMMs, the partial buffering slightly increases both power consumption and memory latency.

LRDIMMs use memory buffers to consolidate the electrical loads of the ranks on the LRDIMM to a single electrical load, allowing them to have up to eight ranks on a single DIMM module. The LRDIMM memory buffer reduces the electrical load to the memory controller and allows higher capacity memory to run at three DIMMs per channel. Using LRDIMMs, you can configure systems with the largest possible memory.

LRDIMMs also use more power and have longer latencies compared to the lower capacity RDIMMs. Similar to RDIMMs, LRDIMMs buffer the address and control signals. Unlike RDIMMs, LRDIMMs also buffer the data lines. In RDIMMs, data signals are driven by a controller, limiting performance.

The LRDIMM memory buffer reduces the electrical load to the memory controller and allows higher capacity memory to run at three DIMMs per channel. LRDIMM is ideal for customers who require the maximum memory capacity.

 Note

Although unbuffered DIMMs (UDIMMs) are defined for the DDR4 standard, they no longer offer any performance advantage (in terms of lower latencies) over RDIMMs and LRDIMMs.

Tiers of data storage

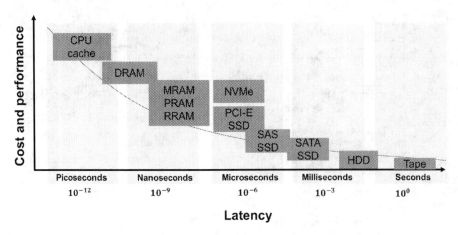

Figure 3-3 Cost, performance, and latency of storage technologies

Figure 3-3 shows the trade-off between latency and cost and performance that customers face when trying to match data storage technologies to workloads. These technologies range from CPU cache to hard drive and tape. Memory-based storage technologies lie along this curve.

NVM Express (NVMe) refers to a nonvolatile interface designed for SSDs. NVMe bypasses the need for an HBA to connect directly to flash-based devices through the PCIe bus, thereby reducing latency and power consumption. HPE NVMe devices attach directly to select ProLiant Gen9 servers and are ideal for consolidating enterprise databases to reduce licensing costs.

Other persistent storage on the IO bus includes:

- **Magnetoresistive random access memory (MRAM)**—MRAM has much lower power consumption than DRAM. It is faster than flash memory and does not degrade over time. With MRAM, data is stored by magnetic storage elements.

- **Phase-change random access memory (PRAM)**—PRAM is a type of nonvolatile, random-access memory with inherent scalability, which is what makes it attractive. It is based on the same storage mechanism technology as CDs and DVDs. Phase change is a thermally driven process rather than an electronic process. There are still great challenges with PRAM, most notably threshold resistance and voltage drift.

- **Resistive Random Access Memory (RRAM)**—RRAM changes resistance across a material often referred to as a *memristor*. RRAM has the potential to replace flash memory, but it is still in the development stage

Technology pyramid

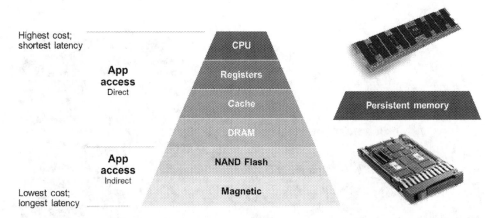

Figure 3-4 Memory and storage hierarchy

Figure 3-4 represents a pyramid of memory and storage convergence. Technology at the top of the pyramid has the shortest latency (best performance) but has a higher cost relative to the technologies at the bottom of the pyramid. These are the technology layers comprising DRAM (memory) and the CPU cache and registers. All of these components are accessed directly by the application—also known as *load/storage access*.

Technology at the bottom of the pyramid—represented by magnetic media (HDDs and tape) and NAND flash (represented by SSDs and PCIe Workload Accelerators)—have longer latency and lower costs relative to the technology at the top of the pyramid. These technology components have block access meaning data is typically communicated in blocks of data and the applications are not accessed directly.

Nonvolatile memory is not shown in this pyramid. This layer sits between NAND flash and DRAM, providing faster performance relative to NAND flash but also providing the nonvolatility not typically found in traditional memory offerings. This technology layer provides the performance of memory with the persistence of traditional storage.

Storage tiering on HPE ProLiant servers—Without persistent memory

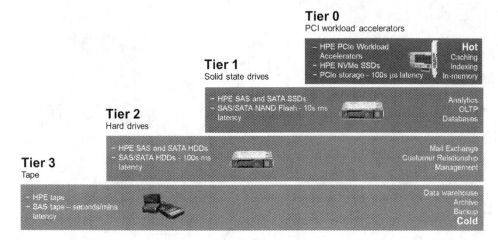

Figure 3-5 Storage tiering on HPE ProLiant servers

Within the concept of storage tiering, Tier 0 is the fastest storage tier and Tier 3 has the lowest performance requirements. On the hot (frequently accessed) and cold (less frequently accessed) data spectrum, performance, capacity, and cost all must be considered. Typically, hotter data has a greater cost per GB and lower capacity requirements relative to colder data, which usually has a lower cost per GB and greater capacity requirements. Typical workloads fall along the hot and cold spectrum, with a large proportion of data residing in the warm tier (Tier 1) represented by analytics, online transaction processing (OLTP), and databases.

Figure 3-5 illustrates a traditional storage tier represented by the following devices:

- **Tier 0**—PCIe SSDs and PCIe workload accelerators (NAND flash on the SAS/SATA or PCIe bus); 10s to 100s of microseconds (μs) of latency

- **Tier 1**—SAS and SATA SSDs; 10s of milliseconds (ms) of latency

- **Tier 2**— SAS HDDs (highest-performing rotational media with lower capacities and greater cost per GB relative to SATA); 10s of ms of latency. SATA HDDs (higher capacity relative to SAS and lower cost per GB but lower performance relative to SAS); 100s of ms of latency

- **Tier 3**—Tape or archival medium (lowest cost per GB for customers looking for the lowest cost per GB and highest capacity for cost-effective archiving); seconds to minutes of latency

Storage tiering on HPE ProLiant servers—With persistent memory

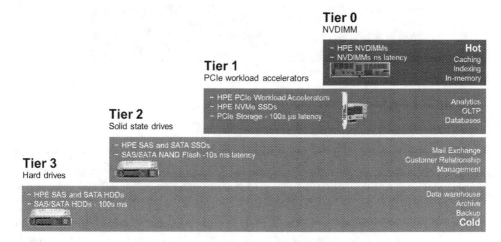

Figure 3-6 Storage tiering with persistent memory

Figure 3-6 looks at emerging storage technologies including persistent memory, compared with the view on the preceding page. NVDIMMs are an example of the type of product in the persistent memory product category. The storage tier hierarchy in Figure 3-6 is represented as follows:

- **Tier 0**—NVDIMMs; nanoseconds (ns) of latency

- **Tier 1**—PCIe SSDs and PCIe workload accelerators (NAND flash on the SAS/SATA or PCIe bus); 10s to 100s of μs of latency

- **Tier 2**—SAS and SATA SSDs; 10s of ms of latency

- **Tier 3**—SAS and SATA HDDs; 10s to 100s of ms of latency

NVDIMMs do not replace NAND flash. NVDIMMs supplanted PCIe NAND flash as the fastest storage tier. And although tape dropped off from this example, you could have another tier for archiving data onto tape. NVDIMM is emerging as the fastest storage tier available in the market and part of an overall tiering strategy that includes HPE PCIe Workload Accelerators, SSDs, HDDs, and tape.

HPE persistent memory

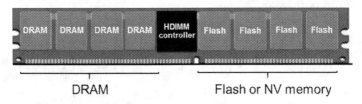

DRAM   Flash or NV memory

Figure 3-7 NVDIMM

Persistent, nonvolatile DIMM (NVDIMM) technology combines the speed and long life of DRAM with the persistent storage of flash memory, resulting in increased system performance and reliability. Also called *hybrid DIMMs* (HDIMMs), NVDIMMs provide performance and cost advantages for a range of enterprise-class server and storage applications. NVDIMMs can be used for broad application acceleration and are ideal for hyperscale computing environments focused on cloud computing, big data analytics, high-performance databases, and low-latency applications such as high-frequency trading.

Figure 3-7 illustrates how NVDIMM technology moves storage closer to the memory bus. The chips on the left side of the DIMM use DRAM technology to store data in standard memory. The memory on the right side of the DIMM stores data in persistent (flash) memory, meaning the data is retained even after the server is powered down. The chip in the middle is the controller that enables the transition between the two types of memory.

DRAM is fast and reliable but cannot retain data when power is lost. In addition, DRAM costs more than traditional storage media such as NAND flash. NAND flash technology, which is used in SSDs and other storage devices, retains data without power and is more cost-effective than DRAM but has lower performance and endurance relative to DRAM.

When persistent memory is used as a traditional storage device, persistence is achieved by combining the speed and endurance of DRAM with the nonvolatility of NAND flash. HPE persistent memory is nonvolatile memory that delivers the performance of memory with the persistence of traditional storage.

DDR4 NVDIMMs are supported with a special controller that is the main interface to the processor. They also have a separate flash controller. If power is lost, the NVDIMM controller moves data from the DRAM to its own onboard flash. The HPE Smart Storage Battery provides a backup power source in the event of a surprise power loss allowing any data in flight on the DRAM to be moved to the nonvolatile NAND Flash memory.

 Note

There is a trade-off between the advantages of persistent memory and total RAM size, as a result of some of the physical space on the memory module being taken up by flash memory, rather than DRAM.

The HPE 8 GB NVDIMM module is the first offering in the HPE persistent memory product category.

HPE persistent memory offerings are not just new hardware technology but a complete software ecosystem designed to work with today's applications and workloads, including databases and analytics workloads. HPE is working actively with companies such as Microsoft, Red Hat, SuSE, and Hortonworks to enable software drivers and software development kits (SDKs) as well as workload optimization around persistent memory.

 Note

For more information on HPE persistent memory, scan this QR code or enter the URL into your browser.

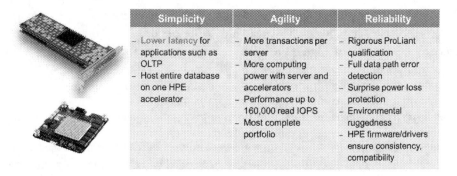

http://www8.hp.com/us/en/products/servers/qualified-options/persistent.html

---

HPE PCIe Workload and IO Accelerators

Simplicity	Agility	Reliability
– Lower latency for applications such as OLTP – Host entire database on one HPE accelerator	– More transactions per server – More computing power with server and accelerators – Performance up to 160,000 read IOPS – Most complete portfolio	– Rigorous ProLiant qualification – Full data path error detection – Surprise power loss protection – Environmental ruggedness – HPE firmware/drivers ensure consistency, compatibility

Figure 3-8 HPE PCIe Workload and IO Accelerators

As shown in Figure 3-8, HPE PCIe Workload and IO Accelerators for ProLiant servers are PCIe card-based direct-attach solid-state storage solutions offering high performance with low latency and enterprise reliability. They are ideal for enterprise workloads that require real-time data access with high transaction rates and low latency.

PCIe Workload Accelerators are available for ProLiant ML and DL servers and increase application performance by moving storage closer to the CPU and have a latency of about 30 µs compared to 100 ms for array controllers.

These workload accelerators greatly increase performance for the following applications:

- Enterprise resource planning (ERP)
- Microsoft Exchange
- Microsoft SharePoint
- Business intelligence and data warehousing
- Virtualization
- Multimedia
- Medical imaging

The HP IO Accelerator for BladeSystem c-Class is part of the comprehensive HPE solid-state storage portfolio. This device is targeted for markets and applications requiring high transaction rates and real-time data access that will benefit from application performance enhancement. It offers the equivalent storage performance of multiple disks. The HP IO Accelerator brings high IO performance and low latency access to storage, with the reliability of solid state.

Customers will benefit from high speed, augmenting existing NAS/SAN HDD-based solutions offered by the HPE IO Accelerator.

The HPE IO Accelerator brings high IO performance and low latency access to storage, with the reliability of solid state. It provides high read and write performance and accelerated application performance for customers with high IO requirements who buy large arrays just to get the performance from a large number of drives.

Storage features of ProLiant servers

As data storage and accessibility requirements grow, customers need solutions that can help overcome performance bottlenecks. In addition to the aforementioned workload and IO accelerators, storage options for ProLiant Gen9 servers include HDDs, SSDs, and Smart Array controllers. These offerings provide customers hassle-free performance, outstanding reliability, and exceptional quality. Backed by more than 2.4 million hours of the industry's most rigorous testing and qualification programs, there is a solution to fit any application workload.

Customer storage challenges

Making IT a strategic enabler of the business has never been more challenging. Data storage requirements are growing exponentially, along with government regulations for protecting sensitive data. This means that storage solutions need to meet a variety of needs. As an example, every minute:

- 217 new mobile users are online

- More than 168 million emails are sent

- 1879 TB data is created

- 698,455 Google searches are performed

- 11 million instant messages are sent

It is currently being predicted that IT systems will create a new yottabyte (10^{24} bytes) of data a year by 2020. It is being predicted that systems will create a new brontobyte (1000 yottabytes) of data a year by 2025.

Customers have several requirements regarding how to manage their complex and expanding storage infrastructure. They need to:

- Manage growing storage with fixed IT resources

- Proactively identify and resolve storage bottlenecks

- Have a single console for deployment, configuration, performance monitoring, and maintenance

---

 Note

For many years, the capacity of storage devices has been described using megabytes (MB), gigabytes (GB), and terabytes (TB). Similarly, throughput has been historically measured in megabits (Mb) and gigibits (Gb) per second. As the scale of devices expands, the difference between the generally used terms (MB, GB, and TB) and their engineering equivalents mebibyte (MiB), gibibyte (GiB), and tebibyte (TiB) grows at an alarming pace. To more accurately represent throughput and storage capacity, the use of mibibit (Mib), gibibit (Gib), tebibyte (TiB), MiB, GiB, and TiB is accepted.

HPE ProLiant Gen9 server drive technologies

SAS SSD   SATA HDD   MicroSD Mainstream Flash Media Kit   SATA DVD-RW

Figure 3-9 HPE ProLiant Gen9 server drive technologies

With some storage requirements escalating and others becoming more complex, factors such as flexibility, performance, increased reliability, greater density, security, scalability, and accessibility are more critical than ever. Today's organizations consist of different kinds of environments. Enterprise data centers must be online 24 x 7, fulfill requests from numerous users simultaneously, and allow for constant growth and expansion while in operation. Other customer environments require high capacity storage and high data availability for low IO environments. The HPE portfolio of drives meets these demands.

All HPE drives pass a rigorous qualification process, which certify that every HPE drive is proven to perform in a ProLiant server environment. As shown in Figure 3-9, internal drive options for ProLiant Gen9 servers include:

- **Solid state drives**—HPE SSDs deliver exceptional performance and endurance and reduce power consumption for customers with applications requiring high random read and write IOPs performance. HPE SSDs are categorized as read intensive, mixed use, and write intensive so that you can choose the SSD tailored to the demands of the workload. Available as SFF and LFF hot-plug devices, nonhot plug SFF devices, and SFF quick release devices, these drives deliver better performance, better latency, and more power-efficient solutions when compared with traditional rotating media.

  - **Write intensive solid state drives**—HPE write intensive 12G SAS and 6G SATA SSDs provide high write performance and endurance. They are best suited for mission-critical enterprise environments with workloads high in both reads and writes. Workloads best suited for these WI SSDs include online transaction processing (OLTP), virtual desktop infrastructure (VDI), business intelligence, and Big Data analytics.

  - **Mixed use solid state drives**—HPE mixed use 12G SAS and 6G SATA SSDs are best suited for high IO applications with workloads balanced between reads and writes. The SAS and SATA SSDs provide the workload-optimized performance required for demanding IO-intensive applications. When paired with ProLiant servers, these SSDs help meet the challenges of Big Data. They achieve twice the performance and endurance of previous generations of HPE SAS and SATA SSDs.

- **Read intensive solid state drives**—HPE read intensive 12G SAS and 6G SATA SSDs deliver enterprise features for a low price in ProLiant server systems. This entry-level pricing is fueling rapid SSD adoption for read-intensive workloads because the cost per IOPS compares very favorably to HDDs. Read intensive SSDs deliver great performance for workloads high in reads such as boot/swap, web servers, and read caching. The HPE M.2 Solid State Enablement Kit is the newest addition to the HPE Read Intensive solid state drive family and is best suited for boot/swap. The M.2 Solid State Enablement Kit is available in 64 GB and 120 GB capacities with ProLiant Gen9 server blades. The 120 GB and 340 GB capacities supported with ProLiant ML/DL servers are also available.

- **Hard disk drives**—HPE SAS and SATA hard drives are available in both 3.5-inch large form factor (LFF) and 2.5-inch small form factor (SFF) and ship with a standard one-year warranty.

  - **SATA HDDs**—HPE SATA hard drives are built for reliability and larger capacity needs for today's nonmission-critical server applications and storage environments. These high-capacity drives provide the lowest cost per GB, and the best price advantage for nonmission critical applications with low workload duty cycles of 40% or less.

  - **SAS HDDs**—HPE SAS drives satisfy the data center requirements of scalability, performance, reliability, and manageability, and provide a storage infrastructure for both enterprise SAS drives and SATA disk drives. SAS midline drives provide the lowest cost per gigabyte and economical reliability and performance. The SAS interface is compatible with SATA devices. This compatibility provides users with unprecedented choices for server and storage subsystem deployment.

- **Flash media drives**—HPE offers high-performance flash media kits for customers requiring boot-from-flash for integrated hypervisors and first tier operating systems. With high data retention and read write cycles, HPE flash media devices are available in both SD and MicroSD form factors.

- **Optical drives**—Optical drives for ProLiant servers feature an industry-standard SATA interface and are supported on most major operating systems.

  - **SATA DVD ROM optical drives**—The DVD ROM drive is designed to read not only CD ROM and CD R/RW discs but also DVD ROM, DVD RAM, DVD +R/RW, and DVD R/RW discs. HPE optical drives are available in half-height, slim, and super slim form factors.

  - **SATA DVD RW optical drives**—The DVD RW drive can read DVD 4.7 GB through 8.5 GB media, as well as standard stamped, CD-R, and CD-RW media. This drive supports writing to CD-R, CD-RW, DVD +R/RW, and DVD -R/RW media via software utilities. For Microsoft operating systems, this is available by installing the included Roxio software disk. For other operating systems, an operating system-specific software utility is required for writing to media.

HPE Smart Storage solutions for ProLiant Gen9 servers

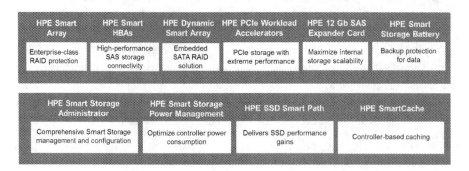

Figure 3-10 HPE Smart Storage solutions

HPE Smart Storage solutions for ProLiant Gen9 servers are built to align IT with business requirements. These solutions help to streamline operations, contain costs, accelerate the delivery of new products and services, and optimize application performance. Smart Storage solutions improve storage utilization while delivering the scalability, reliability, and accessibility required to compete.

Smart Storage encompasses all of the storage technologies built into ProLiant Gen9 servers. As illustrated in Figure 3-10, these technologies include:

- **HPE Smart Array controllers**—HPE offers a complete portfolio of enterprise-class RAID controllers with fault tolerance for ProLiant-attached storage. Designed to enhance server uptime and maintain flexibility for future growth, Smart Array controllers blend the reliability of SCSI with the performance advantages of serial architecture. Providing industry-leading performance with unmatched data protection, these controllers are ideal for companies with direct-attached SAS storage. With support for more than 576 TB of total storage, Smart Array controllers can help customers meet the requirements of a broad range of applications. Moreover, by providing extensive choices for server and storage deployment, these controllers provide high levels of flexibility and return on investment (ROI).

- **HPE Smart HBAs**—Perfect for environments that require fast access, Smart HBAs provide cost-effective and reliable high-performance SAS connectivity to direct-attached storage, shared storage, and tape drives for ProLiant servers running Hadoop, Database Availability Group, and VMware vSAN. Smart HBAs provide a conduit for deploying software-defined storage as a means to manage the IT storage pool. For greater flexibility, Smart HBAs are capable of running in either HBA or simple RAID mode.

- **HPE Dynamic Smart Array**—The Dynamic Smart Array controller provides an embedded serial ATA (SATA) RAID solution for ProLiant Gen9 servers. It eliminates most of the hardware RAID controller components and relocates the RAID algorithms from a hardware-based controller into device driver software. This lowers the total solution cost, while still maintaining comparable RAID protection and full compatibility with Smart Array disk format, configuration utilities, and management/monitoring software. The common metadata format on the drives allows disks to migrate from Dynamic Smart Array to Smart Array or Smart HBA (when running in RAID mode) if needed. This capability helps to achieve higher performance, capacity, and availability. This controller is ideal for supporting boot devices and applications that do not generate massive IO workload.

- **HPE PCIe Workload Accelerators**—PCIe Workload Accelerators for ProLiant servers are PCIe card-based direct-attach solutions. PCIe storage devices provide performance, reliability, and very low latency. With enterprise-class endurance and capacity points up to 6.4 TB, these solutions are ideal for applications and workloads requiring maximized performance.

- **HPE 12 Gb SAS Expander Card**—The SAS expander card allows any of the ProLiant Gen9 DL servers to be configured with their maximum number of drives. The expander card is ideal for users who want to configure RAID for more than eight internal HDDs or add an additional internal drive cage and configure RAID across all internal drives.

- **HPE Smart Storage Battery**—In ProLiant Gen9 servers, a single Smart Storage Battery connected to the system board provides the backup battery power to all of the Smart Array controllers in the system that use flash-backed write cache (FBWC). Each 96 watt Smart Storage Battery in ProLiant ML/DL servers is capable of providing enough power to back up the larger cache sizes (4 GB) found in Gen9 Smart Array controllers.

- **HPE Smart Storage Administrator (SSA)**—SSA is the comprehensive management and configuration application for Smart Storage products and solutions. Available as a stand-alone application, this utility provides advanced scripting and diagnostics capability to simplify and streamline array configuration and management.

- **HPE Smart Storage Power Management**—Optimizes controller power consumption based on both array configuration and workload. Smart Storage Power Management can save several watts on storage controller power consumption without greatly impacting overall storage performance.

- **HPE SSD Smart Path**—With SSD Smart Path, the Smart Storage device drivers analyze each IO request to decide whether it can be executed more quickly through the driver itself or whether it should be passed to the Smart Array firmware for execution as normal IO. It is designed specifically to deliver performance gains for logical drives using SSDs on Smart Array controllers.

- **HPE SmartCache**—In direct-attached storage (DAS) environments, SmartCache uses one or more SSDs as dedicated caching devices for other volumes, increasing storage performance by copying the most frequently accessed data to the low latency SSDs for quicker access that is completely transparent to host applications.

Dynamic Smart Array B140i

All ProLiant Gen9 servers can be purchased with a SATA-only option supporting up to 10 ports. This is delivered in the form of the HPE Dynamic Smart Array B140i, which is the standard embedded storage controller for all ProLiant Gen9 rack servers. It replaces the B120i and B320i controllers that were embedded in ProLiant Gen8 servers.

The Dynamic Smart Array B140i is the entry-level storage controller for ProLiant Gen9 servers, delivering a basic level of storage functionality and performance. It is ideal for supporting operating system boot devices or providing basic protection for data that does not require significant storage performance.

 Note

> The Dynamic Smart Array B140i acts as the default controller only if you do not install the more powerful Smart SAS HBA controller or the Smart Array controller in the system.

The Dynamic Smart Array B140i controller provides SATA HDD, SATA SSD, and SATA M.2 drive support. M.2 SATA and USB-attached, mirrored microSD are devices that can be used to boot the operating system or the hypervisor without having to go out to a drive cage.

The Dynamic Smart Array B140i supports only 6 Gb/s SATA drives using the embedded SATA ports that are part of each ProLiant Gen9 server. On ProLiant Gen9 ML/DL systems, it can support the maximum of 10 SATA drives—an increase from the ProLiant Gen8 embedded controllers.

The Dynamic Smart Array B140i controller embedded on the system board of all ProLiant Gen9 rack servers is upgradeable with an HPE Flexible Smart Array or Smart HBA on ProLiant DL380 and DL360 Gen9 servers (without consuming a PCIe slot) and with a standup PCI adapter on all ProLiant Gen9 rack servers.

Smart Array controllers use the PCIe 3.0 host interface and 12 Gb/s SAS storage interfaces. These controllers also provide active health logging and predictive spare activation and use an embedded RAID-on-Chip (ROC).

 Note

> For more information on the Dynamic Smart Array B140i controller, scan this QR code or enter the URL into your browser.

http://h20564.www2.hpe.com/hpsc/doc/public/display?docId=emr_na-c04406959

Software RAID with the Smart Array B140i

In addition to supporting individually attached SATA drives as standard Advanced Host Controller Interface (AHCI) devices, the B140i also supports RAID operation. Using the SSA, you can configure the B140i and attached SATA drives for RAID 0, RAID 1, RAID 10, or RAID 5 operation.

Unlike the more advanced Smart SAS HBA or Smart Array controllers, the B140i uses driver-based software RAID. All of the RAID functions and calculations are performed by the operating system drivers using server CPU and memory resources. The B140i does not feature write cache capability, and it uses a read cache that it carves from system memory.

The B140i is also dependent on the UEFI and will not operate in Legacy BIOS mode. Using a UEFI driver included in the system ROM, the B140i supports bootable RAID volumes.

Many ProLiant Gen8 servers included a Smart Array card embedded into the system board. For customers who have no need for this Smart Array card, the optional daughter cards provide more flexibility and eliminate the cost of the embedded controller. Customers can choose between P440ar or H240ar controllers on select ProLiant Gen9 rack servers or select P244br or H244br on ProLiant Gen9 server blades.

HPE P-Series controllers

P-Series controllers include support for:

- User configurable power mode

    - **Maximum performance**—Default mode with no power savings

    - **Minimum power**—Maximum power savings and potential for a significant impact on performance

    - **Balanced**—Optimal settings based on configuration and minimal performance impact

- Smart cache

    - Write-back capability for improved write performance

    - Critical for many applications, such as virtual desktop infrastructure (VDI)

    - RAID 0, RAID 1, and RAID 5 support for cache logical unit number (LUN)

    - Added RAID 1 support for cache LUN in write-through mode

    - Seamless encryption support and enabled transformation for noncache volumes

- HBA mode

    - No logical drives

    - Presents all raw drives to applications

    - Benefits customers who choose not to use the protection of RAID in favor of better performance

12 Gb SAS Expander Card

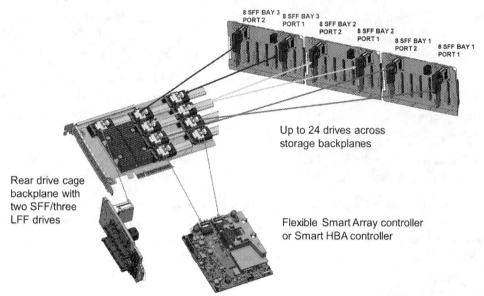

8 SFF BAY 3
PORT 2

8 SFF BAY 3
PORT 1

8 SFF BAY 2
PORT 2

8 SFF BAY 2
PORT 1

8 SFF BAY 1
PORT 2

8 SFF BAY 1
PORT 1

Up to 24 drives across
storage backplanes

Rear drive cage
backplane with
two SFF/three
LFF drives

Flexible Smart Array controller
or Smart HBA controller

Figure 3-11 12 Gb SAS Expander Card

The HPE 12 Gb SAS Expander Card provides internal storage expansion within ProLiant Gen9 DL/ML servers by allowing support for more than eight internal HDDs when connected to a Gen9 supported Smart Array or Smart HBA controller. This full-height card supports 12 Gb/s SAS connectivity and features nine internal ports with a maximum of 26 physical links.

Figure 3-11 shows the internal architecture for a ProLiant DL380 Gen9 server with the 12 Gb SAS Expander Card installed. The DL380 server supports up to 26 SFF drives internally—24 on the storage backplane and two more connected to the rear drive cage backplane.

This expander card is ideal for ProLiant DL380, DL180, and ML350 Gen9 server customers who want to use RAID with more than eight internal HDDs or those who want to add additional internal drive cages and use RAID across all internal drives connected to the controller. It is also well-suited for midrange to enterprise-level customers looking to add internal direct-attached SAS storage for file, messaging, and database applications.

Key features include:

- 12 Gb/s SAS technology delivers high performance and data bandwidth up to 1200 MB/s per physical link and contains full compatibility with 6 Gb/s SATA technology. Mix-and-match SAS and SATA hard drives enable deployment of drive technology as needed to fit in the customer's computing environment.

- Support for up to 26 internal drive bays supports three SFF drive backplanes (with each holding eight drive bays, plus two SFF drives in the back of the server).

- Edge buffering—6 Gb/s drives use buffering to maintain 12 Gb/s back-end speed.

HPE Smart Storage Battery

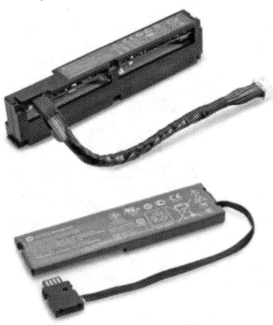

Figure 3-12 HPE Smart Storage Battery

In ProLiant Gen9 servers, a single HPE Smart Storage Battery connected to the system board provides the backup battery power to all of the Smart Array controllers in the system that use FBWC. As shown in Figure 3-12, the Smart Storage Battery for ProLiant ML/DL/SL servers is a single lithium-ion battery pack capable of supporting multiple devices. This battery replaces the individual supercaps used with each controller in Gen8 systems.

This approach for supporting FBWC has several advantages over the use of individual supercaps. The Smart Storage Battery delivers more power for backups. Each 96 watt Smart Storage Battery in ML/DL servers can provide enough power to support backing up the larger cache sizes (4 GB) found in Gen9 Smart Array controllers. Backing up 4 GB of cache to the flash modules can take up to one minute, which is too close to the maximum capabilities of supercaps used with each controller in Gen8 systems. The Smart Storage Battery can also support up to 24 separate devices in the system.

The Smart Storage Battery also simplifies cabling by delivering its power to the system board using a single connection. Daughterboard-based Smart Array controllers that use FBWC draw their battery power directly though their connection to the system board. For P-Series controllers, a single cable from the riser card to the controller provides connection to the battery.

The Smart Storage battery carries a longer life expectancy than earlier solutions. HPE positioned the battery pack in an area of the server that will keep it at a temperature below 50° Celsius, ultimately maximizing the lifespan of the Smart Storage Battery.

Note

For more information on the Smart Storage Battery, scan this QR code or enter the URL into your browser and watch the video.

https://www.youtube.com/watch?v=wQzu7s-hFU4

HPE Smart Storage Administrator

For ProLiant Gen9 servers, SSA continues to evolve as the single control point for configuration, management, and monitoring of Smart Storage. SSA offers a simple and intuitive interface and functionality. Key features of SSA include:

- Array diagnostics

- Support for the HPE SmartSSD Wear Gauge across all Gen9 storage controllers (including AHCI attached drives)

- Configuration and management of advanced Smart Storage functionality—including SmartCache, HPE Secure Encryption, and power management

- Comprehensive management for Smart Storage products

- Simplified and intuitive interface and functionality

 Note

For more information on Smart Storage Administrator, scan this QR code or enter the URL into your browser.

http://www8.hp.com/us/en/products/server-software/product-detail.html?oid=5409020

HPE SmartCache

SSDs have much lower latency and higher performance when compared with traditional rotational hard disk drives (HDDs), but the historical prices of these drives have prevented widespread adoption. As solid-state technology has dropped in price, customers have started investing more into this technology. However, the prospect of an all-SSD solution might still be too big an investment for some customers. SmartCache is the ideal solution for customers looking to invest in SSD technology at a controlled pace to get the benefits of lower latency SSDs without moving to all SSDs.

SmartCache combines different technologies and device types and enables SSDs to be used as caching devices for hard drive media. In the SmartCache architecture, a copy of the data resides on the HDD, as well as on the lower-latency SSD used for caching. The SmartCache architecture includes the following three elements:

- Bulk storage—Any supported HDD or SSD that can be attached to a Smart Array controller.

- Accelerator—A lower-latency SSD device that caches data; the capacity of the accelerator is less than that of the bulk storage device.

- Metadata—Information held in a relatively small storage area of the Flash-Backed Write Cache (FBWC) memory that maps to the location of information residing on the accelerator and bulk storage devices.

SmartCache offers the following benefits:

- Accelerates application performance

- Provides lower latency for transactions in applications

- Supports all operating systems where Smart Array Gen9 controllers are supported, without the need for changes to operating system, driver, or applications

- Offers a choice of write-through or write-back cache (Gen9 servers only)

- Provides seamless integration into a data center

- Simplifies deployment and management using SSA, the same management and configuration application used to manage HPE storage arrays

The ideal environment for SmartCache is a ProLiant Gen8 or later server running a read-intensive workload with repetitive data requests. One example is a database application that frequently accesses specific files such as page files. That type of data is tagged as hot data and sent over to the SSD so that the application can read from the faster-performing SSD drive instead of rotational hard drives.

The amount of data tagged as hot data and sent over to the SSD is called the *cache hit rate*. A cache hit rate of 80% means that 80% of the workload is being read off the SSD. The higher the cache hit rate, the better the performance.

HPE SSD Smart Path

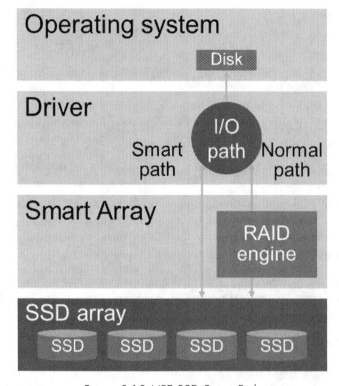

Figure 3-13 HPE SSD Smart Path

The SSD Smart Path feature included in the Smart Array software stack is designed to improve SSD read performance. SSD Smart Path enables an optimized data path to high-performance SSDs. As illustrated by Figure 3-13, the optimized path bypasses the controller's RAID engine and sends IO directly to the drives.

With up to 3.5 times better SSD read performance, SSD Smart Path chooses the optimum path to the SSD and accelerates reads for all RAID levels and RAID 0 writes. SSD Smart Path is ideal for read-intensive workloads using more than six SSDs and is included with Smart Array P-Series controllers.

The following operating systems are supported by SSD Smart Path:

- Microsoft Windows Server 2008
- Microsoft Windows Server 2008 R2
- Microsoft Windows Server 2012
- Microsoft Windows Server 2012 R2
- Red Hat Enterprise Linux 6.1, 6.2, 6.3, 6.4, 6.5, and 7.0

- SUSE Linux Enterprise Server 11 (SP1, SP2, SP3) and 12

- VMware ESXi 5.0 update 3

- VMware vSphere 5.1 update 2 and vSphere 5.5

HPE Secure Encryption

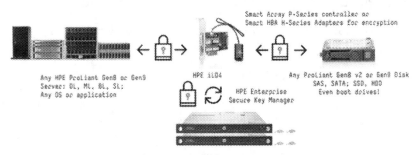

Figure 3-14 HPE Secure Encryption

Today, good business practices and industry regulations require organizations to protect sensitive and private information from unauthorized disclosure or theft. HPE Secure Encryption with HPE Enterprise Secure Key Manager (ESKM) is a simple, controller-based disk data encryption solution that protects sensitive data at rest on any bulk storage attached to a P-Series controller or Smart HBA H-Series Adapter, as illustrated by Figure 3-14. It supports any HDD or SSD in the HPE SmartDrive portfolio for ProLiant Gen8 and Gen9 servers or supported storage enclosures. The solution is available for both local and remote key management deployments.

This helps customers comply with government regulations such as the Health Insurance Portability and Accountability Act of 1996 (HIPAA) and Sarbanes-Oxley Act of 2002, which both have data privacy requirements.

Local key management provides the simplest approach to implementing Secure Encryption and is intended for a limited number of servers. Remote key management deployment includes:

- High-availability clustering and failover with Federal Information Processing Standards (FIPS)-compliant key server appliances that include a separate, secure key database

- Key generation and retrieval services

- Identity and access management for administrators and for data encryption devices

- Secure backup and recovery

- A local certificate authority

- Strong audit logging for compliance validation

HPE ESKM provides a complete solution for unifying and automating an organization's encryption controls by securely creating, protecting, serving, controlling, and auditing access to business- and compliance-critical encryption keys. ESKM supports a growing range of HPE server and storage products, partners, and solutions for data protection. ESKM manages all encryption deployments from just a few servers to thousands of servers and two million keys per ESKM cluster.

Activity: Choosing a storage technology to fit a workload

Consider the following case study that represents challenges faced by many companies. To complete this activity, read the case study and answer the questions that follow.

Customer profile

Derwent Ltd. manufactures and distributes components for the automotive industry. Their compute environment consists of ProLiant DL360p Gen8 and ProLiant DL380 Gen9 servers with internal HDD storage. To achieve acceptable levels of performance from their core business applications, they have had to overprovision the storage. They are now running into capacity and performance limitations. They use many applications to support their business, and the two mission-critical applications that need immediate attention are:

- Online transaction processing (OLTP) for product and order management
- Web-based distribution management system (web servers)

The general manager Jim McDonald has expressed the need to improve performance and address the capacity limitations of the existing storage solution. He wants to reduce power consumption and has asked for your advice. His immediate need is to refresh the internal storage, and in the near future, he wants to investigate an entry-level, low-cost, easy-to-manage shared storage solution so that storage can be consolidated. There is no Fibre Channel network in place, and Jim considers that to install and manage a Fibre Channel solution would be outside of the company's budget and skill levels.

Questions

Use the information at these websites to answer the following questions:

- http://www8.hp.com/h20195/v2/GetPDF.aspx%2F4AA4-7186ENW.pdf
- http://www8.hp.com/h20195/v2/GetDocument.aspx?docname=4AA3-0132ENW
- https://www.hpe.com/uk/en/storage/entry-level.html
- http://www8.hp.com/us/en/products/servers/qualified-options/storage.html

1. Which devices would you recommend for the OLTP systems, and why did you choose these? What is the approximate guideline cost per device?

   _____

   _____

   _____

2. Which devices would you recommend for the web servers, and why did you choose these? What is the approximate guideline cost per device?

   _____

   _____

   _____

3. Which shared storage solution would you recommend, and why did you choose it?

   _____

   _____

   _____

Answers

The following recommendations should not be considered to be the "correct" answers, but should give you an indication of where to begin.

1. Which devices would you recommend for the OLTP systems, and why did you choose these? What is the approximate guideline cost per device?

   HPE Write Intensive 12G SAS and 6G SATA SSDs provide high write performance and endurance.

   Workloads best suited for these SSDs include online transaction processing (OLTP), virtual desktop infrastructure (VDI), business intelligence, and Big Data Analytics. Solid state devices have a significantly better performance-to-power rating than traditional rotating HDDs.

   List price for the 800GB 12G SAS SFF drive is $3429. The 1.6TB 12G SAS SFF drive is $6689.

2. Which devices would you recommend for the web servers, and why did you choose these? What is the approximate guideline cost per device?

   HPE Read Intensive 12G SAS and 6G SATA SSDs deliver enterprise features for a low price in HPE ProLiant server systems. The cost per IOPS compares very favorably to HDDs.

Read Intensive SSDs deliver great performance for workloads high in reads such as boot/swap and Web servers.

List price for the 480GB 12G SAS SFF drive is $1149. The 960GB 12G SAS SFF drive is $2379. The 1.92TB 12G SAS SFF drive is $4509. The 3.84TB 12G SAS SFF drive is $8569.

3. Which shared storage solution would you recommend, and why did you choose it?

HPE MSA 1040 Storage is a simple, fast shared storage system with direct SAS connection for up to four servers with no SAN Infrastructure required. It has a low entry price and also has options to support iSCSI and Fibre Channel connectivity as well as flexibility for SSDs and small and large form factor HDDs.

Networking features of ProLiant servers

The combination of ProLiant servers and HPE server networking brings new levels of network performance, reliability, and efficiency in the data center:

- **Performance**—Engineered to improve networking bandwidth and lower latency across the HPE server networking portfolio.

- **Reliability**—Rigorous qualification and testing that eliminates downtime and works seamlessly with HPE servers.

- **Efficiency**—Workload optimized with HPE software defined features, from virtualization to network partitioning to meet application requirements.

What is RDMA?

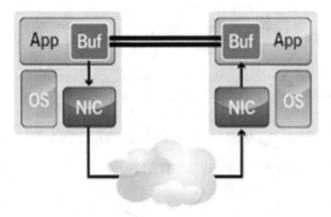

Figure 3-15 RDMA

The latest networking technologies can help businesses boost reliability and productivity, eliminate silos and complexity, and provide better services faster. HPE advancements in networking technologies can address typical data center and server challenges.

As illustrated in Figure 3-15, RDMA enables direct memory access from the memory of one computer to that of another without involving the operating system of either system.

Windows Server 2012 R2 and Windows Server 2012 include SMB Direct, which supports the use of network adapters that have remote direct memory access (RDMA) capability. RDMA network adapters can function at full speed with very low latency and use very little CPU. For workloads such as Microsoft Hyper-V or SQL Server, this enables a remote file server to resemble local storage. SMB Direct provides:

- **Increased throughput**—Leverages the full throughput of high-speed networks where the network adapters coordinate the transfer of large amounts of data at line speed

- **Low latency**—Provides extremely fast responses to network requests, and as a result, makes remote file storage assume that it is directly attached block storage

- **Low CPU utilization**—Uses fewer CPU cycles when transferring data over the network, which leaves more power available to server applications

RDMA and RoCE

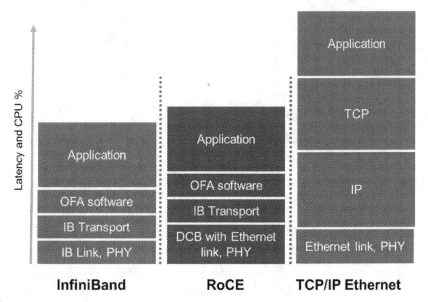

Figure 3-16 RDMA and RoCE

RDMA allows data to move between memory in different servers without any OS involvement. RDMA over Converged Ethernet (RoCE) provides this data transfer with very low latencies on lossless Ethernet networks and is ideal for network-intensive applications such as networked storage, cluster computing, live migration, and Microsoft SMB Direct environments. RoCE is an accelerated IO delivery mechanism that allows data to be transferred directly from the memory of the source server to the memory of the destination server, bypassing the operating system kernel. RoCE benefits from reduced latency and CPU utilization when compared with TCP/IP, and can approach the IO performance of InfiniBand, as illustrated in Figure 3-16.

 Note

> Data center bridging (DCB) is a set of IEEE standards that create a lossless fabric on top of Ethernet. RoCE works best when the underlying fabric implements lossless technologies and requires RoCE/DCB support on the network switches and adapters.

Increase virtualization performance with RoCE

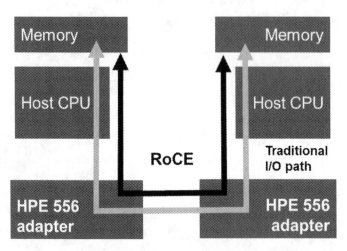

Figure 3-17 Increase virtualization performance with RoCE

Application and user growth are impacting network traffic and driving increased virtual machine (VM) deployment, straining system IO and CPU resources. Customers need agile and efficient infrastructure to maximize use of virtualized workloads throughout their lifecycle.

Converged LAN and SAN can provide cost savings in infrastructure and simplified management. However, low latency and higher efficiency are extremely important and need further improvement in clustered, grid, and utility computing. Every microsecond delay in data transfer, algorithmic execution, or transaction derivation can result in millions of dollars in losses. RoCE reduces CPU utilization and helps maximize host VM density and server efficiency.

With RoCE, the RDMA data transfer is performed by the direct memory access engine on the adapter's network processor as illustrated by Figure 3-17. This means the CPU is not used for the data movement, freeing it to perform other tasks such has hosting more virtual workloads (increased VM density, data mining, and computational databases.

RDMA also bypasses the host TCP/IP stack in favor of upper-layer InfiniBand protocols implemented in the adapter's network processor. Bypassing the TCP/IP stack and removing a data copy step reduce overall latency to deliver accelerated performance for applications such as Hyper-V Live Migration, Microsoft SQL, and Microsoft SharePoint with SMB Direct. For example, Hyper-V Live Migration is much faster using SMB Direct with RoCE than using TCP/IP. Additional workloads that would benefit from low latency include clusters with app-to-app RDMA transfers from one node to another.

RoCE is a key feature of HPE FlexFabric 556 (for HPE rack servers) and FlexFabric 650 series (for HPE server blades) adapters. The adapters provide tunnel offload for efficient overlay networking to increase VM migration flexibility and network scale with minimal impact to server performance. Network switches such as the HPE FlexFabric 59xx and 129xx provide full support for DCB.

What are overlay networks?

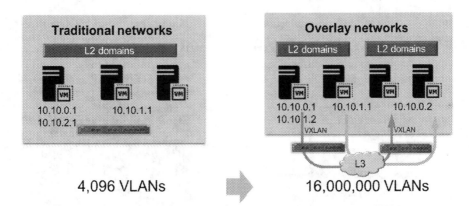

Figure 3-18 Overlay networks

The evolution to cloud data centers requires deployment at scale of tens of thousands of secure, private networks for tenants. Traditional technologies limit these data centers in the areas of speed, scalability, and manageability of application deployments. Current virtual LAN (VLAN) technology is limited to 4096 VLAN IDs, allowing for a relatively small number of isolated private networks.

Two overlay networking technologies address these challenges: network virtualization using generic routing encapsulation (NVGRE) and virtual extensible LAN (VXLAN). A network overlay is a virtual network that runs independently on top of another one. Interconnected nodes share an underlying physical network, allowing applications that require specific network topologies to be deployed without needing to modify the underlying network.

In both NVGRE and VXLAN, a virtual Layer 2 overlay network (tunnel) is automatically created on top of a Layer 3 network. VM-to-VM communications traffic traverses this virtual network, and a VM can now be freely migrated across the data center over an overlay network without reconfiguration, saving time.

Overlay networks can be used in data centers to support the following use cases:

- **Multitenancy at scale**—Provide scalable Layer 2 networks for a multitenant cloud that extends beyond current limitations. VXLAN uses an identifier that is 24 bits long, compared with the 12-bit VLAN ID that provides for only 4094 usable segments. As a result, and as illustrated in Figure 3-18, up to 16 million VXLAN segments can be used to support network segmentation at the scale required by cloud builders with large numbers of tenants.

- **Simplified traffic management**—Shift the network complexity from the physical network to the overlay network with software and provide network resources from a single management point without changing the physical network.

- **Hybrid cloud capabilities**—Incorporate bare-metal servers anywhere and move the workload as needed, with public and private cloud working in sync.

Boost server efficiency for overlay networking

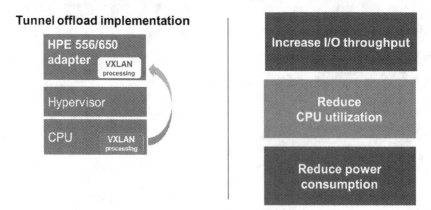

Figure 3-19 Tunnel offload

Overlay network tunneling technologies (VXLAN and NVGRE) help address the issues of traditional Layer 2 networks. However, these technologies can significantly impact performance of data center compute resources. More specifically, they cause significant increases of CPU utilization, reduction in network throughput, and increased power consumption as indicated by Figure 3-19.

Note

Inserting the VXLAN/NVGRE header on an Ethernet frame, as well as calculating the new checksum value, creates a tremendous burden on throughput, host CPU utilization, and power consumption. This limits the number of VMs per physical server platform.

The FlexFabric 556 series (for ProLiant ML and DL servers) and FlexFabric 650 series (for ProLiant BL servers) adapters minimize the impact of overlay networking on host performance with tunnel offload support for VXLAN and NVGRE. By offloading packet processing to adapters, customers can use overlay networking to increase VM migration flexibility and network scale with minimal impact to performance. Tunnel offloading increases IO throughput, reduces CPU utilization, and lowers power consumption. These adapters are the first in the industry to support VXLAN, NVGRE, and RoCE.

HPE Virtual Connect and flex adapter hardware

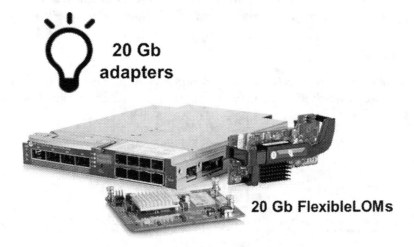

Figure 3-20 FlexFabric network hardware

HPE 20 Gb FlexFabric converged network adapters (CNAs) for BladeSystems, shown in Figure 3-20, remove the 10 Gb bandwidth restrictions imposed by previous generations of adapters. HPE delivers native 20 Gb performance per port or aggregate 40 Gb performance in a dual-port adapter. Industry-available solutions deliver performance through aggregate or multiple 1/10 Gb ports on single or multiple adapters in a teamed fashion. HPE adapters can stream converged 10 GbE and 8 Gb storage simultaneously over a 20 Gb port.

Previously, 10 Gb FlexNIC (CNA) implementations were limited to partitioning 10 Gb into one 8/4 Gb Fibre Channel and multiple GbE physical functions or as a single 10 GbE (no Fibre Channel or bandwidth for other physical functions). Additionally, 20 Gb ports can be partitioned into a full-rate 10 Gb Ethernet and a full-rate 8 Gb Fibre Channel over Ethernet (FCoE), with increased additional bandwidth remaining for other functions, provisioned in 100 Mbps increments.

Traditional Layer 2 networks limit mobility and scale of virtualized and multitenant workloads. When connected to HPE 6125XLG Ethernet Blade Switch, it also supports RoCE, which increases host efficiency and lowers latency. When connected to HPE Virtual Connect FlexFabric-20/40 F8 Modules, the FlexFabric 20Gb 2-port 650FLB Adapter provides 20 GbE performance and it can also provide 10 GbE performance when connected to the HPE 6125XLG Ethernet Blade Switch.

The HPE FlexFabric 10Gb 2-port 556FLR-SFP+ Adapter is installed in the FlexibleLOM socket in ProLiant Gen9 rack servers. The adapter offers 10 GbE FlexibleLOM with RDMA and tunnel offload features.

The FlexFabric 20Gb 2-port 650FLB Adapter and the FlexFabric 10Gb 2-port 556FLR-SFP+ Adapter provide fast Hyper-V Live Migration (using SMB with RoCE instead of TCP/IP), reduce CPU consumption, reduce latency, and improve host VM density and server efficiency using RoCE.

 Note

For more information on HPE server networking, scan this QR code or enter the URL into your browser and watch this video. You might need to scroll down to find the video.

http://www8.hp.com/us/en/products/servers/qualified-options/networking.html

---

Learning check

1. What are some reasons why customers choose ProLiant Gen9 solutions?

   _____

   _____

   _____

   _____

   _____

2. Which technology provides improved reliability to DDR4 memory and analyzes all correctable errors?

   a. HPE SmartCache

   b. HPE Advanced Memory Error Detection

   c. HPE Single Device Data Correction

   d. HPE Control Bus Parity Protection

3. Which storage feature is best suited for customers looking to invest in SSD technology at a controlled pace?

   A. Dynamic Smart Array B140i

   B. HPE Smart Storage Administrator

   C. HPE SmartCache

   D. HPE PCIe Workload and IO Accelerators

4. RDMA allows data to move between application memory in different servers without any CPU involvement.

   ☐ True

   ☐ False

5. What characteristics must an application have in order to take full advantage of a Xeon Phi coprocessor for optimized workload performance?

   _____

   _____

   _____

   _____

Learning check answers

1. What are some reasons why customers choose ProLiant Gen9 solutions?

    - **HPE has the broadest server portfolio in the market**

    - **Each ProLiant server—from the entry-level server to the most scalable server—is engineered to provide meaningful, cutting-edge benefits to customers for their increasingly complex environments**

    - **ProLiant servers provide the right compute resources, for the right workload, at the right economics**

2. Which technology provides improved reliability to DDR4 memory and analyzes all correctable errors?

    A. HPE SmartCache

    B. HPE Advanced Memory Error Detection

    C. HPE Single Device Data Correction

    D. HPE Control Bus Parity Protection

3. Which storage feature is best suited for customers looking to invest in SSD technology at a controlled pace?

    A. Dynamic Smart Array B140i

    B. HPE Smart Storage Administrator

    C. HPE SmartCache

    D. HPE PCIe Workload and IO Accelerators

4. RDMA allows data to move between application memory in different servers without any CPU involvement.

    ☐ **True**

    ☐ False

5. What characteristics must an application have in order to take full advantage of a Xeon Phi coprocessor for optimized workload performance?

    - **An application must scale well to more than one hundred threads**

    - **An application must make extensive use of vectors or efficiently use more local memory bandwidth than is available on a Xeon processor**

Summary

- ProLiant Gen9 servers are an integral part of the HPE portfolio, offering convergence, cloud-ready design, and workload optimization. To meet growing business demands, ProLiant Gen9 rack servers redefine compute economics by delivering more compute and storage capacity, along with less compute energy and floor space consumption.

- The Xeon processor E3 family is designed to deliver the best combination of performance, built-in capabilities, and cost-effectiveness in file and print, messaging, collaboration workloads and cloud environments. The Xeon processor E5 family is designed to deliver agile services for cloud and traditional applications and workloads with versatility across diverse workloads. The Xeon processor E7 family is designed for the scalable performance demands of complex, data-demanding workloads such as in-memory databases and real-time business analytics. You can use Xeon processors and Phi coprocessors together to optimize performance for almost any workload. Phi coprocessors complement the performance and energy-efficiency of the Xeon processor E5 family to make dramatic performance gains.

- Choosing the right memory is crucial to ensuring high reliability and delivering a faster return on IT investment. Memory features of ProLiant servers include DDR4 SmartMemory, RDIMMs, LRDIMMs, and NVDIMMs.

- Data storage requirements are growing exponentially, along with government regulations for protecting sensitive data. HPE Smart Storage solutions for ProLiant Gen9 servers are built to align IT with business requirements, streamline operations, contain costs, accelerate the delivery of new products and services, and optimize application performance.

- The latest networking technologies can help businesses boost reliability and productivity, eliminate silos and complexity, and provide better services faster. Networking features of ProLiant server options include RDMA, RoCE, and overlay network capabilities.

4 HPE ProLiant Rack Server Solutions

WHAT IS IN THIS CHAPTER FOR YOU?

After completing this chapter, you should be able to:

✓ Describe Hewlett Packard Enterprise (HPE) ProLiant DL rack-mounted server models and the workloads they target

✓ Name the HPE options for ProLiant Gen9 DL servers

✓ Describe the power and cooling features and options available for HPE rack-mounted servers

✓ Explain how to use HPE QuickSpecs

OPENING CONSIDERATIONS

Before proceeding with this section, answer the following questions to assess your existing knowledge of the topics covered in this chapter. Record your answers in the space provided here.

1. With which HPE ProLiant Gen9 rack server models are you familiar?

---

2. How familiar are you with storage, networking, and security innovations for ProLiant Gen9 servers?

_____
_____
_____
_____
_____
_____

3. What experience have you had using QuickSpecs? Which other HPE tools and resources do you use frequently in your job?

_____
_____
_____
_____
_____
_____
_____
_____

HPE ProLiant Gen9 rack servers

With the broadest server portfolio in the industry, HPE offers ProLiant Gen9 servers that focus on the needs of all customer segments, including small and medium business (SMB), enterprise, and high-performance computing (HPC).

The ProLiant Gen9 rack portfolio delivers flexible, reliable, secure, and performance-optimized server solutions for a range of workloads and budgets. It features versatile and flexible designs along with improved energy efficiencies to help reduce total cost of ownership (TCO).

The ProLiant rack portfolio is performance optimized for multi-application workloads to significantly increase the speed of IT operations and enable IT to respond to business needs faster. Integrated with a simplified yet comprehensive management suite and industry-leading support, the ProLiant Gen9 rack server portfolio enables customers to accelerate business results with faster compute, memory, and IO performance, coupled with increased storage and networking performance—including lower latency.

HPE ProLiant rack servers offer the following advantages:

- **Flexible choices to redefine compute economics**—Right-sized computing with flexible choices across multiple workloads delivers better operational efficiencies and lower total cost of ownership (TCO). ProLiant rack servers offer three times the compute per watt, based on an HPE internal comparison between ProLiant DL380 Gen9 and DL380p Gen8 servers with Intel Sandy Bridge processors.

 Note

The source for system wattage was the IDC Qualified Performance Indicator. Performance taken from SPECint_rate_base2006 industry benchmark. Calculation: Performance/ Watt. August 2014.

Scan this QR code or enter the URL into your browser for more information on the IDC Qualified Performance Indicator.

http://www.idc.com/QPI/index.jsp

- **Reliable infrastructure to accelerate service delivery**—The ProLiant Gen9 rack portfolio delivers a reliable, fast, and secure infrastructure solution, helps increase IT staff productivity, and accelerates service delivery. According to an IDC whitepaper sponsored by HPE, ProLiant rack servers offer 66x faster service delivery for competitive advantage. According to the study, a customer was able to reduce the time to build and deploy infrastructure for 12 call centers from 66 days to 1. A total of 2000 servers were deployed.

 Note

Scan this QR code or enter the URL into your browser to download the whitepaper "Achieving Organizational Transformation with HP Converged Infrastructure Solutions for SDDC" (January 2014, IDC #246385).

http://www.hp.com/go/oneviewROI

- **Optimized compute to boost business performance**—Optimized computing, storage, and networking capabilities deliver faster workload performance to transform business results.

In addition to these advantages, ProLiant Gen9 servers also provide several infrastructure management tools that accelerate IT service delivery. These include:

- HPE OneView for automation simplicity across servers, storage, and networking

- Online personalized dashboard for converged infrastructure health monitoring and support management with HPE Insight Online

- Unified Extensible Firmware Interface (UEFI) boot mode configuration

- Embedded management to deploy, monitor, and support the server remotely, out of band with HPE iLO

- Driver and firmware update management with HPE Smart Update

ProLiant Gen9 rack series positioning

Table 4-1 ProLiant Gen9 rack series positioning

Features and Segments	Essential rack servers		Performance rack servers	
	New to servers 10 series	New IT growth 100 series	Traditional IT 300 series	Scale up 500 series
Customers	SMB	SMB, enterprise	SMB, enterprise, HPC	Enterprise, HPC
Use cases, apps, and workloads	File and print Messaging Infrastructure core apps	NoSQL, Hadoop, Map/Reduce Virtualization (low VM density)	Mission critical apps Academic and research Virtualization (medium-to-high VM density)	Large databases Unix alternative Monolithic apps
Value proposition	Most cost-efficient servers to run the new style of IT, web, collaboration, and business workloads		Most flexible and best overall performance systems to run compute-intensive workloads	

Table 4-1 shows the positioning of the ProLiant Gen9 rack series. The rack portfolio meets the needs of customers who are either new to servers or who are positioned to grow and expand their business.

- For businesses that are new to servers, HPE recommends the ProLiant 10 series of rack servers, which are easy-to-implement, affordable solutions for the relatively low workloads of an SMB environment. They are designed for SMBs that need servers for first-time workload deployment.

The ProLiant 10 series is ideally suited to performing back office tasks, running core infrastructure applications such as Microsoft Office programs, and performing messaging, printing, and web services.

- For growing businesses or new IT growth customers, HPE recommends the ProLiant 100 series rack servers, which are optimized with the right balance of storage, performance, efficiency, and manageability to address multiple workloads for growing SMB and enterprise businesses. For SMBs and enterprise businesses with higher workloads, the ProLiant 100 series offers virtualization solutions for environments where low density is sufficient and for Hadoop-style applications that require multiple low-cost servers.

The ProLiant Gen9 server portfolio also meets the needs of businesses that are more mature in their lifecycle and require greater levels of performance to address their compute-intensive or scale-up workloads.

- For SMB, enterprise, and HPC customers that use traditional IT and require a mission-critical environment, HPE recommends the ProLiant 300 series of rack servers. The ProLiant 300 series meets the needs of businesses with compute-intensive workloads and applications that require medium- to high-density virtualization or run large databases.

- For customers requiring the most demanding scale-up workloads, HPE offers the ProLiant 500 series, which delivers unparalleled scalability, reliability, and availability. For enterprises and HPC environments with large databases or monolithic applications, the ProLiant 500 series offers high scalability and performance.

ProLiant Gen8 to Gen9 rack server model transitions

Table 4-2 Gen8 to Gen9 rack server model transitions

Gen 8 Rack Server	Gen9 Rack Server
HP ProLiant DL320e Gen8 v2	HPE ProLiant DL20 Gen9
—	HPE ProLiant DL60 Gen9
—	HPE ProLiant DL80 Gen9
—	HPE ProLiant DL120 Gen9
HP ProLiant DL360e Gen8	HPE ProLiant DL160 Gen9
HP ProLiant DL380e Gen8	HPE ProLiant DL180 Gen9
HP ProLiant DL360p Gen8	HPE ProLiant DL360 Gen9
HP ProLiant DL380p Gen8	HPE ProLiant DL380 Gen9
HP ProLiant DL560 Gen8	HPE ProLiant DL560 Gen9
HP ProLiant DL580 Gen8	HPE ProLiant DL580 Gen9

Table 4-2 shows the model transitions from Gen8 to Gen9 rack server models.

In the ProLiant Gen8 family, the "e" (essential) and "p" (performance) designations were assigned to various server models. The portfolio has expanded with common Intel architecture in ProLiant Gen9 servers. The Gen8 naming convention of "e" and "p" is no longer used. The new naming better represents the features and positioning for Gen9 rack servers.

HPE ProLiant DL20 Gen9 Server

Figure 4-1 HPE ProLiant DL20 Gen9 Server

The single-processor 1U HPE ProLiant DL20 Gen9 Server powered by Intel Pentium, Intel Pentium Core i3, and Intel Xeon E3-1200 v5 processors provides a unique blend of enterprise-class capabilities at a great value, making it an ideal rack server platform for growing businesses and service providers. It offers outstanding configuration flexibility to cater to a wide variety of business requirements at an affordable price point. The ProLiant DL20 Gen9 server, shown in Figure 4-1, is compact, versatile, and efficient.

The ProLiant DL20 Gen9 server offers a range of HPE qualified options to fit most needs, such as affordable drives for light workloads or solid state drives (SSDs) for more demanding requirements. Customers can choose from a single nonhot-plug power supply or efficient HPE hot-plug redundant power supplies, and from multiple storage controllers and HPE FlexibleLOM networking cards.

Customers can protect against data loss and downtime with the enhanced error handing of HPE DDR4 ECC memory options while also improving workload performance and power efficiency. Trusted and proven HPE Smart Array technology provides businesses with protection for critical data through RAID mirroring and striping capability and, with flash-backed write cache (FBWC) up to 4 GB, the ProLiant DL20 can capture and hold data in the event of a power loss or equipment failure.

HPE ProLiant DL60 Gen9 Server

Figure 4-2 HPE ProLiant DL60 Gen9 Server

The HPE ProLiant DL60 Gen9 Server (Figure 4-2) is powered by one or two Intel Xeon E5-2600 v3 or v4 processors and provides performance for budget-constrained environments. These servers feature eight DIMM slots of DDR4 SmartMemory and create a powerful and efficient rack server in a compact 1U chassis. It is an ideal rack server platform for file and print, messaging, and collaboration workloads as well as cloud environments.

The ProLiant DL60 offers outstanding configuration flexibility through a range of HPE qualified options to satisfy a wide variety of business requirements. ProLiant DL60 features and options include:

- Standard with 2x1 GbE, up to three PCIe slots, and choice of HPE FlexibleLOM for flexibility of networking bandwidth and fabric

- Embedded SATA HPE Dynamic Smart Array B140i Controller for boot and data, with a choice of optional Smart Array controllers such as 12 Gb/s SAS technology or HPE Smart host bus adapters (HBAs) with reliable high-performance SAS connectivity capable of running HBA mode or simple RAID mode

- 80 PLUS gold certified entry-level HPE power supplies (up to 92% efficient)

HPE ProLiant DL80 Gen9 Server

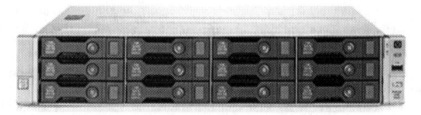

Figure 4-3 HPE ProLiant DL80 Gen9 Server

As shown in Figure 4-3, the HPE ProLiant DL80 Gen9 Server provides ample storage capacity with up to 12 large form factor (LFF) HPE SmartDrives, delivering optimal performance, capacity, and reliability to meet various customer segments and workload requirements at the right economics. Powered by one or two Xeon E5-2600 v3 or v4 processors, the DL80 is an ideal rack server platform for running both basic IT infrastructure workloads and storage capacity intensive applications such as cold storage and backup.

The ProLiant DL80 Gen9 offers outstanding configuration flexibility through a range of HPE qualified options to satisfy a wide variety of business requirements. DL80 features and options include:

- Standard with 2x1 GbE, up to six PCIe slots, and choice of HPE FlexibleLOM for flexibility of networking bandwidth and fabric

- Embedded Dynamic Smart Array B140i controller for boot and data, with a choice of optional Smart Array controllers such as 12 Gb/s SAS technology or HPE Smart HBAs with reliable high-performance SAS connectivity capable of running HBA mode or simple RAID mode

- SmartMemory with built-in intelligence reduces downtime and energy costs, resulting in better throughput performance and lower power consumption with DDR4 memory

- 80 PLUS gold certified entry-level HPE power supplies (up to 92% efficient)

HPE ProLiant DL120 Gen9 Server

Figure 4-4 HPE ProLiant DL120 Gen9 Server

The HPE ProLiant DL120 Gen9 (Figure 4-4) is an enterprise-class design packed in a dense form factor. It offers a combination of performance, redundancy, and expandability as compared with the traditional 1P servers. The ProLiant DL120 Gen9 server meets the growing needs of running virtualization and general-purpose workloads. It helps reduce virtualization licensing costs with its 1U/1-socket cost optimized design for SMBs and service providers.

The DL120 Gen9 server is powered by one Xeon E5-1600 v3/v4 or E5-6600 v3/v4 processor. DDR4 SmartMemory prevents data loss and downtime with enhanced error handling. HPE Smart Drives deliver optimal performance, capacity, and reliability to meet various customer segments and workload requirements at the right economics. DL120 server features and options include:

- Up to eight small form factor (SFF) or four large form factor (LFF) slots support hard disk drive (HDD) and SSD storage

- Embedded with 2x1 GbE, up to three PCIe slots, and choice of FlexibleLOM or PCIe standup 1 GbE–25 GbE or InfiniBand adapters provide flexibility of networking bandwidth and fabric

- 80 PLUS gold certified entry-level HPE power supplies (up to 92% efficient)

HPE ProLiant DL160 Gen9 Server

Figure 4-5 HPE ProLiant DL160 Gen9 Server

The HPE ProLiant DL160 Gen9 Server, shown in Figure 4-5, delivers the right balance of performance, storage, reliability, manageability, and efficiency in a dense and compact chassis. It meets the needs of a variety of customers, including SMBs and service providers with different workloads, from general purpose IT to emerging IT deployments, including cloud and big data in distributed computing environments.

The ProLiant DL160 Gen9 supports up to two Xeon E5-2600 v3 or v4 processors. It has a total memory capability of up to 1 TB using 16 x 64GB DDR4 memory DIMMs. Up to eight SFF or four LFF slots support HDD and SSD storage.

The DL160 Gen9 server offers outstanding configuration flexibility through a range of HPE qualified options to satisfy a wide variety of business requirements. DL160 features and options include:

- Standard with 2x1 GbE, up to three PCIe slots, and choice of HPE FlexibleLOM or PCIe standup 1 GbE–25 GbE or InfiniBand adapters provide flexibility of networking bandwidth and fabric

- Embedded Dynamic Smart Array B140i controller for boot and data, with a choice of Smart Array controllers and HPE Smart HBAs with reliable high-performance SAS connectivity capable of running in HBA mode or simple RAID mode

- 80 PLUS gold certified entry-level HPE power supplies

HPE ProLiant DL180 Gen9 Server

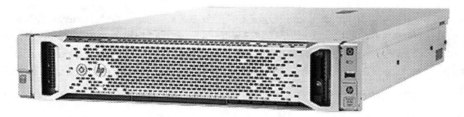

Figure 4-6 HPE ProLiant DL180 Gen9 Server

As shown in Figure 4-6, the HPE ProLiant DL180 Gen9 Server, powered by up to two Xeon E5-2600 v3 or v4 processors, is a 2U server, designed with the right balance of expandability, performance, reliability, and manageability in a new compact chassis. It is the ideal platform for customers who require flexibility in a growing data center, including SMBs and enterprises working with applications such as Hadoop, Cloud, and mobility.

Memory options offer up to 1 TB total RAM using 16 x 64GB DDR4 DIMMs. Internal storage is accomplished with up to 16 SFF or 12 LFF slots for SmartDrives that support HDD and SSD storage.

The ProLiant DL180 Gen9 server offers the following features and options:

- Standard with 2 x 1 GbE and choice of HPE FlexibleLOM or PCIe standup 1 GbE–25 GbE or InfiniBand adapters provide flexibility of networking bandwidth and fabric

- Embedded Dynamic Smart Array B140i controller for boot and data with a choice of Smart Array controllers with 12 Gb/s SAS technology or HPE Smart HBAs with reliable high-performance SAS connectivity capable of running in HBA mode or simple RAID mode

- 80 PLUS gold certified entry-level HPE power supplies

HPE ProLiant DL360 Gen9 Server

Figure 4-7 HPE ProLiant DL360 Gen9 Server

The HPE ProLiant DL360 Gen9 Server, shown in Figure 4-7, is the leading HPE server for dense general-purpose computing. Powered by up to two Xeon E5-2600 v3 or v4 series processors with up to 22 cores each, the ProLiant DL360 Gen9 Server delivers high performance with large memory and IO expandability packed in a 1U rack design.

The ProLiant DL360 Gen9 server uses DDR4 DIMMs for increased performance. Advanced error checking and correcting (ECC) and online spare memory improve uptime and reliability. The memory on this system can be expanded to 3.0 TB using 24 x 128GB DDR4 LRDIMM memory modules. It is also possible to configure the server with up to 128 GB of persistent memory using 16 x 8GB NVDIMM.

Note

NVDIMMs cannot be used with LRDIMMs. A maximum of 16 NVDIMMs can be supported. 128GB DIMMs cannot be mixed with any other size DIMMs.

Internal storage options offer up to ten SFF and four LFF HDDs or SSDs, along with Nonvolatile Memory Express (NVMe) PCIe options delivering optimal performance, capacity, and reliability to meet various customer segments and workload requirements at the right economics.

The ProLiant DL360 Gen9 server offers configuration flexibility through a range of HPE qualified features and options, including:

- A choice of Embedded 4 x 1 GbE, HPE FlexibleLOM, or PCIe standup 1 GbE–25 GbE or InfiniBand adapters which provides flexibility of networking bandwidth and fabric

- A choice of Dynamic Smart Array B140i controller, Smart Array adapter, or Smart HBAs

- 80 PLUS gold certified entry-level HPE power supplies

HPE ProLiant DL380 Gen9 Server

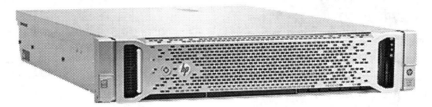

Figure 4-8 HPE ProLiant DL380 Gen9 Server

The data center standard for general-purpose computing, the HPE ProLiant DL380 Gen9 Server, shown in Figure 4-8, delivers the best performance and expandability in the HPE 2P rack portfolio. Powered by up to two Xeon E5-2600 v3/v4 series processors with up to 22 cores each, the HPE ProLiant DL380 Gen9 Server delivers performance with large memory and IO expandability in a 2U form factor.

The HPE ProLiant DL380 Gen9 Server has a flexible redesigned chassis, including new HPE Universal Media Bay configuration options with 8–24 SFF and 4 or 12 LFF drive options along with NVMe options and additional rear drive support for expandability and investment protection.

The DL380 Gen9 server offers the following features and options:

- A choice of Embedded 4 x 1 GbE, HPE FlexibleLOM, or PCIe standup 1 GbE–25 GbE or InfiniBand adapters which provides flexibility of networking bandwidth and fabric

- A choice of HPE Dynamic Smart Array B140i Controller, Smart Array adapter, or Smart HBAs

- 80 PLUS gold certified entry-level HPE power supplies

HPE ProLiant DL560 Gen9 Server

Figure 4-9 HPE ProLiant DL560 Gen9 Server

As shown in Figure 4-9, the HPE ProLiant DL560 Gen9 server is the high-density four-socket (4S) server with balanced performance, increased scalability and reliability, all in a 2U chassis. Powered by up to four Xeon E5-4600 v3 or v4 series processors with up to 22 cores each, the ProLiant DL560 Gen9 server delivers processing power along with significant storage capacity.

The ProLiant DL560 Gen9 server is the ideal server for virtualization, server consolidation, database, business processing, and general four-processor (4P) data-intensive applications where data center space and price/performance are important.

The ProLiant DL560 Gen9 server has HPE Flexible Smart Array and HPE Smart SAS HBA controllers, with support for up to 4 GB FBWC, which provide the flexibility to choose the optimal 12 Gb/s controller most suited to the customer's environment. The DL560 also has an embedded SATA HPE Dynamic Smart Array B140i Controller for boot, data, and media needs.

DDR4 SmartMemory improves workload performance and power efficiency while preventing data loss and downtime with enhanced error handling. The ProLiant DL560 Gen9 server supports up to 3 TB DDR4 max capacity through its 48 DIMM slots.

The ProLiant DL560 Gen9 server has a redesigned chassis that can include up to 24 SFF SmartDrives and NVMe options, along with an optional HPE Universal Media Bay that delivers optimal performance, capacity, and reliability to meet various customer workload requirements at the right price.

The ProLiant DL560 Gen9 provides a choice of Embedded 4 x 1 GbE, HPE FlexibleLOM, or PCIe standup 1 GbE–25 GbE or InfiniBand adapters for flexibility in networking bandwidth and fabric. It can support up to 2x 1200 W or 1500 W HPE Platinum Plus hot plug power supplies (up to 94% efficient) for maximum redundancy and to reduce unnecessary downtime.

HPE ProLiant DL580 Gen9 Server

Figure 4-10 HPE ProLiant DL580 Gen9 Server

The HPE ProLiant DL580 Gen9 Server, as shown in Figure 4-10, is the HPE four socket enterprise standard x86 server offering commanding performance, rock-solid reliability and availability, and compelling consolidation and virtualization efficiencies. Powered by up to four Xeon E7-4800/8800 v3 or v4 series processors with up to 24 cores each, the ProLiant DL580 Gen9 server delivers enhanced processor power.

The ProLiant DL580 Gen9 server has security and data protection features for system resiliency that businesses can depend on. It is ideal for mission-critical enterprise, business intelligence, and database applications.

The ProLiant DL580 Gen9 server ships standard with nine PCIe 3.0 slots and a choice of FlexibleLOM or PCIe standup 1 GbE, 10 GbE, 25 GbE or InfiniBand adapters provide flexibility of networking bandwidth and fabric. HPE Flexible Smart Array controllers, with support for up to 4 GB FBWC, provide the flexibility to choose the optimal 12 Gb/s controller most suited to the customer's environment. The controllers also feature HPE Secure Encryption, Advanced Data Mirroring, and SmartDrive technology.

HPE DDR4 SmartMemory improves workload performance and power efficiency and prevents data loss and downtime with enhanced error handling. The ProLiant DL580 Gen9 server supports up to 6 TB DDR4 max capacity through its 96 DIMM slots.

The ProLiant DL580 Gen9 server can include up to 10 SFF SmartDrives. The DL580 also supports HPE NVMe mixed use and write intensive PCIe Workload Accelerators, ideal for database and virtual desktop infrastructure (VDI) workloads or online transaction processing (OLTP) and Business Intelligence workloads.

The DL580 Gen9 server can support up to 4x 1500W HPE Platinum Plus power supplies for maximum redundancy and to reduce unnecessary downtime. It provides increased system availability and less need for service with advanced error recovery, error diagnosis, and built-in redundancy.

HPE ProLiant DL use cases

Redstone Federal Credit Union, Technicolor, Bally Technologies, and BCDVideo are among the customers experiencing the benefits of HPE Gen9 technologies.

Bally's goal is to build a world-leading gaming platform that continually exceeds internal and user expectations in the entertainment industry. Bally engaged with HPE to evaluate ProLiant Gen9 servers in its business environment. The company discovered that ProLiant Gen9 servers:

- Deliver 30% faster provisioning and installation of customer deployments

- Increase computational density by 20% over previous solution

- Run more than 100 virtual machines (VMs) without stressing the CPU

- Reduce the total technology footprint by 50%–70%, boosting the overall environmental profile

- Offer superior system visibility and management with built-in software tools

Jeff Burgess, President of BCDVideo, commented, "Building on the HPE portfolio and global brand, BCDVideo's business is exploding around the world. Working together with HPE, there is not an environment or budget that we cannot accommodate with an innovative solution."

As a manufacturer, BCDVideo relies exclusively on HPE servers, storage, and networking. As part of the HPE OEM program, BCDVideo can match customer requirements to just the right HPE platform, whether an HPE Z230 Workstation for smaller deployments or the latest HPE ProLiant DL380 Gen9 server for large enterprise installations.

Although ProLiant Gen8 servers have been the standard for BCDVideo's global deployments, the company is moving to ProLiant Gen9 for even greater performance and scalability. The ProLiant DL80 Gen9 server based on the Xeon Processor E5-2600 v3 product family, as optimized for video by BCDVideo, was certified by the leading video software companies to run twice as many cameras as the Gen8-based server—1000 compared with 500. In addition, BCDVideo's BCD524 video server, based on the ProLiant ML350 Gen9 server, provides up to 144 TB of SAS storage per server, increasing the system's internal storage capacity by 150%.

BCDVideo's solution approach includes storage, networking, and client viewing stations, all built on HPE products and technologies and supported by HPE Care Pack Services. The HPE Active Health System is an added differentiator for BCDVideo. It provides continuous, proactive monitoring of more than 1600 system parameters on every ProLiant Gen9 server.

Options for ProLiant Gen9 rack-mounted servers

HPE Smart Array controllers

HPE Smart Array controllers have been a part of the ProLiant story since the launch of the ProLiant server portfolio. They have evolved to provide functionality and features that help maximize workload performance, availability, and reliability. The current generation of Smart Array controllers are available in many form factors, including embedded on the system board, as a stand-up card in a PCIe slot, or as a mezzanine card or FlexibleLOM for HPE server blades.

Smart Array controllers are designed for ProLiant customers who need fault tolerance for their direct attached storage environments. They provide reliable RAID fault tolerance for ProLiant storage to maximize system uptime. They also support advanced features such as RAID 10 Advanced Data Mirroring (ADM), which is a feature exclusive to HPE that provides a three-drive mirror.

Smart Array controllers are optimized to meet the performance needs of workloads and use the fastest PCI 3.0 technology. The FBWC module is used for write caching and as a metadata store for HPE SmartCache. As write commands are sent across the Smart Array controller to be written to disk, they are often sent faster than the actual write to disk. When this happens, the FBWC acts as a holding area, which helps smooth out the performance and efficiency of write commands. Because the FBWC is nonvolatile, data is retained indefinitely if the server goes down. SmartCache caches hot data on lower-latency SSDs to accelerate application workloads and performance dynamically.

HPE SSD Smart Path enhances SSD read performance by bypassing Smart Array firmware for the optimal path to the SSD.

HPE also provides comprehensive and simplified management of the entire Smart Array ecosystem with HPE Smart Storage Administrator (SSA). You can easily expand capacity on an existing volume or create new logical volumes using SSA. Each Smart Array controller is right-sized for the specific ProLiant model and is designed to grow with the business.

Smart Array controllers support SAS and SATA technology for HDDs and SSDs. HPE has a comprehensive portfolio of HPE Smart Drives available in 2.5-inch SFF or 3.5-inch LFF with various capacity points for HDDs and SSDs.

Understanding letters and numbers in the Smart Array controllers

The letters and numbers signify the family, series, generation, ports, and media.

- The first character represents the family (B-series, H-series, or P-series controller)

- The second character represents the series (where 2xx is base/entry, 4xx is scale/mainstream, 7xx is premium/performance for HPE BladeSystem, and 8xx is premium/performance for HPE ProLiant ML/DL/SL family servers)

- The third character represents the generation (0 is 3 Gb SAS, 1 is 6 Gb SAS, 2 is 6 Gbps second generation, 3 is 12 Gb SAS with HPE Secure Encryption, 4 is 12 Gb SAS for Gen9)

- The fourth character represents the ports (0 is internal, 1 is external, 2 is internal and external, 4 is specific to the BL460c, and 6 is specific to the BL660)

- The fifth character represents the media ("i" is integrated, "m" is mezzanine, "br" is flexible controller for server blades, "ar" is flexible controller for rack servers)

As an example, take the SmartArray P440ar. This is a P-series (P), scale/mainstream (4), 12 Gb SAS (4), internal port (0), flexible controller for rack servers (ar).

HPE Smart Array controller comparison

Table 4-3 HPE Smart Array controller comparison (1 of 2)

	Dynamic Smart Array B140i	Smart Array P440ar	Smart Array P440	Smart Array P441
PCI bus	Embedded	Flexible Smart Array	Low-profile PCIe 3.0 x8 card	Low-profile PCIe 3.0 x8 card
Memory bus speed	64-bit	DDR3-1866 MHz 72-bit	DDR3-1866 MHz 72-bit	DDR3-1866 MHz 72-bit
Memory options	Zero Memory RAID has no cache memory	2 GB FBWC	4 GB FBWC	4 GB FBWC
SAS/SATA connectivity	10 SATA ports	Two internal x4 mini-SAS connectors with expander support	One internal x8 mini-SAS double-wide with expander support	Two external x4 mini-SAS connectors

Table 4-3 HPE Smart Array controller comparison (1 of 2) Continued.

	Dynamic Smart Array B140i	Smart Array P440ar	Smart Array P440	Smart Array P441
Max drives	Up to 10 SATA/ NGFF drives	Up to 26 (with expander card)	Up to 60 depending on server model	Up to 200 physical drives
Software management	SSA, SPP, HPE OneView	SSA, SPP, HPE OneView	SSA, SPP, HPE OneView	SSA, SPP, Intelligent Provisioning, HPE OneView
RAID support	RAID 0	RAID 6	RAID 6	RAID 6
	RAID 1	RAID 60	RAID 60	RAID 60
	RAID 1 + 0	RAID 5	RAID 5	RAID 5
	RAID 5	RAID 50	RAID 50	RAID 50
		RAID 1 and 10	RAID 1 and 10	RAID 1 and 10
		RAID 1 ADM and 10 ADM	RAID 1 ADM and 10 ADM	RAID 1 ADM and 10 ADM

Table 4-4 HPE Smart Array controller comparison (2 of 2)

	Smart HBA H240ar	Smart Array P840	Smart Array P841
PCI bus	Flexible PCIe 3.0 x8 card	Full-height, half-length PCIe 3.0 x8 card	Full-height, half-length PCIe 3.0 x8 card
Memory bus speed	64-bit	DDR3-1866 MHz 72-bit	DDR3-1866 MHz 72-bit
Memory options	n/a	4 GB FBWC	4 GB FBWC
SAS/SATA connectivity	Two external x4 mini-SAS connectors	Two internal x8 double-wide connectors with expander support	Two external x8 connectors
Max drives	Up to 90 TB SAS and 90 TB SATA	Up to 16 without expander

Up to 48 with two expanders | Up to 200 physical drives |
Software management	Secure Encryption, HPE OneView	SSA, SPP, Intelligent Provisioning, HPE OneView	SSA, SPP, Intelligent Provisioning, HPE OneView
RAID support	RAID 5	RAID 6	RAID 6
	RAID 1	RAID 60	RAID 60
	RAID 0	RAID 5	RAID 5
		RAID 50	RAID 50
		RAID 1 and 10	RAID 1 and 10
		RAID 1 ADM and 10 ADM	RAID 1 ADM and 10 ADM
		RAID 0	

Several storage controller options are available for ProLiant rack series servers, including the Dynamic Smart Array B140i, Smart Array P440ar, Smart Array P440/P441, and Smart Array P840/841 controllers. Tables 4-3 and 4-4 provide a high-level comparison of the basic functionality of these controllers.

With these controllers, HPE is providing key differentiators in the areas of flexibility, performance, and simplicity. For example, with select ProLiant Gen9 servers, users can choose the controller to suit their workload. Using the HPE Flexible Smart Array and Smart SAS HBA, customers can choose between the Smart Array Controller P440ar, the Smart SAS HBA H240ar, or the Dynamic Smart Array B140i in a rack server without having to consume a PCIe slot.

The ProLiant Gen9 family of controllers are supported by the HPE Smart Storage Battery. With the Smart Storage Battery, multiple Smart Array controllers are supported without needing additional backup power sources for each controller.

 Important

The Smart Array P840/4GB FBWC and Smart Array P841/4GB FBWC controller option kits do not include the backup power source necessary to protect FBWC data. A Smart Storage Battery must be purchased separately if the P840 or P841 is the first P-series Smart Array controller installed in a ProLiant Gen9 server.

HPE has also added new features for power management. ProLiant Gen9 users can use SSA to manage the power level of the Smart Array controller and the Smart SAS HBA to meet the workload. The different power levels are:

- **Performance**—The controller provides the highest performance without any power savings

- **Balance**—The controller will dynamically balance between performance and power

- **Power savings**—The controller saves power when performance is not critical, for example, for boot devices or cold storage

As hard drives increase in size, rebuild times can take many hours. HPE has enhanced the rapid rebuild feature in Gen9 controllers so that rebuilds of a failed drive are faster, helping to prevent the data loss that can occur during slower rebuilds.

HPE Dynamic Smart Array B140i Controller

Figure 4-11 HPE Dynamic Smart Array B140i Controller

The Dynamic Smart Array B140i (Figure 4-11) is the standard embedded storage controller for all ProLiant Gen9 servers, replacing the B120i and B320i controllers that were in Gen8 servers. The B140i is the entry-level storage controller for HP Gen9 servers, delivering a basic level of storage functionality and performance. It is ideal for use to support operating system boot devices or to provide basic protection for data that does not require significant storage performance.

The B140i acts as the default controller only if you do not install the more powerful Smart SAS HBA controller or the Smart Array controller in the system.

Eliminating most of the hardware RAID controller components and relocating advanced RAID algorithms from a hardware-based controller into device driver software lowers the total solution cost, yet still maintains comparable RAID protection and full compatibility with Smart Array disk format, configuration utilities, and management/monitoring software. Dynamic Smart Array controllers share the same easy and consistent UI with Smart Array controllers, thus making storage management and deployment easy.

The Dynamic Smart Array B140i provides customers with greater choice and higher levels of data protection than previous embedded controllers. In the past, these data protection levels were only available on stand-up controllers.

 Note

The B140i supports only 6 Gb/s SATA drives. SAS is not supported with the Dynamic Smart Array B140i.

The Dynamic Smart Array B140i is a 6 Gb/s SATA controller with 10 SATA ports that supports up to 10 SATA drives with RAID 0, RAID 1, RAID 1+0, and RAID 5 capabilities. The controller allows

RAID volumes across all 10 drives in a single RAID set. Given the increasing need for high performance and rapid capacity expansion, Smart Array controllers offer:

- Up to 40 TB of total internal storage with 10 x 4 TB LFF SATA hard drives

- Up to 10 TB of total internal storage 10 x 1.0 TB SFF SATA hard drives

---

 **Important**

The Dynamic Smart Array B140i Controller supports UEFI Boot Mode only. It does not support Legacy BIOS Boot Mode.

HPE Smart HBA H240ar Controller

Figure 4-12 HPE Smart HBA H240ar Controller

The HPE Smart HBA H240ar (Figure 4-12) is a 12 Gb/s controller that provides increased performance over the previous generation. The H240ar controller is designed for ProLiant DL360 and DL380 Gen9 servers and allows expansion of storage density within the server. The controller also offers basic RAID mode allowing RAID 0, 1, and 5 configurations.

The H240ar provides high levels of reliability, storage, and performance. It can provide HPE Secure Encryption for ProLiant Gen9 servers through its support of the latest Smart Array technology and advanced firmware optimizations.

Important

Unlike the Smart Array controllers, the H240ar does not support cache modules for acceleration.

The H240ar Smart HBA is ideal for internal connectivity to HDDs and SSDs with support for 12 Gb/s SAS and 6 Gb/s SATA HDDs. The H240ar Smart HBA can also provide high performance for SSDs that do not require data protection. It is ideal for driving cost-effective and reliable scalability in today's data centers. It provides midrange to enterprise-level direct attached SAS storage performance for file server, application server, messaging server, and database applications.

HPE Smart Array P440ar Controller

Figure 4-13 HPE Smart Array P440ar Controller

The HPE Smart Array P440ar Controller (Figure 4-13) is a Flexible Smart Array 12 Gb/s SAS RAID controller that provides enterprise-class storage performance, increased internal scalability with the optional HPE Smart SAS Expander Card, and data protection for ProLiant Gen9 rack servers. It features eight internal physical links. The controller delivers increased server uptime by providing advanced storage functionality, including online RAID level migration between any RAID levels with FBWC, global online spare, and prefailure warning.

When connected to 12 Gb/s internal storage devices the Smart Array P440ar delivers increased performance in messaging, database, or general server applications. It offers integrated 2 GB DDR3-1866 FBWC providing up to 14.9 GB/s cache bandwidth to reduce latency in write-intensive applications that require heavy logging activities. Optional SmartCache provides read and write acceleration for workloads such as databases or web pages.

The Smart Array P440ar offers optional HPE Secure Encryption capability that protects data at rest on any bulk storage attached to the controller. Other features include long-term data retention with integrated 2 GB FBWC for improved data reliability and predictive spare activation for improved rebuild and increased uptime even in recovery mode.

The Smart Array P440ar controller and the attached storage devices are configured and managed by the easy-to-use HPE SSA software, which is included in the server intelligent provisioning. Data compatibility throughout the Smart Array controller family allows simple migration whenever there is a need for higher performance, capacity, or availability, such as converting an existing storage volume to more advanced disk array configuration.

RAID support includes:

- **RAID 6 Advanced Data Guarding (ADG)**—Supported with a minimum of four drives. This allocates two sets of parity data across drives. This level of fault tolerance can withstand a double drive failure without downtime or data loss.

- **RAID 60**—Supported with a minimum of eight drives. This volume is composed of two or more RAID 6 sub-volumes (parity groups) where data is striped across each parity group as if it were a single physical drive. Each RAID 6 parity group can sustain up to two drive failures without incurring data loss.

- **RAID 5 (Distributed Data Guarding)**—Supported with a minimum of three drives. This allocates one set of parity data across drives. This level of fault tolerance can withstand a single drive failure without downtime or data loss.

- **RAID 50**—Supported with a minimum of six drives. This volume is composed of two or more RAID 5 sub-volumes (parity groups) where data is striped across each parity group as if it were a single physical drive. Each RAID 5 parity group can sustain a single drive failure without incurring data loss.

- **RAID 1 and 10 (Drive Mirroring)**—Supported with a minimum of two drives. This allocates half of the drive array to the data and the other half to the mirrored data, providing two copies of the data.

- **RAID 1 ADM and RAID 10 ADM**—Supported with a minimum of three drives. RAID 1 ADM creates redundant copies of the data using three drives. RAID 10 ADM stripes data across two or more sets of RAID 1 ADM volumes. This level of fault tolerance can withstand a double drive failure within a RAID 1 ADM volume without downtime or data loss.

The P440ar is ideal for midrange to enterprise-level direct attached SAS storage performance for file server, application server, messaging server, and database applications. It is also suited for environments that require encryption of sensitive data.

HPE Smart Array P840/841 Controllers

Figure 4-14 Designed for direct attached storage

The HPE Smart Array P840 (Figure 4-14) and the HPE Smart Array P841 controllers are designed for ProLiant Gen9 rack-mount servers for direct attached storage. They are 12 Gb/s controllers that support FBWC to maximize data retention in case of a power failure and supports HPE Secure Encryption. SmartCache ships standard with both controllers, providing both read and write acceleration for workloads such as databases or web pages.

The Smart Array P840 allows customers to expand internal storage from a single drive cage to the second drive cage while managing volumes and data from a single controller. The controller offers transportable (data in the cache can be migrated to a new controller) 4 GB DDR3-1866 FBWC that provides up to 14.9 GB/s cache bandwidth to reduce latency in write-intensive applications that require heavy logging. This controller supports up to 16 drives without need for the SAS expander card, thus providing low-latency point-to-point connectivity to SSDs.

The P840 is ideal for midrange to enterprise-level direct attached SAS storage performance for file server, application server, messaging server, and database applications. It is also suited for environments that require encryption of sensitive data.

The Smart Array P841 controller delivers 12 Gb/s SAS connectivity on ProLiant Gen9 servers when connected to 12 Gb/s external storage devices for increased performance in messaging, database, or general server applications. Its high port count delivers bandwidth for high performance and attached devices without impacting performance. External storage configurations can be configured for high availability using dual-port connectivity.

The P841 is ideal for customers that need to add additional storage but do not have room in their servers for an internal controller. It features 16 external physical links and delivers increased server uptime by providing advanced storage functionality, including online RAID level migration (between any RAID levels) with FBWC, global online spare, and prefailure warning.

HPE SmartDrives

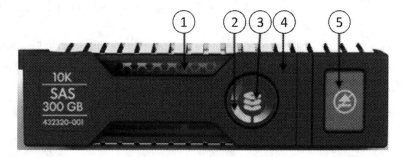

Figure 4-15 HPE SmartDrive

The SmartDrive product portfolio consists of 2.5-inch SFF and 3.5-inch LFF SATA and SAS hard drives and SSDs and features a unique hot plug carrier. Figure 4-15 shows the 300 GB hot-plug SAS hard disk drive—10K RPM, 6 Gb/sec transfer rate, 2.5-inch SFF SmartDrive.

SmartDrive carriers for ProLiant servers feature an enhanced display that clearly communicates the status of each drive so you do not have to translate the multiple combinations of blinking LEDs. This intuitive design for communicating drive activity and location helps save time and avoid data loss.

SmartDrives have fault and activity indicators and a microcontroller to monitor and store information about drive operation and status. An icon-based display reports the drive's status.

Table 4-5 SmartDrive indicators

Item	Description	Status
1	Locate backlight	The blue "locate" backlight allows for easy identification/ location of a specific drive or set of drives, critical for saving time in an environment with lots of drives in a confined space.
2	Activity ring (Spinner)	Either hidden or spins clockwise anytime HDD has one or more outstanding commands.
3	Hard drive status indicator	• Off—Not configured • Solid green—Configured • Blinking green—Rebuilding • Blinking green\amber—Configured and predictive failure • Blinking amber—Predictive failure • Solid amber—Failed
4	Handle	Drive cage handle comes out when eject button is pressed.
5	Do Not Remove (DNR) indicator is either white or hidden	Do not remove the drive when the indicator is white. Ejecting the drive results in a logical drive failure.

Table 4-5 lists the meaning of the SmartDrive status indicators. The item number listed in the first column of the table refers to the numbered labels shown in Figure 4-15.

Activity: Choosing a rack server

Consider the following customer profile that represents challenges faced by a fictional customer. Read the profile and answer the questions that follow.

Customer profile

Jenner PLC is a large professional services company headquartered in the UK. As a multi-disciplinary group, Jenner offers the full range of services demanded by clients in the property and construction industry, ranging from strategy advice to building audits and surveys. They employ 3000 people in more than 80 branch offices across Europe.

The main data center in the UK headquarters has been refreshed recently and features several HPE c7000 BladeSystems with ProLiant Gen9 server blades and HPE 3PAR storage.

The branch offices each house three ProLiant rack servers to cope with local processing needs. The remote office server base has not been refreshed over the last three years because of budget restrictions, and consists mainly of ProLiant G6 and G7 servers.

The central facility manager has been tasked with reducing energy consumption and carbon emissions by 20% within the next six months, and also needs to be able to control power consumption on a per-server basis.

Each branch office uses three main business applications with static requirements:

- Architectural design. This application requires:
    - At least 36 CPU cores
    - At least 2 TB RAM
    - At least four PCIe IO expansion slots
    - 8 TB internal storage

- Building regulations compliance. This application requires:
    - At least 16 CPU cores
    - At least 2 TB RAM
    - At least two PCIe expansion slots
    - 4 TB internal storage

- Customer and supplier database. This application requires:
    - At least 12 CPU cores
    - At least 2 TB RAM
    - At least two PCIe expansion slots
    - 6 TB internal storage

The customer has experienced several power outages in certain branch offices over the last few months and has lost sensitive business data as a result. They have expressed an interest in alternative forms of persistent memory to complement HDD and SSD storage technologies.

The computer room space is very limited and the total rack space available for new servers in each office is 6U.

Questions

Use the information in this book and in the HPE ProLiant rack and tower servers whitepaper at http://www8.hp.com/h20195/v2/GetDocument.aspx?docname=4AA3-0132ENW to answer the following questions.

1. What else do you need to know before you can recommend the most appropriate solution?

    _____

    _____

    _____

    _____

2. What are the customer's key technology challenges?

_____

_____

_____

_____

3. What products and services should you promote to the customer?

_____

_____

_____

_____

_____

_____

_____

_____

_____

Answers

The following recommendations should not be considered to be the "correct" answers, but should give you an indication of where to begin.

1. What else do you need to know before you can recommend the most appropriate solution?

   There are many questions that will help guide your choice of servers. These include:

   - What operating system is required for each application (Windows, Linux, and so forth)?

   - Does the customer make use of virtualization technologies such as VMware ESX or Microsoft Hyper-V?

   - What level of application availability is expected? What are the specifics regarding hours of downtime per year, duration of downtime, acceptable performance degradation in a failure situation, protection against failures (server, disk subsystem, network, and power), and 24 x 7 availability?

   - Are the applications designed to run on separate servers, or can they be run in a shared or virtualized environment?

   - Do the servers need to be located at the branch offices or could they be relocated to the main datacenter?

- Are there spare datacenter resources that could accommodate the workloads from the branch offices?

- Does the customer require IT services or solutions to be up and running in a short time?

- What are the specifications for number of concurrent users, benchmark results, response times, or any other relevant metrics?

- How many users do the applications need to accommodate—now and in the future? What other IT resources do applications require? Are there applications that can be consolidated?

- Does the customer have budget allocated for the project?

2. What are the customer's key technology challenges?

- Reduce energy consumption and carbon emissions by 20% within the next six months

- Several power outages have resulted in sensitive business data being lost

- Total rack space available for new servers in each office is 6U

- Power consumption needs to be controlled on a per-server basis.

3. What products and services should you promote to the customer?

If the applications must run on separate servers, a combination of DL380 Gen9 (for the architectural design application) and DL360 Gen9 (for the building regulations application and the customer and supplier database) should be considered. HPE Persistent Memory Non-Volatile DIMM (NVDIMM) could help with the customer's need to retain important business data in the event of a power failure.

If the applications can run in a shared or virtualized environment, the DL580 Gen9 could be considered.

If the servers could be relocated to the main datacenter, the applications could be consolidated and hosted on existing spare Gen9 blade servers or additional blade servers could be purchased.

Power and cooling features of ProLiant servers

Customers who are building new data center facilities or upgrading existing facilities must be aware of the constantly changing power and cooling requirements of computer hardware. To keep pace with the growing demands for power and cooling, data center infrastructure designs should consider the most current practices. Power distribution practices that support older equipment do not necessarily deliver the power density necessary for the newest enterprise IT equipment.

Recently, commissioned data centers were designed for 50 W–75 W per square foot (500 W–750 W per square meter), but loads are reaching 150 W–200 W per square foot (1500 W–2000 W per square meter). Design criteria based on average wattage per square foot (or square meter) and British

Thermal Units per hour (BTU/hr) assumed that power and cooling requirements were equal across the entire data center. Data centers are populated by racks of scalable computing systems that require enormous amounts of electricity and produce tremendous amounts of heat. Average (per unit area) design criteria do not encompass the specific power and cooling requirements of high-density solutions. Consequently, data centers can no longer be designed by using average wattage and BTU criteria.

Additionally, cooling patterns that support older equipment styles are not compatible with the most current enterprise IT equipment. Although the actual power and heat densities that customers should plan for depend on the equipment specification and IT strategy and hardware adoption rates, HPE best practices maximize effectiveness in high-density data centers.

Thermal management

For HPE servers, managing thermal output is an internal and external process. Internally, HPE server fans draw cool air over the heated components. HPE engineers carefully consider airflow when they determine where to place components within a server. Many designs include baffles and heat sinks to help keep components cool.

Thermal management in the data center is expressed in tons of cooling. Many heating, ventilation, and air conditioning (HVAC) units meet cooling tonnage requirements, but it is essential to get cool air where it is needed in the data center.

 Important

> Whether designing a new data center or retrofitting an existing one, you should work with knowledgeable HVAC engineers to ensure adequate cooling.

Increasing availability through power protection

Businesses cannot rely on utility power as a source of continuous power for critical equipment. HPE has developed a full line of power management products that protect and manage systems ranging from individual workstations to distributed enterprises:

- Automated energy optimization with HPE Intelligent Power Discovery

- HPE Common Slot (CS) Power Supplies

- HPE Intelligent Power Distribution Units (iPDUs)

- HPE uninterruptible power supplies (UPSs)

- HPE Intelligent Power Discovery Services

- HPE Thermal Discovery Services

Intelligent Power Discovery Services combine an iPDU and CS Platinum/Platinum Plus Power Supplies with HPE software to create an automated, energy-aware network between IT systems and facilities. Intelligent Power Discovery Services with iPDUs automatically track power usage and document configurations to increase system uptime and reduce the risk of outages.

Intelligent Power Discovery provides automated server discovery on a network through power line communication technology that is embedded in CS Platinum Power Supplies. Power line communication is a feature that allows the power supply to communicate with the iPDU. The communication between the power supply and iPDU helps:

- Automatically discover the server when it is plugged into a power source

- Map the server to the individual outlet on the iPDU

Thermal Discovery Services increase compute capacity by intelligently placing the workload where customers have the most power and cooling capacity. This feature helps businesses both reduce energy consumption and maximize usage of data center power and cooling capacity.

 Note

Scan this QR code or enter the URL into your browser for more information on rack, power, and cooling solutions from HPE.

https://www.hpe.com/us/en/integrated-systems/rack-power-cooling.html

---

HPE iPDUs

Figure 4-16 HPE iPDUs

The key element of Power Discovery Services is the iPDU, shown in Figure 4-16, which is a power distribution unit with full remote outlet control, outlet-by-outlet power tracking, and automated documentation of power configuration. HPE iPDUs track outlet power usage at 99% accuracy, showing system-by-system power usage and available power. The iPDU records server ID information by outlet and forwards this information to HPE Insight Control/HPE OneView, saving hours of manual spreadsheet data-entry time and eliminating human wiring and documentation errors.

When combined with the HPE line of Platinum-level high-efficiency power supplies, the iPDU actually communicates with the attached servers to collect asset information for the automatic mapping of the power topology inside a rack. This capability greatly reduces the risk of human errors that can cause power outages.

Using the popular core-and-stick architecture of the HPE modular PDU line, the iPDU monitors power consumption at the core, load segment, stick, and outlet level with unmatched precision and accuracy. Remote management is built into the PDU core for power cycle ability of individual outlets on the extension bars and high-precision monitoring of current, voltage, wattage, and power that is 99% accurate across the C19 outlets.

The iPDU can help users track and control power other PDUs cannot monitor. Information is gathered from all monitoring points at half-second intervals to ensure the highest precision. The iPDU can detect a new server even before it is powered on.

An iPDU is ideal for enterprise customers who want to speed power configuration, improve monitoring to reclaim stranded power, and extend data center life. An iPDU enables you to:

- Determine the exact power consumption for every component in the rack with less than 1% variation in accuracy

- Monitor and control outlets individually for the ultimate control of power distribution within the rack

- Save valuable space in the rack with dense rack-mountable form factors

HPE iPDUs have a patented modular architecture that improves their flexibility. This building block concept consists of two main parts, the iPDU core unit and extension bars. The Intelligent PDUs ship with one core unit, the mountable LED display, and mounting hardware. The six-outlet core unit is zero-U or 1U rack mountable. The 12-outlet core unit is 2U rack-mountable. This architecture:

- Communicates with the attached servers to collect asset information for the automatic mapping of the power topology inside a rack

- Speeds implementation time

- Greatly reduces the risk of human errors that can cause power outages

HPE power distribution units

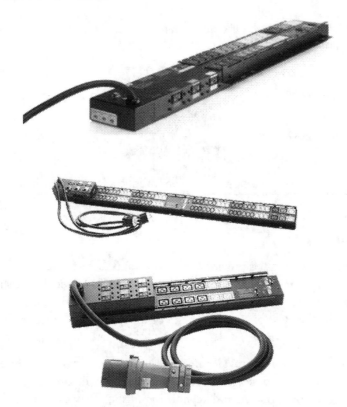

Figure 4-17 HPE power distribution units

HPE PDUs, some of which are shown in Figure 4-17, provide power to multiple objects from a single source. In a rack, the PDU distributes power to the servers, storage units, and other peripherals.

PDU systems:

- Address issues of power distribution to components within the computer cabinet
- Reduce the number of power cables coming into the cabinet
- Provide a level of power protection through a series of circuit breakers

---

 Note

Scan this QR code or enter the URL into your browser for more information on the HPE power distribution unit portfolio.

http://www8.hp.com/us/en/products/pdu/index.html

HPE Metered PDUs

The HPE metered rack-mount PDUs provide both single- and three-phase monitored power, as well as full-rack power utility ranging from 8.3 kVA to 19.9 kVA. HPE Metered PDUs provide rack-level power monitoring over the network. With billing grade accuracy, these PDUs are well suited for data centers wanting to make the most of circuit capacity or for co-location facilities needing to precisely track power usage for billing purposes.

The low-profile, single-piece design provides multiple mounting options and ease of access to rear devices for maintenance. They are designed for HPE racks but will also work in a large variety of third-party racks using standard button and keyhole mounting. They can be installed on either side of the rack with outlets facing the back of the rack for easy access and improved clearance. For greater power density, they can be installed side by side on both sides of the rack with the outlets facing in toward the center of the rack.

Available metered PDUs include:

- Full-height (42U), mid-height (36U), and half-height (22U) vertical versions as well as horizontal (1U) versions
- Three-phase models with C19 receptacles
- Single-phase models with C13 and C19 receptacles

HPE Horizontal/Modular PDUs

HPE Horizontal/Modular PDUs have a unique modular architecture designed specifically for data center customers who want to maximize power distribution and space efficiencies in the rack.

Modular PDUs consist of two building blocks: the core unit and extension bars. The core unit can be mounted in either 1U or 0U locations and houses the output receptacles and circuit breakers for each load segment. Extension bars plug into the core unit, extending the outlet receptacles along the length of the rack. Extension bars reduce the need for long power cords and confusing clutter. They fit easily into the frame of the rack. Available models range from 16A to 48A current ratings, with output connections ranging from 2 to 32 outlets.

Benefits of HPE Modular PDUs include:

- Increased number of outlet receptacles

- Modular design

- Superior cable management

- Flexible 1U/0U rack mounting options

- Easy accessibility to outlets

- Limited three-year warranty

ProLiant Gen9 platform power architecture

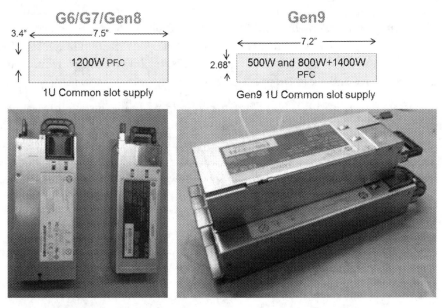

Figure 4-18 Gen9 power supply compared to previous generations

The ProLiant Gen9 power supply is 25% smaller by volume than the supplies of previous server generations, as shown in Figure 4-18. The large reduction in volume and 25% improvement in power density (from 40 W per inch to 50 W per inch) provide more server power in a smaller space. This large reduction in volume enables the HPE server design teams to add more internal devices:

- 1U platforms can add one PCI slot

- 2U platforms can add four SSDs (and potentially one more PCI slot)

HPE Flex Slot Power Supplies share a common electrical and physical design that allows for hot-plug, tool-less installation into ProLiant Gen9 servers. Flex Slot Power Supplies are rated for Platinum-level certification with efficiency of up to 94% and Titanium-level certification with efficiency of up to 96%. Support for Power Discovery Services, through embedded PLC technology, is available with the 800 W Titanium and 1400 W Platinum Plus models. This feature enables each server to communicate identification, location, and power-related data to the Intelligent PDU in the rack; this information can then be shared with HPE Insight Control/HPE OneView to manage power usage and efficiency in the data center.

These power supplies support the HPE portfolio of AC, DC, and high voltage (HVAC/HVDC) power supply options. Flex Slot Power Supplies are certified for high-efficiency operation and offer multiple power output options, allowing users to "right-size" a power supply for specific server configurations. They support both low-line and high-line AC input voltages providing additional flexibility to operate in multiple IT environments (500 W and 800 W Platinum only). Input voltages of –48VDC, 277VAC, and 380VDC are also available. This flexibility helps to reduce power waste, lower overall energy costs, and avoid "trapped" power capacity in the data center.

An HPE innovation, the HPE 750 W Flex Slot Battery Backup Module installs into a single Flex Slot Power Supply bay to free up the space needed for rack mount UPS without compromising server uptime. Daisy-chained operation between two battery backup modules allows paralleled connection for pass-through power sharing between two ProLiant DL300 Gen9 series servers. In the event of a power outage, the module can provide up to 750 W of power for up to 60 seconds, or up to 500 W when in paralleled configuration.

HPE uninterruptible power systems

Figure 4-19 HPE uninterruptible power systems

HPE UPSs, shown in Figure 4-19, are supported with ProLiant Gen8 and Gen9 servers. They provide high-efficiency power protection for all environments, from workstation to data center.

Features include:

- Highest wattage per U space saves valuable room in the rack

- True sine wave output protects valuable equipment

- More than 97% efficiency prevents wasted power

- Enhanced battery management (EBM) yields up to two times the usable battery life

- Hot-swappable batteries save time

- There is a standard three-year warranty, even on the batteries

- Prefailure battery notification prevents outages

HPE Intelligent Series Racks with Location Discovery Services

HPE Intelligent Series Racks with HPE Location Discovery Services enable ProLiant Gen8 and Gen9 servers to self-identify and inventory to optimize workload placement. Location Discovery

Services is a solution that merges the physical rack with IT equipment to provide automated location awareness of installed servers for advanced asset management and workload placement.

HPE Intelligent Series Racks equipped with the Location Discovery Services option provide detailed location information to ProLiant Gen8 and Gen9 servers to track new installations and equipment moves. Upon installation, the server's iLO queries and records the rack identifier as well as the exact location of the server in the rack.

After the server identifies exactly where it is located, this information can be forwarded to other systems such as HPE OneView to update new rack configurations automatically. This saves time in configuring hardware and software to manage the data center. It also eliminates constant manual updates and the associated human errors that can slow resolution of issues and even result in downtime or loss of business.

 Note

For more information on Location Discovery Services, scan this QR code or enter the URL into your browser to read the QuickSpecs.

http://www8.hp.com/h20195/v2/GetPDF.aspx/c04123199.pdf

Elevated temperature support

The ASHRAE guidelines for data center operation suggest thermal ranges of operation for IT equipment that are commonly accepted by many data center administrators and facility managers. ProLiant Gen9 products support A3 and A4 guidelines, which means that certain ProLiant Gen9 servers can operate at higher temperatures than previous generations. This provides more choices for air-cooled solutions and the potential to reduce cooling costs.

The allowable ranges for classes A3 and A4 are intended to remove obstacles to data center cooling strategies such as free-cooling methods. Free-cooling takes advantage of a facility's local climate by using outside air to cool IT equipment directly—without the use of mechanical refrigeration (chillers or air conditioners) whenever possible.

A variety of implementations for free-cooling are possible. For example, filtered outside air can be drawn directly into the data center. Other techniques keep the outside air isolated from the data center but still transfer the data center heat directly to the outside air without refrigerating it. Careful application of the new ASHRAE guidelines might enable free-cooling in more climates or allow for the data center to be cooled without refrigeration more days of the year. Reducing the use of refrigerated cooling lowers the operating expenses for the data center. In some cases, the refrigeration equipment can be eliminated or significantly reduced in size, saving capital expense as well.

The traditional maximum server inlet air temperature is 35°C. For ProLiant Gen9 servers, 40°C (ASHRAE A3) support is available on most platforms with configuration limitations. For select ProLiant Gen9 platforms, 45°C (ASHRAE A4) support is available with configuration limitations.

 Note

Check the QuickSpecs for specific servers to find ASHRAE support information.

HPE Modular Cooling System 200/100

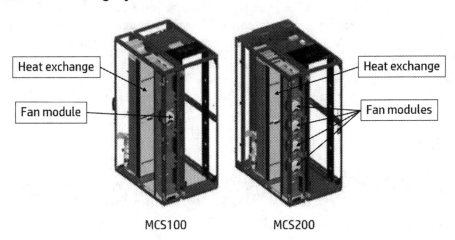

Figure 4-20 Airflow for MCS-100 (single-rack configuration)

The Modular Cooling System (MCS) is a portfolio of modular, rack-based cooling solutions that remove the high levels of heat generated by current server, mass storage, and core networking systems. Equipped with rack infrastructure, cooling, and IT power distribution, the MCS provides a standardized high-performance cooling approach for servers and other IT equipment installed in server racks. HPE Bi-Directional cooling technology allows simultaneous cooling of two racks when the MCS Expansion Rack is utilized. The MCS racks can cool as little as 5 kW, depending on configuration, and as high as 50 kW of server capacity, either all on one rack or split between two racks.

MCS is designed to complement any new or existing data center by enabling the scaling of computing power without adding to the heat load in the data center. In addition, by packing up to 10 times the kW capacity of a standard rack, the MCS will extend the life of the data center. The horizontal air flow of the MCS fully supports industry-standard front-to-back cooling designs. With the variable-speed fan system, all devices receive adequate and evenly distributed cool air, regardless of the mounting position or workload. MCS-100/200 models are available in four configurations:

- MCS-100 single-rack configuration—10 kW capacity

- MCS-100 dual-rack configuration— Expandable to 30 kW with additional fan modules

- MCS-200 single-rack configuration—30 kW capacity

- MCS-200 dual-rack configuration—Expandable to 50 kW or N + X redundancy with additional fan modules

MCS units ship with the server, storage, and networking infrastructure fully integrated, cabled, and tested. You can choose the standard rack or a combination with the MCS Expansion Rack; both options are 42U tall.

The MCS 100 ships with one fan module and one heat exchanger (HEX) as shown on the left side of Figure 4-20. The MCS 200 ships with four fan modules standard as shown on the right side of Figure 4-20. Each fan module contains a variable-speed circulation fan. The HEX module contains an air-to-water heat transfer device, which discharges cold air to the front of the rack via a side portal. Chilled water for the heat exchanger can be provided by your facility's chilled water system or by a dedicated chilled water unit. As the servers expel warm exhaust air out through the rear of the rack, the MCS fan modules re-direct the warm air into the HEX module. The air is re-cooled and then re-circulated to the front of the rack. Any condensation that forms is collected in the HEX module and sent through a discharge tube to a condensation tray located in the base of the enclosure.

Using QuickSpecs

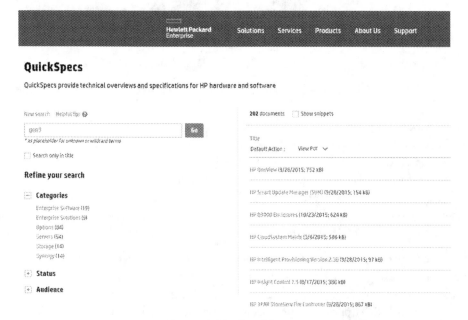

Figure 4-21 QuickSpecs

QuickSpecs is a convenient resource of overviews and technical specifications of HPE hardware and software. It can be accessed through the HPE Marketing Document Library, as shown in Figure 4-21. The library provides an enhanced online and mobile experience for QuickSpecs by offering full text search, faceted navigation, and search results sorted by most popular documents.

QuickSpecs also can be accessed offline by using the HPE Product Bulletin application or through synchronization with standard cloud solutions.

 Note

Scan this QR code or enter the URL into your browser for more information on QuickSpecs.

http://www.hpe.com/info/qs

Learning check

1. List the three defining characteristics of ProLiant Gen9 rack servers.

   _____

   _____

   _____

2. Which ProLiant Gen8 server preceded the HPE ProLiant DL180 Gen9 server?

   A. ProLiant DL320e Gen8

   B. ProLiant DL360e Gen8

   C. ProLiant DL380e Gen8

   D. ProLiant DL380p Gen8

3. What is the leading ProLiant server for dense general-purpose computing environments?

   A. ProLiant DL120 Gen9

   B. ProLiant DL 180 Gen9

   C. ProLiant DL360 Gen9

   D. ProLiant DL580 Gen9

4. Which features are available on ProLiant 580 Gen9 servers? (Select two.)

   A. Xeon E5-2600 v3 processors

   B. 3000W HPE Common Slot Power Supplies

   C. Nine PCIe 3.0 slots

   D. 144 DIMM slots

   E. 6 TB DDR4 RAM

5. What are two features of the Dynamic Smart Array B140i controller? (Select two.)

   A. Embedded in system board of ProLiant DL Gen9 servers

   B. 12 Gb/s SAS connectivity

   C. Support for up to 10 SATA drives

   D. Support for RAID 6

   E. 8 GB FBWC

6. What is an HPE iPDU?

    _____

    _____

    _____

Learning check answers

1. List the three defining characteristics of ProLiant Gen9 rack servers.

    ● **Flexible—"Right-sized" computing with flexible resources**

    ● **Reliable—Reliable, fast, and secure infrastructure coupled with simple automation**

    ● **Performance optimized—Optimized workload performance in compute, storage, and networking**

2. Which ProLiant Gen8 server preceded the HPE ProLiant DL180 Gen9 server?

    A. ProLiant DL320e Gen8

    B. ProLiant DL360e Gen8

    C. ProLiant DL380e Gen8

    D. ProLiant DL380p Gen8

3. What is the leading ProLiant server for dense general-purpose computing environments?

    A. ProLiant DL120 Gen9

    B. ProLiant DL 180 Gen9

    C. ProLiant DL360 Gen9

    D. ProLiant DL580 Gen9

4. Which features are available on ProLiant 580 Gen9 servers? (Select two.)

    A. Xeon E5-2600 v3 processors

    B. 3000W HPE Common Slot Power Supplies

    C. Nine PCIe 3.0 slots

    D. 144 DIMM slots

    E. 6 TB DDR4 RAM

5. What are two features of the Dynamic Smart Array B140i controller? (Select two.)

   A. Embedded in system board of ProLiant DL Gen9 servers

   B. 12 Gb/s SAS connectivity

   C. Support for up to 10 SATA drives

   D. Support for RAID 6

   E. 8 GB FBWC

6. What is an HPE iPDU?

   - **An HPE iPDU is a power distribution unit with full remote outlet control, outlet-by-outlet power tracking, and automated documentation of power configuration.**

   - **HPE iPDUs:**
     - **Track outlet power usage at 99% accuracy, showing system-by-system power usage and available power**
     - **Record server ID information by outlet and forward this information to HPE Insight Control, saving hours of manual spreadsheet data-entry time and eliminating human wiring and documentation errors**

Summary

- ProLiant Gen9 rack-mounted servers meet the needs of customers who are either new to servers or who are positioned to grow and expand their business.

  - For businesses that are new to servers, HPE recommends the ProLiant 10 series.

  - For growing businesses or new IT growth customers, HPE recommends the ProLiant 100 series.

  - For SMB, enterprise, and HPC customers that use traditional IT and require a mission-critical environment, HPE recommends the 300 series.

  - For customers requiring the most demanding scale-up workloads, HPE offers the 500 series.

- HPE offers optional Smart Array controllers and SmartDrives for ProLiant DL servers.

- Customers who are building new data center facilities or upgrading existing facilities must be aware of the constantly changing power and cooling requirements of computer hardware. HPE has developed a full line of power and cooling management products that protect and manage computer service systems ranging from individual workstations to distributed enterprises.

- QuickSpecs is a convenient online and offline resource of overviews and technical specifications of HPE hardware and software.

5 HPE BladeSystem Server Solutions

After completing this chapter, you should be able to:

✓ Explain how the Hewlett Packard Enterprise (HPE) BladeSystem Gen9 portfolio provides solutions in the compute era

✓ List the steps to build a BladeSystem solution

    a. Select the operating environment

    b. Select the BladeSystem enclosure

    c. Select the interconnects and adapters

    d. Select the server blades

    e. Select the storage infrastructure

    f. Select the infrastructure management

    g. Select the power and cooling configurations

    h. Select the services

OPENING CONSIDERATIONS

Before proceeding with this section, answer the following questions to assess your existing knowledge of the topics covered in this chapter. Record your answers in the space provided here.

1. What is your experience working with BladeSystem solutions? What have you learned from designing BladeSystem solutions?

_____

_____

_____

_____

2. Describe your experience with HPE server interconnects and adapters such as HPE Virtual Connect, blade switches, and server adapters.

_____

_____

_____

_____

_____

_____

3. Describe your experience with HPE infrastructure management tools such as HPE Onboard Administrator, HPE integrated Lights-Out (iLO), Insight Online, HPE OneView, and Insight Online for HPE ProLiant BladeSystem solutions.

_____

_____

_____

_____

_____

_____

_____

_____

HPE BladeSystem Gen9 solutions in the compute era

With the growth of data fueling today's global environment, the ability of a business to adapt to change quickly is becoming critical for survival. IT managers are under enormous pressure to deliver applications and services that innovate and transform the business at a lower cost.

Rather than adding more technology silos, IT organizations need an agile and reliable converged infrastructure platform that is purpose-built for enterprise workloads such as virtualization and cloud computing, ready to deliver industry-leading total cost of ownership (TCO), and able to increase IT staff productivity.

To help address these ever-changing business needs, HPE BladeSystem enables customers to shift investment from routine maintenance to innovation, maximize availability by reducing unplanned downtime, and accelerate enterprise workload deployment, such as virtualization and cloud computing much faster.

BladeSystem is a modular infrastructure platform that converges servers, storage, and network fabrics to accelerate operations and speed delivery of applications and services running in physical, virtual, and cloud-computing environments. Because the core infrastructure is shared, capital costs can be significantly lower. Blades share power, cooling, network, and storage infrastructure at the BladeSystem enclosure level, resulting in a dramatic reduction in power distribution units, power cables, LAN and SAN switches, connectors, adapters, and cables.

BladeSystem delivers the Power of One

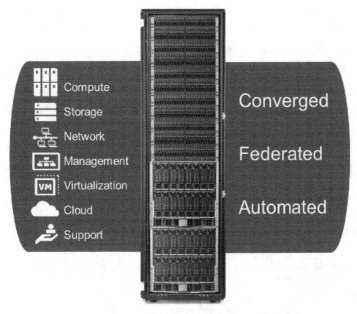

Figure 5-1 The Power of One—One infrastructure, one management platform

BladeSystem is founded on the principles of having a single infrastructure and one management platform. BladeSystem with HPE OneView delivers a new experience for IT with the Power of One—one infrastructure, and one management platform to speed the delivery of services. BladeSystem can lower data center costs by 68%[1] so that you can shift investment from routine maintenance to innovation, maximize availability by reducing downtime up to 90%[1], and dramatically accelerate enterprise workload deployment such as virtualization and cloud computing.

 Note

1. IDC white paper sponsored by HPE, *Business Value of Blade Infrastructures,* #227508R2.

HPE OneView combines server, storage, and networking with control of the data center environment into a single, integrated management platform designed to deliver life cycle management for the complete Converged Infrastructure. With HPE Onboard Administrator, HPE iLO remote management, and HPE OneView, BladeSystem server blades can be managed and controlled regardless of the state of the server operating system.

BladeSystem delivers the Power of One by having one management platform and a single infrastructure that is converged, federated, and automated. As illustrated by Figure 5-1, it provides a single experience that offers key advantages:

- **Converged**—Compute, network, storage, virtualization, and management are included in one infrastructure so that costs can be reduced.

- **Federated**—With BladeSystem and HPE OneView, the chassis becomes a single system. With a federated system, HPE OneView makes it possible to move server profiles across multiple chassis, so any compute blade is available for any appropriate workload in the whole system. A valuable benefit is the ability to automate the environment. Because it is federated and not hierarchal, the environment is safer. Failure domains are smaller, and a failure in any one part does not impact the other parts.

- **Automated**—Day-to-day lifecycle management is made easier with automated provisioning, proactive health monitoring, and virtual machine (VM) failover.

Improve IT services at lower cost with HPE BladeSystem

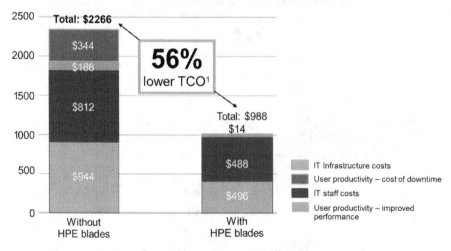

Figure 5-2 Costs for IT infrastructure, IT staff, and user productivity

As shown in Figure 5-2, IDC research shows that organizations migrating to BladeSystem from traditional rack servers or upgrading their BladeSystem servers make substantial user productivity gains and reduce the cost of delivering computer services by an average of 56%[1].

Note

1. IDC white paper sponsored by HPE, *Business Value of Blade Infrastructures*, #227508R2.

- BladeSystems enhance user productivity and support the business. HPE customers told IDC that users are more productive with BladeSystem thanks to improved performance and availability of important business applications. IDC projects that employee productivity gains are worth an average of $469 per year per user over three years.

- BladeSystems deliver IT infrastructure cost savings. HPE customers reported that they have consolidated their server footprints and deployed more virtual machines, which has helped them reduce related costs, including server hardware, network infrastructure, power, and facilities-related costs. Over three years, IDC calculates that these cost savings have an average value of $424 per user.

- BladeSystems enable IT staff efficiencies. HPE customers explained that their IT staff has become more efficient by saving time and making productivity gains thanks to the integrated nature of the blade platform. IT staff spends less time monitoring and managing server environments and needs substantially less time to deploy a server blade. IDC puts the value of these efficiencies and time savings at an average of $307 per user per year over three years.

The cumulative result of these efficiencies is that these organizations are able to provide improved IT services at a substantially lower cost to their users with BladeSystems.

BladeSystem converged architectural advantages

The converged architecture in BladeSystem provides several key advantages including simplified IT operations, workload optimization, and an optimized infrastructure.

- Simplified IT operations

  - **Unified management**—One console manages compute, networking, and storage

  - **Seamless server configuration**—Readily provisioned templates enable profile mobility for rapid scalability and failover operations

  - **Change-ready networking**—HPE Virtual Connect and FlexFabric connections can be provisioned in HPE OneView

  - **Managed data services**—Shared experience for device and data service management with HPE 3PAR storage systems

- Workload-optimized

  - **Seamlessly virtualized**—In a virtual desktop infrastructure (VDI) with VMware vMotion and virtual volumes (VVols), the entire infrastructure to transform business operations and productivity is available

- **Data-driven optimized**—Better-together innovation reduces costs and consolidates operational and application silos

- Optimized infrastructure

  - **Wire-once, simplified connectivity**—Converge IO fabrics to reduce sprawl while virtualizing the edge including Flat SAN connectivity to HPE 3PAR StoreServ storage systems

  - **Flash-optimized storage**—Tier 1 HPE 3PAR storage delivers quality of service (QoS), extreme performance, and built-in protection

BladeSystem is the bridge to a Composable Infrastructure

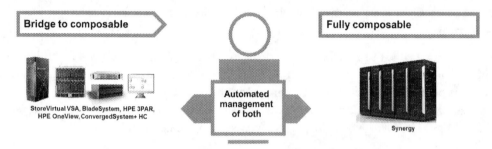

Figure 5-3 Bridge to Composable Infrastructure

HPE Composable Infrastructure is a new architecture that lets IT administrators and developers use infrastructure as code to control their internal environments. The HPE Synergy platform is the world's first platform built from the ground up to bridge traditional and new IT with the agility, speed, and continuous delivery needed for today's applications.

The bridge to Composable Infrastructure is through BladeSystem with HPE OneView, as illustrated by Figure 5-3. BladeSystem with ProLiant Gen9 servers already takes advantage of some of the attributes of composability through HPE OneView.

With HPE OneView, customers can already start to treat resources as blocks of compute, storage, and fabric. They can use the software defined intelligence through HPE OneView and can leverage the unified API ecosystem today and can carry forward those partnerships through to Composable Infrastructure. Customers who are using HPE 3PAR or leveraging HPE StoreVirtual VSA or other software assets can experience composable attributes.

Building a BladeSystem solution

The HPE global community of business technology experts and partners is committed to helping build solutions and support plans that are right for a customer's business needs. HPE integrates the infrastructure essentials inside the BladeSystem so that it arrives at the customer site ready to deliver the best business results.

Building a BladeSystem infrastructure solution begins with eight steps:

1. Select the operating environment.

2. Select the BladeSystem enclosure.

3. Select the interconnects and adapters.

4. Select the server blades.

5. Select the storage infrastructure.

6. Select the infrastructure management.

7. Select the power and cooling configurations.

8. Select the services.

 Note

These eight steps are a connected sequence, rather than a linear, one-time progression. It might be necessary to repeat the process to arrive at the most appropriate solution to meet the customer's business needs. For example, if a BladeSystem c3000 enclosure is selected during the first iteration and it later becomes evident that more than eight server blades are required, it will be necessary to review the enclosure selection and either add another c3000 enclosure, or select a different enclosure, such as the c7000.

Step 1: Select the operating environment

Customers can purchase their entire operating environment from HPE. HPE resells and provides full service and support for the following supported operating systems and virtualization software:

- Canonical Ubuntu—http://h17007.www1.hpe.com/us/en/enterprise/servers/supportmatrix/ubuntu.aspx

- CentOS—http://h17007.www1.hpe.com/us/en/enterprise/servers/supportmatrix/cent_os.aspx

- Citrix—http://h17007.www1.hpe.com/us/en/enterprise/servers/supportmatrix/xenserver.aspx

- Microsoft Windows Server with Hyper-V—http://h17007.www1.hpe.com/us/en/enterprise/servers/supportmatrix/windows.aspx

- Oracle Linux—http://h17007.www1.hpe.com/us/en/enterprise/servers/supportmatrix/oracle_linux.aspx

- Oracle Solaris—http://h17007.www1.hpe.com/us/en/enterprise/servers/supportmatrix/solaris.aspx

- Red Hat Enterprise Linux (RHEL)—http://h17007.www1.hpe.com/us/en/enterprise/servers/supportmatrix/redhat_linux.aspx

- SAP Linux—http://h17007.www1.hpe.com/us/en/enterprise/servers/supportmatrix/hplinuxcert-sap.aspx

- SUSE Linux Enterprise Server (SLES)—http://h17007.www1.hpe.com/us/en/enterprise/servers/supportmatrix/suse_linux.aspx

- Wind River Linux—http://h17007.www1.hpe.com/us/en/enterprise/servers/supportmatrix/wind_river.aspx

- VMware—http://h17007.www1.hpe.com/us/en/enterprise/servers/supportmatrix/vmware.aspx

The URLs in this list lead to the HPE Servers Support and Certification Matrices webpages. Navigate to the appropriate URL and select the operating system version to obtain the latest information.

Step 2: Select the BladeSystem enclosure

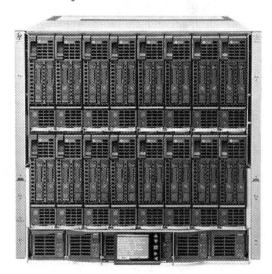

Figure 5-4 BladeSystem c7000 enclosure

Both the BladeSystem c7000, shown in Figure 5-4, and the c3000 enclosures provide all the power, cooling, and IO infrastructure needed to support modular server, storage, interconnect, and power management components. They consolidate the components into a single solution that can be managed as a unified environment.

Both models include a shared 7.1 Tbps high-speed midplane for wire-once connectivity of server blades to network and shared storage. Power is delivered through a pooled power backplane that ensures that the full capacity of the power supplies is available to all blades.

With demanding workloads, the increased power supply wattage and midplane bandwidth aligned with intelligent infrastructure technologies such as Platinum Power Supplies, Intelligent Power Module, and Location Discovery Services enhanced the foundation for a converged infrastructure. HPE Thermal Logic technology helps to minimize power consumption and reduce cooling.

The Onboard Administrator module provides a single point of control for intelligent management of the entire enclosure. An additional Onboard Administrator system management module can be added to provide redundancy. Insight Display, shown at the bottom center of Figure 5-4, is powered by the Onboard Administrator and provides local management through an LCD display conveniently sited on the front of the system.

The BladeSystem c7000 enclosure is 10U high and supports the following components:

- Up to 16 half-height server blades, eight full-height server blades, and eight expansion blades per enclosure (not exceeding 16 total blades)

- Up to four redundant interconnect IO fabrics (Ethernet, Fibre Channel, InfiniBand, iSCSI, and serial attached SCSI [SAS]) supported simultaneously within the enclosure

- Choice of single-phase high-line AC, three-phase high-line AC, single-phase high voltage AC, -48V DC, or high voltage DC power options for power input flexibility

- Up to six hot-plug high-efficiency power supplies per enclosure

- A minimum of four hot-plug HPE Active Cool 200 Fans; for redundancy, additional capacity, and improved power consumption and acoustics, Active Cool 200 Fan kits can be added for a maximum of 10 fans

The BladeSystem c3000 enclosure is 6U high and supports:

- Up to eight half-height server blades, four full-height server blades, and four expansion blades per enclosure (not exceeding eight total blades)

- Up to two redundant IO interconnect fabrics (Ethernet, Fibre Channel, iSCSI, SAS, and so forth) supported simultaneously within the enclosure

- A single-phase power subsystem for connecting to data center power, uninterruptible power supplies (UPSs), or low-line (100VAC to 110VAC) wall outlets

---

 Caution

When connecting directly to wall outlets, determine the maximum amperage of the wall outlet circuit to prevent power overload.

- Up to six hot-plug Active Cool 100 fans

- Redundant hot-plug cooling, redundant hot-plug power supplies, redundant connections, redundant interconnect modules, and optional redundant BladeSystem Onboard Administrator management module

---

 Note

Scan this QR code or enter the URL into your browser for more information on BladeSystem and its components.

https://www.hpe.com/us/en/integrated-systems/bladesystem.html

---

Step 3: Select the interconnects and adapters

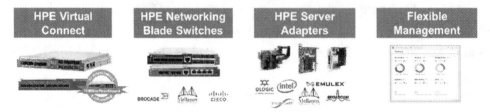

Figure 5-5 BladeSystem interconnects and adapters

HPE has a broad networking portfolio for BladeSystem solutions that provides industry-leading flexibility and performance. As shown in Figure 5-5, HPE has a wide portfolio of Virtual Connect, blade switches, and server adapters, coupled with the flexible management capabilities of HPE OneView.

With its open architecture, BladeSystem eliminates vendor lock-in by providing the industry's most flexible, multi-vendor framework for investment protection. HPE provides superior choice and flexibility based on industry standards, ensuring interoperability.

Virtual Connect

Virtual Connect is an essential building block for any virtualized or cloud-ready environment. This innovative, wire-once connection management technology simplifies server connectivity, making it possible to add, move, and change servers in minutes instead of hours or days. Virtual Connect is the simplest way to connect servers to any network and reduces network sprawl at the edge of the network.

Virtual Connect converges server edge connections, making server changes transparent to storage and networks, and delivering four times the number of connections per physical network link than possible with traditional networking technology. Virtual Connect enables server administrators to dynamically optimize and control bandwidth using fewer physical ports for the same performance and reduces server edge infrastructure (switches, HBAs, NICs, and cables), and costs.

Virtual Connect enables routine infrastructure changes in less time with wire-once connectivity so server administrators can add, replace, and recover server resources from a centralized console and without involving storage and networking administrators.

Additional cost savings can be achieved by consolidating switch connectivity with dual-hop Fibre Channel over Ethernet (FCoE). Virtualization network traffic demands low-latency server-to-server and server-to-storage connectivity. Together with the HPE Intelligent Resilient Framework (IRF), Virtual Connect enables flatter low-latency networks. With more than 10 million ports shipped, Virtual Connect continues to deliver proven simplification of operations.

 Note

Scan this QR code or enter the URL into your browser for more information on the Virtual Connect portfolio.

http://www.hp.com/go/virtualconnect

Switch and Virtual Connect interconnect solutions comparison

Switch interconnect	Virtual Connect interconnect
– Is part of the LAN network – Includes enterprise-class network management and features – Has consistent network architecture from server edge through TOR/EOR – Is directly connected to server and any server change affects the network	– Is part of the server system – Is managed with the servers – Provides abstraction through a layer between servers and network, so network does not see server changes – Allows flexibility and control of system resources

Figure 5-6 Switch and Virtual Connect interconnect solutions comparison

In a BladeSystem solution, a Virtual Connect interconnect module forms a Layer 2 bridge between the server blades and the Ethernet and storage networks. Figure 5-6 shows a comparison between switches and Virtual Connect interconnect solutions.

Typically, the Virtual Connect module is managed as part of the overall server system by the server administrator, who makes use of storage and networking resources that have been provided by the storage and networking administrators. Managing a Virtual Connect module is relatively simple because it is not as complicated as a switch. Therefore, the server administrator can easily handle the configuration tasks without detailed networking knowledge.

The Virtual Connect module pools and shares network connections for the servers within the BladeSystem so that server administrators can upgrade, replace, or move server blades without changes being visible to the external LAN and SAN environments and, therefore, without requiring intervention by LAN and SAN administrators.

Essentially, Virtual Connect is an edge port aggregator that provides Layer 2 networking capabilities. It is not configured, deployed, or managed like a traditional switch, and its uplink ports are termination ports, not transit ports. In other words, Virtual Connect presents itself to the network as an endpoint—like a server instead of a switch—which means it is managed as part of the server system rather than as part of the network.

One of the major technical benefits of using the Virtual Connect modules is that they enable any server NIC to be connected to any data center switch.

Virtual Connect—Performance for virtualized, cloud-ready data centers

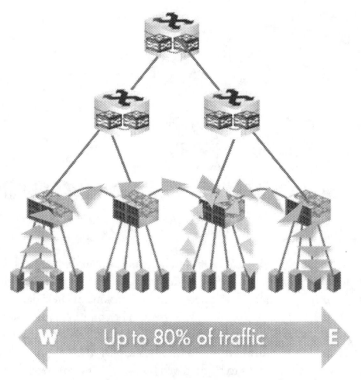

Figure 5-7 Data center traffic flow

Analysts estimate that more than 80% of data center traffic flows between servers, as illustrated by Figure 5-7. VMware vMotion is a key driver of this trend. In these situations, it is not necessary to push all traffic to network core switches as typically practiced in traditional switched networks.

Virtual Connect is optimized for east-west (server-to-server) traffic within a Virtual Connect domain. Workloads on up to 64 servers within a four-enclosure domain can communicate with each other without leaving the domain, thereby minimizing network latency. This unique east-west traffic-flow capability extends to allow traffic between any servers connected to the same Virtual Connect module to communicate over an internal link without oversubscription. By planning server deployment within BladeSystem enclosures, networking can be simplified and performance can be improved.

Validated vMotion acceleration

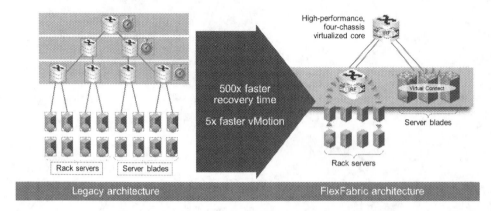

Figure 5-8 Validated vMotion acceleration

Virtual Connect can be combined with HPE Intelligent Resilient Framework (IRF) to improve server-to-server performance significantly. IRF is an innovative switch platform virtualization technology that allows customers to simplify the design and operations of their data center and campus Ethernet networks. With IRF, multiple physical switches can be combined into one virtualized switch known as an *IRF domain,* which is managed and maintained as a single entity. The resulting IRF domain virtualizes switching platforms to dramatically simplify the design and operations of the network fabric. It also enables networks to be flattened by eliminating the need for a dedicated aggregation layer. IRF provides more direct, larger capacity connections between users and network resources.

For virtualized environments, having flat network infrastructure enabled by IRF in core switches and Virtual Connect at the access layer provides a converged data center architecture, which is a key benefit of the Virtual Connect FlexFabric architecture. FlexFabric connects servers and virtual machines to data and storage networks over Ethernet, Fibre Channel, and iSCSI protocols.

Figure 5-8 illustrates this concept. On the left is a legacy architecture. East-west traffic must travel in a north-south direction traversing multiple layers, each introducing latency. With FlexFabric, shown on the right of Figure 5-8, traffic travels in an east-west direction (server-to-server) and does not need to pass through the core switches.

Key components of Virtual Connect solutions

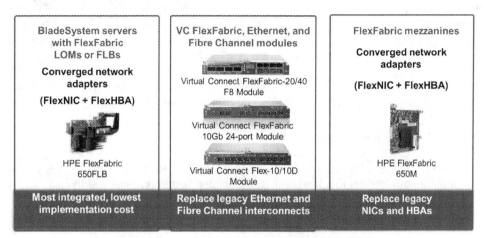

Figure 5-9 Key components of Virtual Connect solutions

As shown in Figure 5-9, the key components of Virtual Connect solutions are:

- BladeSystem servers with FlexFabric LAN-on-motherboard (LOM) or FlexibleLOM Blade (FLB) adapters

---

 Note

LOM architecture refers to servers with a NIC embedded on the system board. FlexibleLOM technology is a variation of LOM architecture that allows you to select ProLiant Gen8/ Gen9 servers with the NIC that best meets your needs without having to embed the NIC on the system board. FlexFabric provides seamless interoperability with existing data center networks and enables HPE networking and security devices to be managed within a single framework.

- Virtual Connect FlexFabric, Ethernet, and Fibre Channel modules
- FlexFabric mezzanine cards

A FlexFabric adapter is more than just a converged network adapter (CNA). It also enables support of interface virtualization, which enables each physical port to be partitioned into multiple FlexNICs, FlexHBA-FCoE, and FlexHBA-iSCSI virtual ports.

Virtual Connect FlexFabric 20/40 F8 module

The HPE Virtual Connect FlexFabric 20/40 F8 module provides 20 Gb server downlinks and 40 Gb uplinks to the data center network.

The Virtual Connect FlexFabric 20/40 F8 Module works with HPE FlexFabric Flex-20 adapters to eliminate network sprawl at the server edge. This module converges traffic inside enclosures and directly connects to external LANs and SANs. Flex-20 technology provides high-speed 20 Gb connections to servers and can achieve a total bandwidth of up to 240 Gbps to a LAN and SAN—a 3x improvement over legacy 10 Gb Virtual Connect modules.

The Virtual Connect FlexFabric 20/40 F8 module reduces the amount of hardware needed to connect servers to networks. Using only two of these modules to connect an enclosure of server blades to both data and storage networks can reduce multiple switches, cards, and cables and provide a 4:1 consolidation of interconnect equipment.

Each Virtual Connect FlexFabric 20/40 module can replace up to three 10 GbE-based switches and one 8 Gb Fibre Channel switch in a BladeSystem enclosure. It reduces networking TCO at the server edge by saving on equipment.

Flex-20 technology enables HPE 20 Gb adapters to stream converged 10 Gb Ethernet and 8 Gb storage simultaneously over a 20 Gb port. Earlier 10 Gb FlexNIC (CNA) implementations were limited to partitioning 10 Gb into one 8/4 Gb Fibre Channel and multiple GbE physical functions (PFs) or as a single 10 GbE (no Fibre Channel or bandwidth for other PFs). Ports that are 20 Gb can be partitioned into a full-rate 10 Gb Ethernet and a full-rate 8 Gb FCoE, with increased additional bandwidth remaining for other functions, provisioned in 100 Mbps increments.

Virtual Connect FlexFabric 10Gb/24-port module

HPE Virtual Connect FlexFabric 10Gb/24-port modules provide a simple, flexible way to connect server blades to data or storage networks. Virtual Connect FlexFabric modules eliminate network sprawl at the server edge with one device that converges traffic inside enclosures and directly connects to external LANs and SANs.

Virtual Connect FlexFabric modules work by first converting FCoE traffic from multiple FlexFabric adapter ports to native Fibre Channel and then aggregating it over a single N-port uplink through the use of N_port_ID virtualization (NPIV). NPIV allows multiple distinguishable identities (multiple port WWNs or port IDs) over a single N-port connection.

The FlexFabric interconnect module separates the converged traffic. Fibre Channel and IP traffic continue beyond the server-network edge using the existing native Ethernet and Fibre Channel infrastructure. Using Flex-10 technology with FCoE and accelerated iSCSI, these modules converge traffic over 10 Gb connections to servers with FlexFabric adapters.

Each redundant pair of Virtual Connect FlexFabric modules provides eight adjustable downlink connections (six Ethernet and two Fibre Channel, or six Ethernet and two iSCSI or eight Ethernet) to dual-port 10 Gb FlexFabric adapters on servers. Up to eight uplinks are available for connection to upstream Ethernet and Fibre Channel switches. Virtual Connect FlexFabric modules avoid the confusion of traditional and other converged network solutions by eliminating the need for multiple Ethernet and Fibre Channel switches, extension modules, cables, and software licenses.

Uplinks support Ethernet, iSCSI, and Fibre Channel (native and FCoE on ports X1 through X4 only).

Virtual Connect Flex-10/10D module

HPE Virtual Connect Flex-10 technology is a hardware-based solution that lets you split a 10 Gb/s server network connection into four variable partitions. It lets you replace multiple lower bandwidth physical NIC ports with a single Flex-10 port. This reduces management requirements, the number of NICs and interconnect modules needed, and power and operational costs.

Flex-10/10D and FlexFabric server ports are subdivided into multiple PCIe PFs called *FlexNICs*. A FlexNIC is a PCIe function that appears to the system ROM, operating system, or hypervisor as a discrete physical NIC with its own driver instance. It is not a virtual NIC contained in a software layer.

Virtual Connect Flex-10/10D and FlexFabric modules support four PFs per adapter port. This provides operating system transparency, because each PF is recognized as an individual NIC or HBA with its own driver by the server operating system. PF2 can be configured as Ethernet, iSCSI, or FCoE. When configured with an iSCSI or FCoE personality it is referred to as a *FlexHBA*.

Virtual Connect supports dynamic bandwidth reallocation among the FlexNICs and FlexHBAs of each FlexFabric adapter port of a server blade. The key benefit of this feature is that it can improve bandwidth utilization and throughput performance of a given server blade's physical adapter port. This feature affects the transmission rates of the egress server port (from the server to the Virtual Connect module). The administrator can control bandwidth utilization and traffic performance from the server blades to the Flex-10/10D or FlexFabric modules.

The Virtual Connect Flex-10/10D module uplinks support Ethernet, iSCSI, and FCoE. The Flex-10/10D module does not support native Fibre Channel. The Flex-10/10D module supports copper direct-attach copper (DAC) cables and short reach (SR), long reach (LR), and long reach multi-mode (LRM) optical fiber transceiver modules.

Virtual Connect Fibre Channel modules

The HPE Virtual Connect Fibre Channel modules for c-Class BladeSystem is a class of blade interconnects that simplifies server connections by cleanly separating the server enclosure from the SAN. It simplifies the process of connecting servers to Fibre Channel networks by reducing cables and the SAN switch management domain. The following Virtual Connect Fibre Channel modules are available:

- Virtual Connect 16Gb 24-port Fibre Channel Module

- Virtual Connect 8Gb 24-Port Fibre Channel Module

- Virtual Connect 8Gb 20-Port Fibre Channel Module

The Virtual Connect Fibre Channel modules offer enhanced Virtual Connect capabilities, allowing a large number of virtual machines running on the same physical server to access separate storage resources. Provisioned storage resource is associated directly to a specific virtual machine, even if the virtual server is relocated within the BladeSystem. Storage management of virtual machines is no longer limited by the single physical HBA on a server blade. SAN administrators can manage virtual HBAs with the same methods and viewpoint of physical HBAs.

You can reduce costs and simplify connections to SANs, consolidate network connections, and enable administrators to add, replace, and recover server resources quickly and easily. Because it is standards-based, a Virtual Connect module looks like a pass-thru device to the Fibre Channel network, yet provides all the key benefits of integrated switching, including high-performance 8 Gb or 16 Gb uplinks to the SAN. Versions of the module provide 8 Gb or 16 Gb server downlinks and either 20 or 24 ports.

The integrated design frees up rack space and reduces power and cooling requirements. It also reduces cables and uses small form-factor pluggable (SFP) transmitters.

Flex-10 for high availability

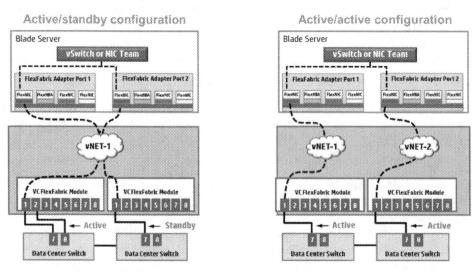

Figure 5-10 Flex-10 high availability

In an **active/standby** configuration, as shown on the left side of Figure 5-10, a single Virtual Connect network is created using ports from two different Virtual Connect FlexFabric modules. Virtual Connect uses the ports from one module or the other but not both simultaneously. If the active Virtual Connect module fails, the standby module takes over.

In an **active/active** configuration, as shown on the right side of Figure 5-10, two different Virtual Connect networks are created, with each network using uplink ports from a single FlexFabric module. Both of these networks can be used simultaneously, but if NIC teaming is set to active/passive or failover, only one NIC receives and transmits, so traffic is limited to the uplinks from the IO slot for that NIC. If one of the active Virtual Connect modules fails, the remaining module continues to operate.

In an **active/active** configuration, failure of an upstream switch or uplink is communicated all the way to the server operating system or hypervisor, and the traffic would be sent through the other NIC in the team instead. In **active/standby**, Virtual Connect routes the traffic through the other module using stacking links and would use the uplinks from the other module. The operating system/hypervisor would not need to know about the upstream switch failure; the failover is handled at the Virtual Connect level.

High levels of north-south traffic are well served by an active/active design because all uplinks are active. High levels of east-west (server-to-server) traffic, such as traffic in multi-tier applications, are well served by an active/standby design because this minimizes the amount of server-to-server traffic leaving the enclosure.

Dual-hop FCoE support in Virtual Connect

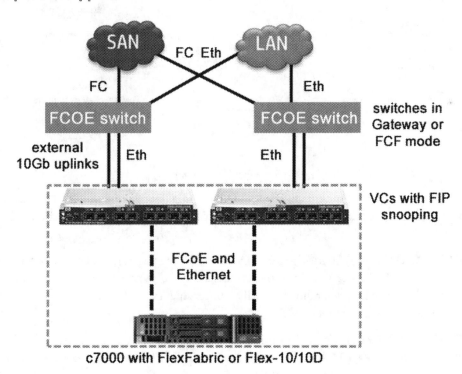

Figure 5-11 BladeSystem c7000 enclosure with FlexFabric or Flex-10/10D

Virtual Connect supports dual-hop FCoE configurations when Virtual Connect FlexFabric or Virtual Connect Flex-10/10D modules are installed in a BladeSystem c7000 enclosure as illustrated by Figure 5-11.

The "D" in "10D" in the Virtual Connect Flex-10/10D module product name stands for "data center bridging," which enables dual-hop FCoE. There are two FCoE hops between the server and the storage. One goes from the server blade HBA to the FlexFabric or Flex-10/10D module. The other goes from the FlexFabric or Flex-10/10D module to the external upstream switch that converts the FCoE traffic to Fibre Channel traffic.

The dual-hop FCoE feature in Virtual Connect employs a technology known as *FCoE Initialization Protocol (FIP) snooping*. Virtual Connect uses this protocol on the external uplinks connected to upstream FCoE switches. FIP snooping enables a device that might not have a native Fibre Channel interface (for example, Flex-10/10D modules) to transfer FCoE traffic to the FCoE upstream switch and thereby provide a path to Fibre Channel storage devices.

HPE 6125 Switch Series

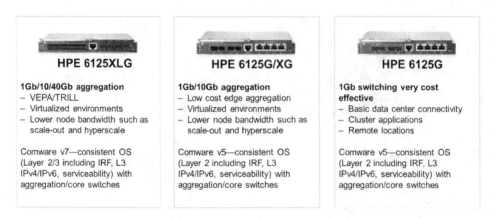

Figure 5-12 HPE 6125 Switch Series

Additional interconnect options for BladeSystem include the HPE 6165 Switch Series, as shown in Figure 5-12.

- **HPE 6125XLG Ethernet Blade Switch**—Designed for the enterprise data center, this switch is built to deliver 880 Gb of switching performance for the most demanding applications. The HPE 6125XLG provides flexibility, versatility, and resiliency, making it the optimal choice for any blade switching environment.

- **HPE 6125G/XG**—With sixteen 1 Gb downlink (server) ports, a combination of 1 Gb and 10 Gb uplink ports, and a 10 Gb cross-link port, the HPE 6125G/XG reduces cost and increases data center efficiency and capability. Switches can be combined at the enclosure, rack, or data center level into a single virtual switch, and managed through a single IP address. It delivers excellent investment protection, flexibility, and scalability for mixed bandwidth applications.

● **HPE 6125G Ethernet Blade Switch**—This switch is ideal for remote office applications or wherever IPv6, full Layer 3 routing, and distributed trunking are required for 1 Gb applications. It appeals to budget-conscious data centers that need 1 Gb switching and routing with the resiliency of IRF, stacking, and the stability of the HPE Comware v7 network operating system.

HPE 6127XLG 20/40GbE Blade Switch

Figure 5-13 HPE 6127XLG 20/40GbE Blade Switch

The HPE 6127XLG Ethernet Blade Switch (Figure 5-13) is designed to support virtualized server environments with exceptional bandwidth of 20 GbE to each server, and provides a mix of 40 GbE and 10 GbE links to the core network. It can provide an aggregate 240 Gb uplink bandwidth and 320 Gb available server side bandwidth. It supports 16 x 1/10/20 Gb downlink ports, 8 x 1/10Gb uplink ports, and 4 x 40 Gb uplink ports. The 6127XLG uses the Comware operating system and supports Layer 2/3 features such as TRILL, VEPA, and FCoE. Combined with hardware support for native VXLAN encapsulation, the HPE 6127XLG is an ideal switch for data center and cloud applications.

The HPE 6127XLG provides a converged fabric solution that supports Ethernet, iSCSI, FCoE, and Fibre Channel Forwarder (FCF) protocols that enables connectivity for multiple storage topologies. The 6127XLG supports the HPE software-defined network (SDN) ecosystem that delivers simple, open, and Enterprise ready benefits to automate the data center network. Using HPE IRF, multiple switches can be virtualized and managed as a single entity with the HPE Intelligent Management Center (IMC).

Capabilities and compatibilities

Table 5-1 Feature comparison

Feature	FlexFabric 20/40 F8 (Virtual Connect 4.20)	Flex-10/10D	HPE 6125XLG
20 G	✓	—	—
10 G	✓	✓	✓
Tunnel offload	✓	✓	✓
FCoE	✓	✓	✓
RoCE	—	—	✓

Table 5-1 shows the features of the Virtual Connect FlexFabric 20/40 F8 module, the Virtual Connect Flex-10/10D module, and the 6125XLG Ethernet Blade Switch.

 Note

The FlexFabric 20/40 F8 module must run Virtual Connect firmware 4.20 or later to support the features in the table.

Other interconnect options

There are several additional interconnect options for ProLiant server blades.

- **Ethernet blade switches**

    - **Mellanox SX1018HP Ethernet Switch**—The highest-performing Ethernet fabric solution in a blade switch form factor, it delivers up to 1.36 Tb/s of nonblocking throughput perfect for high-performance computing (HPC), storage/Hadoop, telecommunications/carrier, oil/gas, service provider, financial services, and cloud/hyperscale environments The SX1018HP is an ultra-low latency switch that is ideally suited as an access switch providing InfiniBand like performance, making this switch the perfect solution for any high-performance Ethernet network.

    - **Cisco Catalyst Blade Switch 3120 Series Switch**—Specifically designed for a server blade-based application infrastructure, this switch enables customers to stack up to nine switches into a single virtual switch. Emulating a redundant top-of-rack (TOR) switch provides an integrated switching solution that optimizes uplinks per rack, reduces the number of switches managed and decreases network complexity.

    - **Cisco Fabric Extender for HPE BladeSystem**—Logically, this switch behaves like a remote line card to a parent Cisco switch, forming a distributed modular system. It forwards traffic to the parent switch over eight 10 Gigabit Ethernet uplinks. Downlinks to each server are auto-negotiating and work with all HPE Ethernet and CNA modules, allowing a choice of Ethernet, FCoE, or iSCSI connections.

- **SAS switches**

  - **HPE 6Gb SAS BL Switch for HPE BladeSystem c-Class enclosures**—An integral part of HPE direct connect SAS storage, it enables a shared SAS storage solution. The SAS architecture combines an HPE Smart Array Controller in each server and the 6Gb SAS BL switches connected to supported HPE storage enclosures for SAS storage. It features an embedded Virtual SAS Manager GUI and command line interface (CLI) used to zone, monitor, and update SAS fabric devices.

- **Fibre Channel switches**

  - **Brocade 16Gb SAN Switch for HPE BladeSystem c-Class**—An easy-to-manage embedded Fibre Channel switch with 16 Gbps Fibre Channel performance, it hot-plugs into the back of the BladeSystem enclosure. Its integrated design frees up rack space, enables shared power and cooling, and reduces cabling. Enhanced trunking support with external switches enhances bandwidth. The switch significantly simplifies the SAN environment, enables easier deployment and management, and delivers the performance required for greater-throughput applications.

  - **Brocade 8Gb SAN Switch for HPE BladeSystem c-Class**—Similar to the Brocade 16Gb SAN Switch for HPE BladeSystem c-Class, this switch delivers 8 Gb/s performance.

- **InfiniBand switches**

  - **HPE BLc 4X QDR InfiniBand Switch**—This Quad Data Rate (QDR) switch has 16 downlink ports to connect up to 16 server blades in a c7000 enclosure, and 16 QSFP uplink ports for inter-switch links or to connect to external servers. All ports are capable of supporting 40 Gbps bandwidth; a subnet manager is required.

  - **HPE BLc 4X FDR InfiniBand G2 Switch**—This Fourteen Data Rate (FDR) switch provides up to 56 Gb/s full bidirectional bandwidth per port in a blade switch. It doubles server throughput, providing 4 Tb/s of nonblocking bandwidth with 165 ns port-to-port latency. Available as an unmanaged or managed switch, it is backward-compatible with QDR InfiniBand and reduces power consumption over previous generations.

- **Pass-thru modules**

  - **HPE 10GbE Pass-Thru Module**—This module is designed for BladeSystem customers who require a nonblocking, one-to-one connection between each server and the network. The pass-thru module provides 16 uplink ports that accept both SFP and SFP+ connectors. It can support 1 Gb and 10 Gb connections on a port-by-port basis. Optical as well as direct attach copper (DAC) cables are supported. Standard Ethernet and Converged Enhanced Ethernet traffic to an FCoE capable switch is possible when using the appropriate NIC or adapter. This solution tends to be the most expensive and cumbersome method of connection, so it is not recommended for common usage.

- **HCA mezzanine cards**

  - **HPE 4X QDR InfiniBand Dual-Port Mezzanine HCA**—This mezzanine card delivers low-latency and up to 40 Gbps bandwidth for performance-driven server and storage clustering applications in HPC and enterprise data centers. It is designed for PCIe 2.0 x8 connectors on BladeSystem server blades.

  - **HPE 4X DDR InfiniBand Dual-Port Mezzanine HCA**—This mezzanine card delivers low latency and up to 20 Gbps bandwidth. Like the QDR HCA, it is based on the Mellanox ConnectX InfiniBand technology. Parallel or distributed applications running on multi-processor multi-core servers benefit from the reliable transport connections and advanced multicast support offered by ConnectX InfiniBand. End-to-end quality of service (QoS) enables partitioning and guaranteed service levels. Hardware-based congestion control prevents hot spots from degrading the effective throughput.

FlexFabric Gen9 adapter innovations improve performance and efficiency

Figure 5-14 HPE FlexFabric 20Gb 630 Series Adapters

HPE 20 Gb FlexFabric CNAs for BladeSystems (Figure 5-14) remove the 10 Gb bandwidth restrictions. One FlexFabric 20 Gb adapter equals one 10 Gb adapter plus one 8 Gb HBA and one extra 2 Gb. It is possible to carve out 8 Gb Fibre Channel bandwidth and have 12 Gb bandwidth available for other FlexNICs. FlexFabric adapters simplify, consolidate, and virtualize the server edge for BladeSystem customers. They provide:

- **Lower cost**—FlexFabric adapters can save considerably on adapter cards, switch ports, and cables needed to support network and storage traffic.

- **Support for converged networks with Virtual Connect**—Virtual Connect provisions LAN and SAN connectivity for BladeSystem server blades through administration of media access control (MAC) and worldwide port names (WWPN) addresses. This allows server administrators to independently manage server blades and their connectivity, maintaining high-availability connections and securely administering MAC addresses and WWPNs for each server.

- **Support for FlexFabric and Flex-10**—FlexFabric adapters support network convergence with Virtual Connect FlexFabric. Each physical port is virtualized as three NIC ports and one FCoE or iSCSI port. They also support Flex-10, which allows each 10 Gb port to be divided into up to four PFs to optimize bandwidth allocation for virtualized servers.

The advantages are:

- Improved performance

  - Greater server efficiency with VMware and Hyper-V by offloading host packet processing

  - High-performance Ethernet networking and FCoE/iSCSI storage IO

  - Balanced system matching networking performance with server CPU performance

  - Advanced storage offload processing, freeing up the server CPU cycles that can be allocated to other application requirements

- Reduced CapEX and OpEX

  - Twice the performance without doubling the budget

  - FCoE or remote direct memory access (RDMA) converged with Ethernet on a single wire

  - More VMs on one link

- Reduced complexity

  - Converged fabric and reduced cabling

  - 10 Gb bandwidth restrictions between server blades and their storage removed

HPE FlexFabric 20Gb 2-port 650 Adapter

Figure 5-15 FlexFabric 650FLB and FlexFabric 650M adapters

HPE FlexFabric 650 adapters (Figure 5-15) were introduced with ProLiant Gen9 server blades. They offer a rich set of offload technologies including overlay network tunneling and storage as well as RDMA over Converged Ethernet (RoCE) capabilities that increase host efficiency and virtualization performance. RoCE is supported with the HPE 6125XLG switch.

In addition to supporting the features of current FlexFabric 20Gb 2-port 630 adapters, FlexFabric 650 series adapters converge FCoE or RoCE traffic with data center LAN traffic. These adapters also support virtual extensible LAN (VXLAN) and network virtualization using Generic Routing Encapsulation (NVGRE) tunnel offload to reduce the CPU load on the server and improve IT service delivery, reduce costs, and deliver greater data center efficiency.

The FlexFabric 650 adapters can be used for both 10 Gb and 20 Gb connections.

Customers choose FlexFabric 650 adapters because the adapters:

- **Can simplify the infrastructure**—FlexFabric 650 adapters deliver faster services by converging FCoE or RoCE with LAN traffic on a single Ethernet wire. Infrastructure can be simplified through eliminating hardware, reducing operational and acquisition costs.

- **Improve performance of overlay networking with tunnel offload on a 20 GbE adapter (VXLAN and NVGRE)**—FlexFabric 650 adapters use tunnel offload to take full advantage of overlay networking investment. They increase the scale of virtual LANs (VLANs) beyond the traditional limit of 4096 and simplify network services provisioning. Using RoCE with Hyper-V Live Migration, customers can reduce migration times and improve server utilization, giving them flexibility to address dynamic workloads. In addition, these adapters offer improvements in input/output operations per second (IOPs) and small packet performance.

- **Deliver IT services efficiently**—CPU and power consumption can be reduced using tunnel offload. In addition, RoCE allows faster data transfer and low latency.

HPE Ethernet 20Gb 2-port 650 Series Adapters are supported with:

- ProLiant BL460c Gen9 server blades
- Virtual Connect FlexFabric-20/40 F8 modules
- Virtual Connect Flex-10/10D modules
- HPE 6125XLG Ethernet Blade Switches

FlexFabric adapters and FlexNICs

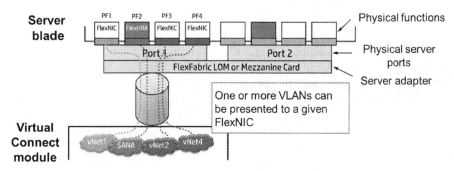

Figure 5-16 FlexFabric adapters and FlexNICs

A FlexFabric adapter presents up to four PFs to an operating system or a hypervisor in a Virtual Connect environment, as illustrated by Figure 5-16. The adapters allow flexible personality definition for networking and storage protocols. All four connections can have their hardware personalities defined as FlexNICs to support only Ethernet traffic. One of the PFs can also be defined as an FCoE or iSCSI adapter. However, only one storage protocol offload can be used at a time. The same protocol must be used on both ports of the adapter.

A single lane (downlink) of 20 GbE for each adapter port serves one to four FlexNICs, or one to three FlexNICs and one FlexHBA. Each server port connects to a different interconnect bay (odd, even) based on the hard-wired BladeSystem design.

FlexNICs and FlexHBA traffic

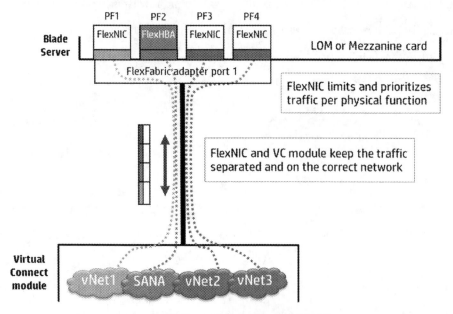

Figure 5-17 FlexNICs and FlexHBA traffic

Although FlexNICs share the same physical port, traffic flow for each is isolated with its own MAC address and FlexNIC VLAN tags. As illustrated by Figure 5-17, each FlexNIC can be mapped to one or more Virtual Connect networks (VLANs) and isolate data traffic by using VLAN tags.

The operating system sees each PF on the FlexFabric adapter as a conventional hardware NIC, Fibre Channel HBA, or iSCSI HBA device. Each PF advertises its VLAN assignment as designated by the server profile. The advertised device type and VLAN assignment steers individual traffic classes to the appropriate PF (PF1, PF2, PF3, or PF4) on the FlexFabric adapter.

Active optical cables

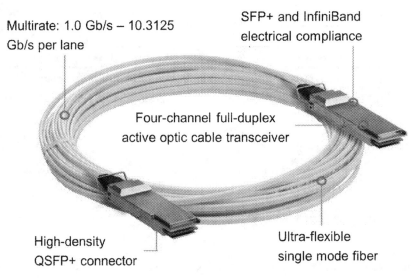

Multirate: 1.0 Gb/s – 10.3125
Gb/s per lane

SFP+ and InfiniBand
electrical compliance

Four-channel full-duplex
active optic cable transceiver

High-density
QSFP+ connector

Ultra-flexible
single mode fiber

Figure 5-18 Active optical cables

With the need for ever-increasing data throughput and increased cable lengths coupled with cost sensitivity, traditional copper cables become less suitable. Active optical cables (AOCs), as shown in Figure 5-18, are supported with the FlexFabric-20/40 F8 module, provide a solution by supporting increased cable lengths, greater bandwidths, more immunity to electromagnetic interference (EMI), and lower cost.

Another advantage is that AOCs are physically smaller than traditional copper cables and are therefore less likely to disrupt airflow in densely packed IT environments. AOCs are a fraction of the cost of buying transceivers and optical cables separately.

The AOCs available from HPE include:

- HPE BladeSystem c-Class 40G QSFP+ to QSFP+ 7m Active Optical Cable

- HPE BladeSystem c-Class 40G QSFP+ to QSFP+ 10m Active Optical Cable

- HPE BladeSystem c-Class 40G QSFP+ to QSFP+ 15m Active Optical Cable

- HPE BladeSystem c-Class 40G QSFP+ to 4x10 7m Active Optical Cable

- HPE BladeSystem c-Class 40G QSFP+ to 4x10 10m Active Optical Cable

- HPE BladeSystem c-Class 40G QSFP+ to 4x10 15m Active Optical Cable

Step 4: Select the blades

Table 5-2 BladeSystem positioning

Features and segments	Performance server blades		
	Entry-level server blade	General-purpose IT	Scale up
Customers	SMB, enterprise	SMB, enterprise, HPC	Enterprise
Use cases	IT infrastructure (file and print)	Virtualization (mid-to-high VM density)	Large databases
			HPC
	Virtualization (lower VM densities)	Cloud foundation	Large virtualization projects
		Application development	Footprint consolidation
Value proposition	Most flexible and best overall performance systems to run compute-intensive workloads		

Table 5-2 illustrates how BladeSystem ProLiant Gen9 server blades are mapped to customer segments, applications, and workloads.

Compared with the previous generation, ProLiant Gen9 server blades present these key features:

- Intel Xeon E5-2600 v3/v4 and E5-4600 v3/v4 processors

- HPE Smart Storage—workload optimized, flexible, embedded storage with 12 Gb SAS controllers

  - Intel Serial Advanced Technology Attachment (SATA) B140i or HPE Smart host bus adapters (HBAs) SATA/SAS

  - Optional HPE Smart Array P244br and Smart Array H244br controllers

 Note

> On the ProLiant BL460c Gen9 server, support for the array controller moved from an embedded RAID controller to the optional Smart Array daughter card.

- Boot devices

  - Slot for two M.2 storage devices

  - Optional redundant microSD card

- HPE SmartMemory DDR4 and support for nonvolatile DIMMs (NVDIMM)

- Improved storage options with support for USB 3.0

- Smart Storage battery with 12W shared backup power

- System management options

  - HPE OneView

ProLiant server blade family—BL400 series

Table 5-3 ProLiant BL400 series

	BL420c Gen8	BL460c Gen9	BL465 Gen8
Number of processors	1 or 2	1 or 2	1 or 2
Max. cores per blade	20	22	32
Processor	Xeon E5-2400 v2 or E5-2400	Xeon E5-2600 v3/v4	Opteron 6200 or 6300 Series
Max. frequency	2.5 GHz	3.5 GHz	3.5 GHz
RAM slots	12	16	16
Max. RAM	384 GB	2 TB	512 GB

Table 5-3 compares the features of the ProLiant BL400 Series server blade family. It includes:

- **HPE ProLiant BL420c Gen8**—Offers breakthrough server blade economics for essential enterprise workloads.

- **HPE ProLiant BL460c Gen9**—Delivers the ideal balance of performance, scalability, and expandability, making it the standard for dense data center computing.

- **HPE ProLiant BL465c Gen8**—Offers unprecedented performance, enhanced flexibility, and simplified management. It is ideal for virtual workloads, and it is flexible enough for any application.

 Note

The number of applications, virtual machines, and users determines the number of server blades needed. Together with HPE channel partners, HPE can help with the choice of the right number of blades with solution-sizing tools and expertise.

Note

Scan this QR code or enter the URL into your browser for more information on the HPE ProLiant BL400 series servers.

https://www.hpe.com/us/en/integrated-systems/bladesystem.html

HPE ProLiant BL460c Gen9 Server Blade

Figure 5-19 HPE ProLiant BL460c Gen9 Server Blade

The HPE ProLiant BL460c Gen9 Server Blade (Figure 5-19) powered by Xeon E5-2600 v3 or v4 series processors delivers performance, scalability, and economics for the converged data center. The Xeon E5-2600 v4 processors provide:

- Up to 21% better performance over previous generations

- Up to 22 cores

Flexible internal storage controller options help customers strike the right balance between performance and price, helping to reduce overall TCO. Customers can choose from the following storage controllers:

- Standard HPE Dynamic Smart Array B140i, which provides low-cost chipset SATA (no SAS)

- HPE H244br 12Gb 2-port Int FIO Smart Host Bus Adapter

- HPE Smart Array P244br/1GB FBWC 12Gb 2-ports Int FIO SAS Controller

- HPE Smart Array P246br/1GB FBWC 12Gb 4-port Int FIO SAS Controller

- HPE Smart Array P741m/4GB FBWC 12Gb 4-port Ext Mezzanine SAS Controller

Additional features of the Smart Array controllers include:

- Improved ease of use with HPE Smart Storage Administrator (SSA), which provides a simple comprehensive utility to manage, configure, and diagnose the attached storage.

- Increased IO speeds with 12Gb/s SAS allowing internal storage scalability while protecting data.

- Maximum reliability with controller-based encryption through HPE Secure Encryption. Data at rest is protected on any bulk storage attached to a Smart Array controller.

- HPE OneView provides a single comprehensive view of the data center, managing hardware, software, firmware, and drivers. HPE iLO software features server life cycle management advancements including iLO Federation, which remotely manages groups of servers at scale with built-in rapid discovery of all iLOs, group configurations, group health status, and ability to determine iLO licenses.

Designed for a wide range of configuration and deployment options, the ProLiant BL460c Gen9 server blade provides customers with the flexibility to enhance core IT applications with right-sized storage for the right workload, which results in lower TCO. The BL460c Gen9 adapts to many demanding blade environments, including virtualization, IT and web infrastructure, collaborative systems, cloud, and HPC.

ProLiant server blade family—BL600 series

Table 5-4 ProLiant BL600 series

	BL660c Gen8	BL660c Gen9
Number of processors	2 or 4	2 or 4
Maximum cores per blade	48	88
Processor	Xeon E5-4600 v2	Xeon E5-4600 v3/v4
Maximum frequency	3.3 GHz	2.9 GHz
RAM slots	32	32
Maximum RAM	1.0 TB	4.0 TB

Table 5-4 compares the features of the ProLiant BL600 Series server blade family. It includes:

- **HPE ProLiant BL660c Gen8**—Offers a four-socket dense form factor without compromising on performance, scalability, or expandability.

- **HPE ProLiant BL660c Gen9**—Provides flexibility, more storage options, faster IO, and more powerful processing to meet any workload needs.

Note

Scan this QR code or enter the URL into your browser for more information on the ProLiant BL660c server blade.

http://www8.hp.com/h20195/v2/GetHtml.aspx?docname=c04543743

HPE ProLiant BL660c Gen9 Server Blade

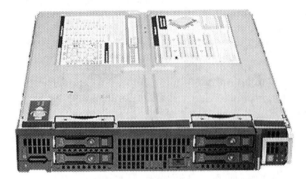

Figure 5-20 HPE ProLiant BL660c Gen9 Server Blade

The HPE ProLiant BL660c Gen9 Server Blade (Figure 5-20), powered by Xeon E5-4600 v3 or v4 series processors, redefines density-optimized four-socket blade technology without compromising on performance. The ProLiant BL660c Gen9 server is ideal for virtualization, database, business processing, and other four-processor applications where fine-tuning of data center space and price/performance is critically important.

HPE DDR4 SmartMemory offers a significant performance increase from the previous generation. Additional support includes tiered storage controller options, 12 Gb/s SAS, 20 Gb FlexibleLOM network interface cards (NICs), additional small form factor (SFF) drives, and support for M.2 and USB 3.0. Both Unified Extensible Firmware Interface (UEFI) and Legacy BIOS modes are available for increased configuration and deployment versatility.

Flexible internal storage controller options with HPE OneView features help customers strike the right balance between performance and price, helping to reduce overall TCO. Customers can choose between the following controllers:

- Standard HPE Dynamic Smart Array B140i, which provides low-cost chipset SATA

- HPE Smart Array P246br for additional performance features including a 1 GB flash-backed write cache (FBWC)

HPE storage blades

Figure 5-21 HPE D2220sb Storage Blade

HPE offers flexible storage solutions designed to fit inside the BladeSystem enclosure, as well as external expansion for virtually unlimited storage capacity. HPE storage blade models include:

- **HPE D2220sb Storage Blade (Figure 5-21)**—Delivers direct-attached storage for ProLiant Gen8 and later server blades with support for up to 12 hot-plug SFF SAS or SATA midline hard disk drives (HDDs) or SAS/SATA solid state drives (SSDs). The enclosure backplane provides a PCIe connection to an adjacent server blade and enables high-performance storage access without any additional cables. The D2220sb features an onboard Smart Array P420i controller with 2 GB FBWC for increased performance and data protection. Up to eight D2220sb storage devices can be supported in a single BladeSystem c7000 enclosure for 192 TB of maximum capacity.

- **HPE X3800sb G2 Network Storage Gateway Blade**—Is used to access Fibre Channel, SAS, or iSCSI SAN storage, translating file data from the server into blocks for storage to provide consolidated file, print, and management hosting services in a package that can be clustered. Built on the ProLiant BL460c server blade, the X3800sb G2 Network Storage Gateway Blade is a ready-to-deploy SAN gateway solution and ships with Microsoft Storage Server 2008, Enterprise x64 Edition preinstalled. The HPE Rapid Start-up Wizard accelerates new deployments by walking IT through basic system settings step-by-step. The X3800sb also includes a Microsoft Cluster Server license and Microsoft iSCSI Software Target.

- **HPE X1800sb G2 Network Storage Blade**—Can be used as a low-cost SAN gateway to provide consolidated file-serving access to Fibre Channel, SAS, or iSCSI SANs. The flexibility of the X1800sb G2 extends to external storage as well.

- **HPE Direct Connect SAS Storage for HPE BladeSystem**—Extends and redefines direct-attached storage for BladeSystem servers. Local storage can be built with zoned storage, or low-cost shared storage can be enabled within the rack using high-performance 3 Gb/s SAS architecture. By combining the simplicity and cost-effectiveness of direct-attached storage with the flexibility and resource utilization of a SAN, Direct Connect SAS Storage gives server administrators a simple in-rack zoned direct attach SAS storage solution that is ideal for growing capacity requirements. The SAS architecture consists of a Smart Array P700m controller in each server, and 3 Gb SAS BL switches connected to HPE 600 Modular Disk System or 2000sa Modular Smart Arrays. The simplicity of SAS results in a very low cost per GB, enabling low-cost zoned or shared storage.

Creating an iSCSI SAN with HPE StoreVirtual VSA on a D2220sb

HPE StoreVirtual VSA turns the D2220sb into a scalable and robust iSCSI SAN for use by all servers in the enclosure and any server on the network. It features storage clustering for scalability, network RAID for storage failover, thin provisioning, snapshots, remote replication, and cloning.

Customers can expand capacity within the same enclosure or to other BladeSystem enclosures by adding more D2220sb storage blades and StoreVirtual VSA licenses. A cost-effective bundle of the D2220sb storage blade and a StoreVirtual VSA license makes purchasing convenient. If storage needs to be increased, customers can add HPE P4300 G2 or P4500 G2 storage systems externally and manage everything through a single console.

HPE Ultrium tape blades

Figure 5-22 Ultrium SB3000c Tape Blade

HPE Ultrium tape blades (Figure 5-22) are ideal for BladeSystem customers who need a complete data protection, disaster recovery, and archiving solution. These half-height tape blades provide direct attached data protection for the adjacent server and network backup protection for all data residing within the enclosure.

Each Ultrium tape blade solution ships standard with HPE Data Protector Express Software Single Server Edition software. In addition, each tape blade supports HPE One-Button Disaster Recovery (OBDR), which allows quick recovery of the operating system, applications, and data from the latest full backup set. Ultrium tape blades are the industry's first tape blades and are developed exclusively for BladeSystem enclosures.

The following models are available:

- HPE SB3000c Tape Blade
- HPE SB1760c Tape Blade

Designed for midrange markets, the HPE LTO Ultrium tape blades set the standards for capacity, performance, manageability, and security. HPE LTO Ultrium tape blades represent five generations of LTO tape drive technology and are capable of storing up to 3 TB per cartridge.

HPE BladeSystem PCI Expansion Blade

Figure 5-23 HPE BladeSystem PCI Expansion Blade

The BladeSystem PCI Expansion Blade (Figure 5-23) fits into a half-height device bay and provides PCI card expansion slots to an adjacent server blade. This blade expansion unit uses the midplane to pass standard PCI signals between adjacent enclosure bays so that customers can add up to two off-the-shelf PCI-X or PCIe cards. The PCI expansion blade and its PCI boards are managed by the adjacent server blade and its operating system.

Customers need one PCI expansion blade for each server blade that requires PCI card expansion. Any third-party PCI card that works in ProLiant ML and ProLiant DL servers should work in this PCI expansion blade.

Note
HPE does not offer any warranty or support for third-party PCI manufactured products.

HPE ProLiant WS460c Gen9 Graphics Server Blade

Figure 5-24 HPE ProLiant WS460c Gen9 Graphics Server Blade

The HPE ProLiant WS460c Gen9 Graphics Server Blade (Figure 5-24) allows the customer to centralize its organization's workstations at a lower cost per seat and with the performance users expect. Locating the graphics server blade in the data center enables it to be more easily, securely, and economically managed. The results are improved uptime and business continuity, enhanced data center security, and reduced IT costs. Features include:

- High-performance NVIDIA Tesla M6 and AMD FirePro S7100X GPUs with virtualized graphics

- Xeon E5-2600 v4 processors, HPE 2400 MT/s DDR4 SmartMemory, and 128 GB DIMMs

- Workload Acceleration with NVM Express (NVMe) SSDs and HPE ProLiant Persistent Memory

- Internal USB 3.0, dual micro-SD, and optional 64/120 GB M.2 support

- Optional HPE Smart Array P244br Controller with 1 GB flash-backed write cache (FBWC) DDR3 at 1866 MHz improves storage performance for demanding workloads

The WS460c Gen9 graphics server blade has a high host density of up to 16 (or 8 expansion) blades and high graphics density of up to 48 GPUs or 512 users per 10U rack space.

It delivers a virtualized, full-fidelity desktop experience including support for 3D graphics applications. It provides hardware accelerated graphics for workstation-class performance and offers up to 21%[1] performance increases with Xeon E5-2600 v4 processors.

HPE DDR4 SmartMemory at 2400 MT/s (up to 2 TB) delivers a 33% increase in throughput compared to competition and 35% lower power consumption than 1.5v DDR3 at the same speed[2].

 Note

1. Intel performance testing, http://www.intel.com/performance, comparing measurements on platform with two E5-2600 v3 vs E5-2600 v4.

2. Based on similar capacity DIMM running on HPE server vs a non HPE server with DDR4.

The WS460c Gen9 graphics server blade provides professional-grade performance and industry-leading GPU density with NVIDIA Tesla M6 or AMD FirePro S7100X, graphics virtualization technology, and the HPE MultiGPU carrier.

Step 5: Select the storage infrastructure

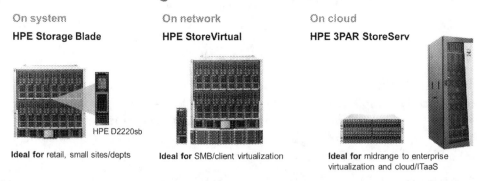

Figure 5-25 Storage solutions

Connect to external HPE SAN, NAS, and backup solutions, or put storage solutions inside the BladeSystem enclosure, side by side with server blades, to quickly expand storage and extend data protection without adding a single cable.

ProLiant server and BladeSystem technology are foundational elements of the HPE Converged Storage architecture. As shown in Figure 5-25, storage infrastructure options include:

- D2220sb storage blade for on-system storage
- HPE P4800 SAN for HPE BladeSystem for on-network storage
- HPE 3PAR StoreServ storage for on-cloud storage

Other storage options include:

- StoreVirtual Storage
- StoreOnce Backup
- StoreAll Storage

Built on modular, industry-standard hardware, scale-out federated software, and integrated management, HPE Converged Storage delivers the simplicity, efficiency, and agility you need to support virtualization, the cloud, and today's proliferation of data. Add further efficiency and IT agility with streamlined storage and networking solutions. Many HPE 3PAR storage solutions can be directly connected to Virtual Connect FlexFabric modules with HPE Flat SAN direct-attach technology, to help reduce infrastructure and multi-tier storage solution complexity.

Converged and collapsed infrastructure with Flat SAN

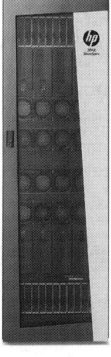

Figure 5-26 HPE 3PAR StoreServ storage and Virtual Connect Flat SAN technology

Virtual Connect Flat SAN technology provides the industry's first direct-attached connection to Fibre Channel storage that does not require dedicated Fibre Channel switches. With Flat SAN, you can connect HPE 3PAR StoreServ storage systems, as shown in Figure 5-26, directly to the Virtual Connect FlexFabric module without an intermediate SAN fabric. The SAN is not eliminated but is created by the switch-on-a-chip technology inside the FlexFabric module and is managed through Virtual Connect Manager.

Storage solutions usually include components such as server HBAs, SAN switches and directors, optical transceivers and cables, and storage systems. The number of components causes concern among customers about management and efficiency. Moreover, different components require different tools, such as SAN fabric management, storage management (for each type of storage), and HBA management.

Because the Flat SAN solution allows you to connect directly to HPE 3PAR storage, it reduces the number of components, lowers latency, and speeds provisioning. It also simplifies connection management by enabling you to wire once, and then add, move, and change network connections to thousands of servers in minutes instead of days from one console without affecting your LAN and SAN.

Flat SAN technology minimizes latency between servers and storage by eliminating the need for multi-tier SANs. Designed for virtual and cloud workloads, this solution reduces storage networking costs and enables faster provisioning compared to competitive offerings.

It also simplifies management by enabling you to use Virtual Connect Manager and HPE OneView through a single console.

Step 6: Select the infrastructure management

BladeSystem infrastructure management is delivered through a complete portfolio of ProLiant server lifecycle management capabilities that can flexibly operate from embedded on-system utilities, on-premises software-defined data centers, and from the cloud. The HPE software-defined approach to infrastructure provisioning leverages a template-based, profile-driven approach that dramatically improves the speed with which customers deploy servers, storage, and network infrastructure.

Managing ProLiant servers with HPE infrastructure management results in increased efficiency and precise control of server infrastructure resources. With a rich set of capabilities that are easy to access and simple to use, HPE infrastructure management covers critical areas such as server deployment and configuration, health and alerting, power and remote management, automated support, and warranty and contract status and control through a cloud-based portal. The core components of HPE infrastructure management are HPE OneView, iLO, and Insight Online.

Onboard Administrator

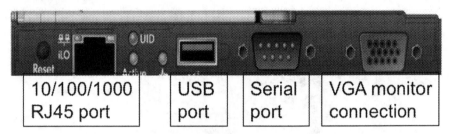

Figure 5-27 Onboard Administrator

The Onboard Administrator for BladeSystem enclosures is the intelligence of the BladeSystem infrastructure. Together with the enclosure's Insight Display, the Onboard Administrator module, as shown in Figure 5-27, has been designed for both local and remote administration of BladeSystem enclosures.

This BladeSystem feature provides wizards for:

- Simple, fast setup, and configuration
- Highly available and secure access to the BladeSystem infrastructure
- Security roles for server, network, and storage administrators
- Agent-less device health and status
- Thermal Logic power and cooling information and control

Each enclosure ships with one Onboard Administrator module. A customer can order a second redundant Onboard Administrator module for each enclosure.

When two Onboard Administrator modules are present in a BladeSystem enclosure, they work in an active/standby mode, assuring full redundancy with integrated management. Either module can be the active module. The other becomes the standby module. Configuration data is constantly replicated from the active Onboard Administrator module to the standby Onboard Administrator module, regardless of the bay in which the active module currently resides.

Insight Display

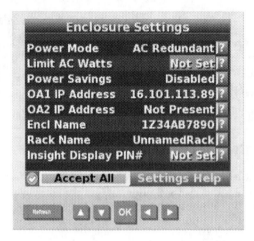

Figure 5-28 Enclosure Settings menu

The LCD panel on the front of the BladeSystem enclosure houses the Insight Display for initial configuration of the enclosure. After the initial configuration has been performed, the Onboard Administrator can be used for ongoing monitoring and management of the enclosure and its components.

The BladeSystem Insight Display panel is designed for local configuring and troubleshooting. It provides a quick visual view of enclosure settings and at-a-glance health status, as shown in Figure 5-28. Green indicates that everything in the enclosure is properly configured and running within specification. It has a keyboard-video-mouse option for local system setup and management

Main Menu

From the Insight Display Main Menu, users can navigate to the submenus, which include:

- Health Summary
- Enclosure Settings
- Enclosure Info
- Blade or Port Info
- Turn Enclosure UID on
- View User Note

Converged management with HPE OneView

To manage an IT infrastructure effectively, customers need one management platform with one approach to the entire infrastructure, including compute, storage, and the network. HPE OneView provides this single management platform.

HPE OneView can be used to automate the deployment and ongoing management of BladeSystem environments. The HPE OneView dashboard provides an easy-to-understand status summary of servers, storage pools, and enclosures. Color-coded icons indicate which systems are functioning properly and which ones need help. Features include:

- One platform manages BladeSystem and HPE 3PAR StoreServ storage.

- Customizable templates that define infrastructure services, enabling the delivery of IT services in a fast, repeatable, and reliable manner, at lower cost and with fewer errors.

- Profiles and groups capture best practices and policies. Profiles and groups can be created once and rolled out to as many enclosures, servers, and storage arrays as needed to increase productivity and ensure compliance and consistency.

- Visualized connections between infrastructure elements enable IT staff to better understand the impact of hardware faults or performance bottlenecks.

HPE OneView creates a closed-loop automation hub with consistent, industry-standard application program interfaces (APIs), a uniform data model, and a state-change message bus. IT staff can automate deployment of multiple enclosures, blades, storage, and networking programmatically—using the industry-standard REST API or a choice of PowerShell and Python language bindings. With HPE OneView automation capabilities, it is possible to:

- Inform multiple management tools or systems administrators of changes to the infrastructure managed by HPE OneView

- Quickly react to environmental changes by deploying or updating resources, updating asset management records, or automatically creating service tickets

- Enable virtualization administrators to automate control of all HPE resources—with no detailed knowledge of each device—through integration with VMware vCenter, VMware vCenter Operations, and Microsoft System Center

Orchestration across the infrastructure

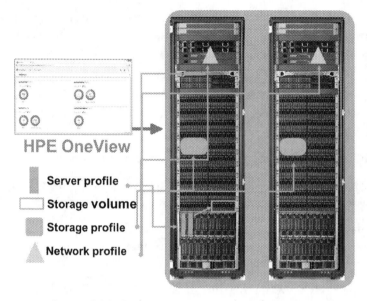

Figure 5-29 Orchestration across the infrastructure

HPE OneView improves operational efficiency by taking the personality of a server and putting it in software. The resulting server profile contains all of the configuration information necessary to instantiate the server, including connections to storage and associated addresses, and connections to networking devices and associated addresses, as illustrated by Figure 5-29.

HPE OneView provides orchestration across the infrastructure by enabling you to:

- Visualize the data center for planning

- Manage and provision hardware, software, and firmware

- Implement changes faster

Insight Online

Figure 5-30 HPE Insight Online

HPE Insight Online (Figure 5-30) is a cloud-based infrastructure management and support portal available through the HPE Support Center and powered by HPE remote support technology, including Insight Online direct connect and Insight Remote Support.

The management capabilities built into ProLiant Gen8 and Gen9 servers seamlessly integrate with Insight Online direct connect and Insight Remote Support 7.0 and later.

Insight Online is designed for IT staff who deploy, manage, and support systems, as well as for HPE Authorized Partners who support IT Infrastructure. Using this portal saves time, reduces complexity, and helps ensure uptime.

Through the HPE Support Center, Insight Online can automatically display devices remotely monitored by HPE. Insight Online provides a personalized dashboard for simplified tracking of IT operations and support information from anywhere, at any time, on any device. Customers can use the Insight Online dashboard to track service events and support cases, view device configurations, and proactively monitor HPE contracts, warranties, and HPE Proactive service credit balances.

Insight Online provides all-in-one secure access to the information needed to support devices in the IT environment with standard warranty and contract services.

After you install HPE remote support tools and register with Insight Online, HPE Proactive Care service assists in proactively supporting infrastructure, providing quick access to support experts, and preventing problems before they occur.

Step 7: Select the power and cooling options

Figure 5-31 Active Cool 200 Fan and BladeSystem c7000 enclosure

The BladeSystem c7000 enclosure provides the power, cooling, and IO infrastructure needed to support modular server, interconnect, and storage components.

Power is delivered through a pooled-power backplane, and power input flexibility is provided with a choice of single-phase high-line AC, three-phase high-line AC, single-phase high-voltage AC, –48V DC, or high-voltage DC. The power supplies are designed to be highly efficient and self-cooling. Single- or three-phase enclosures and N + N or N + 1 redundancy yield the best performance per watt.

HPE Power Regulator gives you a tool to increase server efficiency and free up cooling and power resources. This lets you use power/cooling resources where you need them the most. It lets you manage processor power consumption and system performance to meet business needs. Power-state decisions made in Dynamic Power Savings mode take into account all processor activity. Power Regulator determines CPU utilization by reading a performance event counter residing within the processor.

HPE Thermal Logic technology minimizes power consumption and reduces cooling. HPE Intelligent Infrastructure technology combines energy-efficient design with accurate measurement and control—all without sacrificing performance.

The capacity of the server blades in the data center can be doubled with Dynamic Power Capping delivered through HPE OneView. HPE ProLiant servers enable Dynamic Power Savings mode by default to provide significant, out-of-the-box power savings.

Cooling is provided by a minimum of four Active Cool 200 Fans. For additional capacity, up to six more fans can be added as illustrated by Figure 5-31. These fans provide adaptive flow for maximum power efficiency, air movement, and acoustics. The enclosure architecture is designed to draw air through the interconnect bays. This allows the interconnect modules to be smaller and less complex.

HPE Intelligent Series Racks with the Location Discovery option installed provide detailed location information to ProLiant Gen8 and Gen9 servers and the BladeSystem c7000 Platinum enclosure to track new installations and equipment moves. Upon installation, the server's iLO queries and records the rack identifier as well as the exact U location of the server in the rack.

Step 8: Select the services

HPE Technology Services consulting and support can help customers achieve maximum benefits from their server technology. HPE Technology Services delivers confidence by helping customers prevent problems, reduce risk, solve problems faster, and realize agility and stability as they deploy and operate new technology.

- HPE offers two versions of HPE Proactive Care Services for BladeSystem servers, each with flexible hardware and software coverage windows and response times.

  - Proactive Care leverages innovative HPE remote support technology to help prevent problems and provide rapid access to expertise to stabilize IT.

  - Proactive Care Advanced is designed to support servers running business-critical IT. This service expands on the Proactive Care service by providing localized account managers who work with the customer's IT staff to keep systems in peak performance, manage critical events, and quickly address complex issues

- HPE Foundation Care is an economical alternative that provides hardware and software support with a simplified choice of coverage windows and response times. This support coverage includes collaborative call management for assistance with leading x86 operating system software.

- HPE Datacenter Care offers the most flexible service. It supports the entire IT environment to provide the right mix of enhanced call management, proactive services, and hardware and software assistance to manage a solution holistically for maximum control, performance, and simplicity.

- HPE Education Services helps address the challenge of managing costs and resources while keeping up with the latest technology. HPE Education Services provides IT professionals, enterprise businesses, and end users with the highest quality, most comprehensive technical and business education services and expertise using advanced technologies.

Learning check

1. Building a BladeSystem infrastructure solution begins with eight simple steps. Place the steps in their correct sequence from 1 to 8.

   Select the services

   Select the storage infrastructure

   Select the BladeSystem enclosure

   Select the interconnects and adapters

   Select the operating environment

   Select the power and cooling configurations

   Select the server blades

   Select the infrastructure management

2. What is the capacity in terms of number of server blades of a c7000 enclosure?

   Half-height server blades

   Full-height server blades

   Expansion server blades

3. List at least two benefits of Virtual Connect.

   _____

   _____

   _____

4. What are the key components of Virtual Connect solutions? (Select three.)

   A. Virtual Connect FlexFabric Modules

   B. FlexFabric mezzanine cards

   C. BladeSystem servers with FlexFabric LOMs or FLBs

   D. InfiniBand modules

   E. Blade switch modules

5. Which statement is characteristic of a Flex-10 high availability active/active configuration?

    A. If an upstream switch or uplink fails, Virtual Connect routes the traffic through the other module using stacking links.

    B. Virtual Connect uses the ports from one VC module or the other but not both simultaneously.

    C. Two different Virtual Connect networks are created, with each network using uplink ports from a single Virtual Connect module.

    D. A single Virtual Connect network is created using ports from two different Virtual Connect modules.

6. What is the maximum number of cores possible in a ProLiant 660c Gen9 server blade?

    A. 48

    B. 64

    C. 88

    D. 128

7. Virtual Connect Flat SAN technology provides direct attach connection to Fibre Channel storage that does not require dedicated Fibre Channel switches.

    ☐ True

    ☐ False

8. Which tool could an on-site technician use to visually check the health of server blades in an enclosure?

    A. Insight Online

    B. Insight Display

    C. Smart Storage Administrator

    D. Onboard Support Assistant

9. Your customer wants to be able to monitor HPE contracts, warranties, and service credits. Which tool should you recommend to the customer?

    A. Insight Online

    B. Insight Display

    C. Smart Storage Administrator

    D. Onboard Administrator

Learning check answers

1. Building a BladeSystem infrastructure solution begins with eight simple steps. Place the steps in their correct sequence, from 1 to 8.

Select the services	**8**
Select the storage infrastructure	**5**
Select the BladeSystem enclosure	**2**
Select the interconnects and adapters	**3**
Select the operating environment	**1**
Select the power and cooling configurations	**7**
Select the server blades	**4**
Select the infrastructure management	**6**

2. What is the capacity in terms of number of server blades of the c7000 enclosure?

Half-height server blades	**16**
Full-height server blades	**8**
Expansion server blades	**8**

3. List at least two benefits of Virtual Connect.

   - **Virtual Connect enables routine infrastructure changes in less time with wire-once connectivity**
   - **Virtual Connect simplifies and reduces infrastructure (switches, HBAs, NICs, and cables) and costs**
   - **Virtual Connect provides edge safe networking that provides full network visibility**

4. What are the key components of Virtual Connect solutions? (Select three.)

   A. Virtual Connect FlexFabric Modules

   B. FlexFabric mezzanine cards

   C. BladeSystem servers with FlexFabric LOMs or FLBs

   D. InfiniBand modules

   E. Blade switch modules

5. Which statement is characteristic of a Flex-10 high availability active/active configuration?

    A. If an upstream switch or uplink fails, Virtual Connect routes the traffic through the other module using stacking links

    B. Virtual Connect uses the ports from one VC module or the other but not both simultaneously

    C. Two different Virtual Connect networks are created, with each network using uplink ports from a single Virtual Connect module

    D. A single Virtual Connect network is created using ports from two different Virtual Connect modules

6. What is the maximum number of cores possible in a ProLiant 660c Gen9 server blade?

    A. 48

    B. 64

    C. 88

    D. 128

7. Virtual Connect Flat SAN technology provides direct attach connection to Fibre Channel storage that does not require dedicated Fibre Channel switches.

    ☐ **True**

    ☐ False

8. Which tool could an on-site technician use to visually check the health of server blades in an enclosure?

    A. Insight Online

    B. Insight Display

    C. Smart Storage Administrator

    D. Onboard Support Assistant

9. Your customer wants to be able to monitor HPE contracts, warranties, and service credits. Which tool should you recommend to the customer?

    A. Insight Online

    B. Insight Display

    C. Smart Storage Administrator

    D. Onboard Administrator

Summary

- BladeSystem is a modular infrastructure platform that converges servers, storage, and network fabrics to accelerate operations and speed delivery of applications and services running in physical, virtual, and cloud-computing environments.

- Building a BladeSystem infrastructure solution begins with eight steps:

  1. Select the operating environment.

  2. Select the BladeSystem enclosure.

  3. Select the interconnects and adapters.

  4. Select the server blades.

  5. Select the storage infrastructure.

  6. Select the infrastructure management.

  7. Select the power and cooling configurations.

  8. Select the services.

6 Workload-Optimized Solutions

After completing this chapter, you should be able to:

✓ Explain why high-performance computing (HPC) is important

✓ Describe the features and functions of Hewlett Packard Enterprise (HPE) Apollo systems

✓ Discuss the management options available for HPE Apollo solutions

OPENING CONSIDERATIONS

Before proceeding with this section, answer the following questions to assess your existing knowledge of the topics covered in this chapter. Record your answers in the space provided here.

1. How have HPC business requirements changed recently, and how have these affected the market?

---

2. Which features of HPC solutions such as Apollo do you think are most appealing to customers?

_____

_____

_____

_____

_____

_____

3. What experiences have you had with HPE Apollo systems?

_____

_____

_____

_____

_____

_____

_____

HPC is key to continued business success

High-performance computing customers use IT differently from corporate IT organizations. An HPC customer's expenditure on IT often approaches 90% of the total cost of the business, 10 times that of a corporate IT organization. Additionally, the IT growth rates for a service provider can be many times the IT growth rate of a corporation. For service providers, IT is the business.

HPC customers are focused on how to get the best performance possible with limited resources. HPC is firmly linked to economic competitiveness as well as scientific advances. Governments, academia, and enterprises use HPC to drive advances in their respective fields:

- Governments and academia leverage IT to solve the world's greatest problems (such as curing genetic illnesses, solving climate change, or determining the origin of the universe).

- Researchers are continually trying to solve more and increasingly complex problems in the life and materials sciences industries. HPC solutions increase research agility, lower costs, and allow researchers to process, store, and interpret petabytes of data. HPC enables simulation and analytical solutions to some of the most vexing problems in areas such as nanotechnology, climate change, renewable energy, neuroscience, bioinformatics, computational biology, and astrophysics.

- HPC solutions for upstream oil and gas exploration and production enable the industry to meet the increasing global demand for petroleum products.

- Financial services companies face the most challenging analytics and trading environments in the industry. From risk management to high-frequency trading, IT solutions need to deliver the performance, efficiency, and agility to maximize the ability to add or adapt services quickly as market conditions change.

At one time, HPC was regarded as a specialist area. Today, it is becoming essential to the continued success of businesses requiring optimal computational performance, unprecedented reliability, memory, and storage scalability. Examples of such applications include:

- Computer-aided engineering (CAE)

- Electronic design automation (EDA)

- Research and development

- Life sciences

- Pharmaceutical

- Geophysical sciences

- Energy research and production

- Meteorological sciences

- Entertainment

- Media production

- Visualization and rendering

- Government

- Academia

- Financial services

- Automotive and aerospace design

HPE Apollo systems

HPE Apollo systems provide rack-scale solutions with high density, optimal performance, power efficiency, and low total cost of ownership (TCO). The demand for more compute performance for applications such as EDA, risk modeling, or life sciences is relentless. For customers working with single-threaded application workloads like these, success depends on optimizing performance with maximum efficiency and cost-effectiveness, along with easy management for large-scale deployments.

Apollo systems are designed to deliver compute, storage, networking, power, and cooling solutions for big data, analytics, object storage, and HPC workloads. With rack-scale efficiency, Apollo systems deliver excellent business benefits. These systems:

- Apply just the right amount of scalability, performance, and efficiency with systems that are optimized for specific workloads

- Reduce implementation time from months to days

- Provide architectural flexibility with both scale-up and scale-out solutions

- Provide significant capital and operating expenditure savings

- Leverage complete service and support offerings from HPE

The Apollo portfolio includes:

- Apollo 2000 systems provide a bridge to scale-out architecture for traditional rack-server data centers and deliver hyperscale and general-purpose scale-out computing.

- Apollo 4000 systems allow customers to analyze growing volumes of data in order to turn information into insight and enable faster strategic decision making. The Apollo 4000 series is ideal for big data analytics and object storage needs.

- The air-cooled Apollo 6000 system optimizes rack-scale performance for any budget and makes HPC capabilities accessible to a wide range of enterprise customers.

- The Apollo 8000 system is a supercomputer that combines high levels of processing power with a warm liquid-cooled design for ultra-low energy usage and recycling.

This complete range of offerings makes highly dense server storage, management, and rack-scale efficiency available to organizations of all sizes. Its tiered approach provides a logical starting point for data-driven organizations that want to implement big data, object storage, and HPC solutions.

HPE Apollo 2000

Figure 6-1 HPE Apollo 2000

The Apollo 2000 shown in Figure 6-1 offers all the features of traditional enterprise servers and provides twice the amount of density when compared with standard 1U rack servers. This system increases available data center floor space, improves performance with lower energy consumption, and provides flexible configurations that fit into industry standard racks.

Apollo 2000 systems offer a dense solution with up to four HPE ProLiant XL170r Gen9 or up to two HPE ProLiant XL190r Gen9 server nodes in a standard 2U chassis. Each server node can be serviced individually without impacting the operation of other nodes sharing the same chassis, providing increased server uptime. The ability to combine ProLiant XL170r Gen9 servers and ProLiant XL190r Gen9 servers in the same chassis and the unique drive-mapping flexibility lends itself to optimizing server configurations for many applications. A chassis, or groups of chassis, can be custom-configured to act as affordable, modular, 2U building blocks for specific implementations at scale—and for future growth.

The Apollo 2000 is also compatible with the HPE Advanced Power Manager (APM), which enables aggregate and detailed level of power measurement and control of groups of servers. It also provides static and dynamic capping of power across the nodes.

Apollo 2000 features and benefits

Features of the Apollo 2000 include:

- Redundant fans and power infrastructure with up to two 1400 W power supplies

- Increased storage flexibility with options that support serial-attached SCSI (SAS)/serial advanced technology attachment (SATA)/solid state drives (SSDs)

- Up to four independent, hot-pluggable server nodes in one chassis, delivering twice the compute density when compared with 1U rack-mount servers

- Front hot-pluggable drives and rear serviceable nodes

- Cost-effective configurations for various workloads

- 1U and 2U servers that can be mixed and matched for workload optimization, allowing customers to partially populate the chassis and scale out as they grow

- HPC performance with accelerators, premium (also called *top bin*) processors, and broad range of IO options

- 12 large form factor (LFF) or 24 small form factor (SFF) drive cage options, including an option for an SAS expander to enable flexible allocation of drives per server node

HPE compute nodes and options for Apollo 2000

Table 6-1 Apollo 2000 compute nodes and options

	ProLiant Apollo XL170r: Gen9 1U node	ProLiant Apollo XL190r: Gen9 2U node
Maximum number	**1U** half width—up to **four** per chassis	**2U** half width—up to **two** per chassis
Processor	Dual Intel Xeon E5-2600 v3 or v4 series processors with options for 4–22 cores, 1.6 GHz–3.5 GHz CPU speed, 85–145 W	
Memory	16 x DDR4 up to 2,133 MHz 512 GB maximum	
Network module	2 x 1 Gb Ethernet, Serial RJ45 connector, SUV connector (one serial/two USB/one video), optional FlexibleLOM	
PCIe 3.0 slots	**Two** externally accessible IO options that allow you to choose how the PCIe lanes are used	**Three** externally accessible and **one** internally accessible IO options
Storage	• Up to 24 drives per node • Dual SATA host based M.2 2242 NGFF SSDs (internal) • Hot-plug hard disk drive (HDD) support • Internal USB port • Hard drive mapping feature on r2800 chassis	
Storage controller	• Integrated Smart Array B140i storage controller • Optional PCIe host bus adapters (HBAs) and Smart Array Controllers with advanced array features such as HPE SmartCache and RAID 10 advanced data mirroring	
Supported accelerators	**N/A**	Support for **up to two per server**: NVIDIA Quadro K4200, Tesla K40 or K80 GPUs or Intel Xeon Phi 5110P coprocessors

Table 6-1 Apollo 2000 compute nodes and options (Continued)

	ProLiant Apollo XL170r: Gen9 1U node	ProLiant Apollo XL190r: Gen9 2U node
Management	• HPE integrated Lights-Out (iLO) 4	
	• Advanced Power Manager (optional rack-level management)	
Common workloads	• HPC	• HPC (with GPUs or coprocessors)
	• Cloud server	• Density-optimized general-purpose server
	• Density-optimized general-purpose server	• Computing/storage all-in-one server for SMB, FSI, and EDA
	• Computing/storage all-in-one server for small to midsize businesses (SMBs), financial risk modeling, and engineering design automation	• Server storage gateway controller for SAN, EDA, and HPC cloud server for online gaming

Table 6-1 highlights the differences between the two server models that are compatible with Apollo 2000 systems. The key differences are in bold.

Apollo 2000 system chassis options

Table 6-2 Apollo 2000 chassis options

	Apollo r2200 chassis	Apollo r2600 chassis	Apollo r2800 chassis
Description	Gen9 **12 LFF** disk or SSD chassis	Gen9 **24 SFF** disk or SSD chassis	Gen9 **24 SFF** disk or SSD chassis **with drive mapping capability**
Storage configuration	**12 LFF** hot-plug SAS or SATA HDDs or SSDs, allocated equally across server nodes	**24 SFF** hot-plug SAS or SATA HDDs or SSDs, allocated equally across server nodes	**24 SFF** hot-plug SAS or SATA HDDs or SSDs; **supports flexible drive mapping, enabling custom drive allocations to match workloads**
Power supplies	800 W or 1400 W Platinum Power Supplies, N + 1 redundancy option		
Management	Advanced Power Manager		

Table 6-2 highlights the differences between the Apollo 2000 system chassis options. The key differences are in bold.

Note

To access the QuickSpecs for Apollo 2000, scan this QR code or enter the URL into your browser.

http://www8.hp.com/h20195/v2/GetPDF.aspx/c04542552.pdf

HPE Apollo 4000

Apollo 4200 Gen9

Apollo 4510
(with top panel removed)

Figure 6-2 HPE Apollo 4000

Big data is growing at an exponential rate, and enterprises are seeking to translate big data analytics into a competitive business advantage. Today's general-purpose infrastructure runs into problems when big data workloads move to petabyte scale. The data center can experience capacity constraints, spiraling energy costs, infrastructure complexity, and inefficiencies. To maximize the value of big data, businesses require systems that are purpose-built for big data workloads.

Apollo 4000 systems (Figure 6-2) are designed specifically for Hadoop and other big data analytics and object storage systems. These systems allow customers to manage, monitor, and maintain increasing data volumes at petabyte scale. Businesses can use Apollo 4000 to address data center challenges of space, energy, and time to results.

HPE Apollo 4200 Gen9 server

Apollo 4200 Gen9
servers

Figure 6-3 HPE Apollo 4200 Gen9

As shown in Figure 6-3, the Apollo 4200 Gen9 server is a density-optimized server storage solution designed for traditional enterprise and rack-server data centers. This versatile 2U big data server integrates seamlessly into traditional data centers with the same rack dimensions, cabling, and serviceability, as well as the same administration procedures and tools. It is the ideal bridge system for implementing purpose-built big data server infrastructure, with the capability to scale in affordable increments in the future.

This server is available in LFF and SFF versions:

- The LFF server features up to 224 TB of direct-attached storage per server and 4.48 PB storage capacity per rack[1]. It supports up to 28 hot-swappable LFF SAS or SATA HDDs/SSDs.

- The SFF system features up to 108 terabytes of direct-attached storage per server and supports up to 54 hot-swappable SFF SAS or SATA HDDs/SSDs. The SFF HDD model supports 15k RPM SAS and SSD drives with 12 Gb ports to speed data transfer for analytics workloads.

 Note
[1]Based on 8 TB LFF drives.

Key features and benefits of the Apollo 4200 Gen9 server include:

- Offers a standard size rack, front/side hot-plug disk serviceable, rear aisle cabling, and standard rack server system administration

- Fits in traditional data centers with the same racks, cabling, and servicing accessibility

- Allows enterprises to start and grow object storage solutions in cost-effective 2U increments

- Provides two-processor server configuration options, including:

  - Xeon E5-2600 v3 or v4 series processors with choices from 4 to 22 cores, 1.6 GHz to 3.5 GHz CPU speed, and power ratings between 55 watts and 145 watts

  - Sixteen DIMM slots with up to 1 TB DDR4 memory at up to 2133 MHz, which is ideal for object stores needing high performance with small objects or in-memory data processing for near-real-time analytics software

  - Up to seven PCIe Gen3 slots to meet networking and cluster performance needs in applications requiring very fast IO

 Note

Scan this QR code or enter the URL into your browser to access the Apollo 4200 Gen9 server QuickSpecs.

http://www8.hp.com/h20195/v2/GetDocument.aspx?docname=c04616497

HPE Apollo 4500 systems

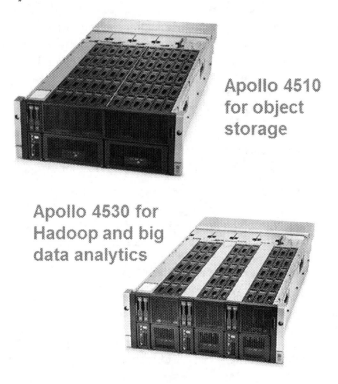

Apollo 4510 for object storage

Apollo 4530 for Hadoop and big data analytics

Figure 6-4 HPE Apollo 4500 systems

Apollo 4500 systems include three purpose-built systems, two of which are shown in Figure 6-4, that address specific needs: the Apollo 4510 for object storage workloads, the Apollo 4520 for clustered storage environments and the Apollo 4530 for Hadoop and big data analytics workloads. These systems provide the following features:

- **Apollo 4510**—This single-server system is ideal for object storage solutions at any scale including collaboration and content distribution, content repositories and active archives, and backup repositories and retention of inactive data (also known as cold storage). This ultra-dense system includes one server in a 4U chassis. It also includes up to 68 hot-plug SAS or SATA HDDs/SSDs with up to 544 terabytes storage capacity per server and up to 5.44 PB of storage per 42U rack.

- **Apollo 4520**—This two-server system is ideal for clustered storage environments. It offers internal cabling for failover, plus massive disk density of 23 LFF drives per node for a total of 368 TB storage per node. With two nodes per chassis this provides a massive 736 TB of storage. This system is an optimal fit for workloads such as Lustre parallel file system and Microsoft Storage Spaces.

- **Apollo 4530**—This three-server system is purpose-built for the wide variety of big data analytics solutions based on parallel Hadoop-based data mining, as well as NoSQL-based big data analytics solutions. It offers three servers per chassis, allowing businesses to house three copies of data in a single system. It provides the performance and storage density needed to develop a complete view of customers to improve marketing cost-effectiveness, boost online sales, and enhance customer retention and satisfaction.

    - This 4U, three-server system has three two-processor ProLiant Gen9 servers, each with 15 LFF hot-plug SAS or SATA HDDs/SSDs per server and up to 45 drives per chassis.

    - Each server has up to 120 TB of capacity providing economical building blocks for efficient implementations at scale with up to 30 servers and 3.6 PB of capacity in a 42U rack.[1]

 Note

[1.] Based on 10 Apollo 4530 systems with 8 TB HDDs.

The Apollo 4510, 4520, and 4530 provide similar features, including:

- Xeon E5-2600 v3 or v4 series processors with choices from 4 to 20 cores, 1.6 GHz–2.6 GHz CPU speed, and power ratings between 55 Watts and 135 Watts

- Sixteen memory DIMM slots with up to 1 TB DDR4 memory at up to 2400 MHz, which is ideal for complex analytics needing fast performance or in-memory data processing analytics applications

- SSDs and high-performance storage controllers to speed data transfer

- Up to four PCIe slots with flexible performance and IO options to match the variety of workload performance and throughput criteria

 Note

Scan this QR code or enter the URL into your browser to access the Apollo 4500 QuickSpecs.

http://www8.hp.com/h20195/v2/GetPDF.aspx/c04616501.pdf

Activity: Technical University of Denmark case study

Watch a video about how Apollo systems helped the Technical University of Denmark (DTU) meet the HPC needs of life sciences researchers.

 Note

To access the video, scan this QR code or enter the URL into your browser.

https://www.youtube.com/watch?v=IjceCDNC4z0

Be prepared to pause the video as needed to answer these questions:

1. What prevented the DTU from using a traditional HPC system?

_____

_____

_____

2. What types of data does the DTU manage?

_____

_____

_____

3. How did DTU use HPE Apollo and other big data systems to drive innovation in medicine?

_____

_____

_____

_____

HPE Apollo 6000

Apollo 6000 chassis

Apollo 6000 Power Shelf

Figure 6-5 HPE Apollo 6000

The Apollo 6000 system shown in Figure 6-5 was designed to provide power, air cooling, and IO infrastructure to support ProLiant XL servers. Each standard rack chassis is 5U, with up to 10 independent server trays. The innovation zone in the back of the chassis holds up to 10 IO modules, with dual FlexibleLOMs supporting 1 GbE or 10 GbE. Power is delivered through an external 1.5U power shelf (14.4 kW or 15.9 kW) supporting up to six chassis depending on power load. Customers can choose from single- or three-phase AC input. An Advanced Power Manager (APM) module helps optimize rack, chassis, and server power.

Benefits of an Apollo 6000 system include:

- Leading performance per watt and per dollar

    - Twenty percent more performance for single threaded applications[1]

    - Forty-six percent less energy per system[2]

    - Four times better performance per dollar per watt[3]

- Rack-scale efficiency

    - Sixty percent less space than a competing blade[2]

    - Simplified, rack-scale administration efficiencies

- Flexibility to tailor the solutions to the workload

    - Innovation zone allows for choice of NIC, FlexibleLOM options to fit workload needs with increased cost savings

    - Capability to scale by chassis or rack with a single modular infrastructure and a selection of compute, storage, and GPU/accelerator trays

Note

1. This percentage was documented using benchmarks from Synopsys, presented at the Synopsys Users Group (SNUG) conference in 2014.

2. This data is based on HPE internal analysis comparing an Apollo 6000 system to a Dell M620.

3. This percentage is based on HPE internal calculations.

HPE Apollo a6000 chassis

Apollo 6000 chassis

Figure 6-6 HPE Apollo a6000 chassis

The Apollo a6000 chassis and the servers it supports are designed for scalability and efficiency at rack scale with projected TCO savings over three years. The Apollo a6000 chassis is designed to help enterprises manage and scale to their business computing demands. It holds various compute servers or accelerator trays to fit specific workloads. Each chassis can hold up to 10 single slot trays or up to 20 servers. Five dual rotor fans share a cooling zone, and power can be managed using an APM option at the server, chassis, or power shelf level.

The 5U Apollo a6000 chassis shown in Figure 6-6:

- Accommodates up to 10 hot-swap server trays

- Fits standard racks with rear-cabled, cold aisle servicing

- Offers rack-level management, networking, and cabling consolidation with chassis-level iLO port aggregation and chassis-to-chassis serial iLO connections

After choosing chassis and power shelf configurations, customers can select a server tray. Options include:

- **HPE ProLiant XL220a Gen8 v2 server tray**—Offers two 1P servers per tray with Xeon E3-1200 v3 series processors with up to four cores. This server tray increases performance per core up to 35% for single threaded applications over a 2P blade.

- **HPE ProLiant XL230a Gen9 server tray**—Includes one 2P server per tray with Xeon E5-2600 v3/v4 series processors. The 16 DIMM slots are ready with DDR4 2133(1024 GB maximum)/2400 (2048 GB maximum) MHz memory.

- **HPE ProLiant XL250a Gen9 server tray**—Provides one 2P server per tray with up to two accelerator cards. This server features the Xeon E5-2600 v3/v4 series processors and 16 DIMM slots with DDR4 2133 (1024 GB maximum)/2400 (2048 GB maximum) MHz memory.

HPE Apollo 6000 Power Shelf

Apollo 6000
Power Shelf

Figure 6-7 HPE Apollo 6000 Power Shelf

The Apollo 6000 Power Shelf (Figure 6-7) optimizes power and efficiency at the rack to accommodate dynamic workload needs. Depending on the power configurations of the trays within a chassis, the power shelf can support two to four fully populated Apollo a6000 chassis with a maximum DC power up to 15.9 kW. With redundant hot-plug power supplies, the Apollo 6000 Power Shelf can also be configured for single- or three-phase input. The 1.5U external power shelf is unique to the Apollo 6000 system.

Customers can use the Apollo 6000 Power Shelf to:

- Simplify their environment with fewer PDU ports and cables

- Deploy in half to full rack increments for maximum energy efficiency

Additional features include:

- Efficient pooled or shared power infrastructure

- Capacity for a maximum of six power supplies

- Support for N, N+1, or N+N redundancy

Note

Scan this QR code or enter the URL into your browser to access the Apollo 6000 QuickSpecs.

http://www8.hp.com/h20195/v2/GetPDF.aspx/c04293373.pdf

HPE Apollo 8000

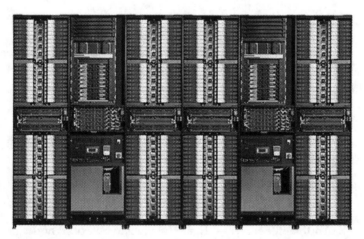

Figure 6-8 HPE Apollo 8000

Supercomputers provide the massive compute power that allows leading research institutions to run the simulations and analytics that are behind breakthroughs in science and technology. To address the barriers of traditional supercomputing, HPE designed the Apollo 8000 system, shown in Figure 6-8.

This product is a high-density, energy-efficient, HPC solution that uses a groundbreaking warm-water liquid cooling system to deliver a rapid, energy-efficient, sustainable infrastructure for research computing workloads. Liquid cooling is 1000 times more efficient than air cooling. The Apollo 8000 design allows extremely dense configurations to deliver hundreds of teraflops of compute power in a very compact space. It enables businesses to use the transferred facilities' heat for a dramatic reduction in costs and carbon footprint. Bringing the heat extraction closer to the processor further enhances computational performance capabilities.

The Apollo f8000 rack holds up to 144 ProLiant servers with CPU, accelerator, memory, and networking options in a standard rack footprint. It powers HPC workloads with three-phase high-voltage AC (HVAC), single stage power rectification, and high-voltage DC distribution. The HVAC power distribution system eliminates conversion steps and exceeds ENERGY STAR Platinum certification levels. Each rack supports an integrated 1 GbE switch with up to 160 ports for simplified cabling and management connectivity. Smart sensors automatically track and dynamically adjust system components for optimum efficiency.

Water coming into the rack is distributed through manifolds that split out to thermal bus bars for heat transfer from the dry-disconnect server trays. Dry-disconnect server trays contain sealed heat pipes that cool server components and keeps them safe. This process separates the water from the electrical components. Water traveling out of the rack features isolation valves and is aggregated for energy recycling. Additionally, a quick-connect, modular plumbing kit can be operational quickly as opposed to multiple days or weeks.

Apollo 8000 components

The Apollo 8000 system is available with a scalable configuration, starting with one HPE Apollo f8000 rack and one HPE Apollo 8000 iCDU rack. This converged system offers various accelerator, PCIe, and throughput options.

Components include:

- **HPE Apollo f8000 rack**—This rack is capable of holding up to 288 processors or up to 144 accelerators and a total of eight InfiniBand switches.

- **HPE Apollo 8000 iCDU rack**—This half-rack solution has 26U available in the top half of the rack and maintains sub-atmospheric water pressure for operational resiliency and serviceability.

- **HPE ProLiant XL730f Gen9 server**— This server includes two 2P nodes per tray support up to 256 GB HPE DDR4 SmartMemory per node, with an optional SSD.

- **HPE ProLiant XL740f Gen9 server**—This server includes two Xeon E5-2600 v3 processors and two Intel Xeon Phi 7120D coprocessors per tray.

- **HPE ProLiant XL750f Gen9 server**—This server includes a single 2P node with two NVIDIA Tesla K40 XL GPUs per tray.

- **HPE InfiniBand Switch for Apollo 8000 system**—Within this 36-port InfiniBand FDR switch, each tray has 18 quad small form-factor pluggable (QSFP) uplinks and 18 downlinks for node connectivity.

Note

Scan this QR code or enter the URL into your browser to access the Apollo f8000 rack QuickSpecs.

http://www8.hp.com/h20195/v2/GetPDF.aspx/c04390756.pdf

Apollo 8000 features

Many innovations differentiate the Apollo 8000 system from other supercomputing systems, including:

- **Primary and secondary water loops**—The primary loop circulates facility water through the cooling distribution unit (CDU) located within the iCDU rack. The closed secondary loop located in the f8000 rack distributes water from the iCDU to the f8000 rack where it flows through the heat exchanger and water wall adjacent to the server tray compartment (there is no liquid in the server tray).

- **Hybrid component cooling**—The direct-contact cooling (server heat pipes) manages the CPU, GPU, and memory temperature with no liquid or liquid connections. Other, less heat-sensitive node components are cooled with the hybrid air/water-cooled component of the Apollo 8000 system.

- **HPE dry disconnect technology**—This technology makes a liquid-cooled system as easy to service as air-cooled. Compute trays remains dry for server maintenance and allow removal without breaking a water connection.

Managing Apollo systems

In addition to iLO 4, HPE also offers additional management options that are ideal for managing Apollo solutions at scale:

- **HPE Advanced Power Manager (APM)**—The APM is an optional rack-level solution for Apollo 8000, Apollo 6000, Apollo 4000, Apollo 2000, ProLiant SL6500 Scalable System, and Moonshot 1500 systems. The APM automatically discovers hardware components and enables bay-level power on and off, server metering, aggregate dynamic power capping, configurable power-up dependencies and sequencing, consolidated Ethernet access to all resident iLOs, and asset management capabilities.

The APM does not replace rack power distribution units (PDUs), but it is designed to enable the utilization of basic, low cost rack PDUs and provides the functionality of switched PDUs. Switched PDUs provide hardware power on and off of individual servers by turning off the AC power to the power supplies of a given server. Because the servers share power supplies to optimize power efficiency, using switched PDUs to turn off all the power supplies in the chassis results in the loss of all server nodes in that chassis. The APM solves this by allowing server node-level hardware power on and off of the DC power to the individual server node motherboards. Additional features include:

- Rack-level event logging

- Remote Authentication Dial-In User Service (RADIUS) authentication

- Integrated serial concentrator

- Up to 11 local user accounts

- Read-only service port

- Supports SNMP, Secure Shell (SSH), Syslogd, and Telnet

- **HPE Insight Cluster Management Utility (CMU)**—The Insight CMU is a hyperscale-optimized management framework that includes software for the provisioning, control, performance, and monitoring of groups of nodes and infrastructure. This collection of tools helps customers manage, install, and monitor a large group of compute nodes, specifically HPC and large Linux clusters. The CMU features a GUI and CLI that can remotely halt, boot, or power off any selection of nodes, and broadcast commands to selected sets of nodes. CMU monitoring makes it possible to see, at a glance, the state of the entire cluster. Businesses can use Insight CMU to lower the TCO of this architecture. Insight CMU is scalable and can be used for any size cluster.

  - The Insight CMU GUI:

    - Monitors all the nodes of the cluster at a glance

    - Configures Insight CMU according to the actual cluster

    - Manages the cluster by sending commands to any number of compute nodes

    - Replicates the disk of a compute node on any number of compute nodes

  - The Insight CMU CLI:

    - Manages the cluster by sending commands to any number of compute nodes

    - Replicates the disk of a compute node on any number of compute nodes

    - Saves and restores the Insight CMU database

Note

Scan this QR code or enter the URL into your browser to access the Insight CMU user guide.

http://h20628.www2.hp.com/km-ext/kmcsdirect/emr_na-c05083084-1.pdf

- **HPE Apollo 8000 System Manager**—With the Apollo 8000 System Manager, users can see and manage shared infrastructure power along with facility and environmental controls from a single console. Customers can avoid spending capital on serial concentrators, adaptors, cables, and switches. This solution is flexible enough to meet workload demands with dynamic power allocation and capping.

  The software provides real-time data from hundreds of smart sensors and environmental controls. It provides email alerts when pre-configured events occur. This solution integrates two products: Insight CMU and APM. Within this integration, Insight CMU cannot be used in its traditional cluster management role. Instead, Insight CMU is used only to provide the back-end monitoring and alerting support.

  The Apollo 8000 System Manager provides a web-based interface for Apollo 8000 rack-level environmental status. Centralized rack management lets users monitor Apollo 8000 shared infrastructure that includes:

  - Air temperature throughout the Apollo f8000 rack and the Apollo 8000 iCDU rack

  - Water temperature and pressure in both the f8000 rack and the iCDU rack

  - Temperature, humidity, and leak detection water sensors for additional safety

  - iCDU pump status and flow rate

Learning check

1. How can researchers benefit from HPC solutions?

_____

_____

_____

_____

2. Match each feature to the correct Apollo system. Note that each system will match more than one feature.

    A.   Apollo 2000   ____ Redundant fans and power infrastructure with up to two 1400 W power supplies

    B.   Apollo 4000   ____ A design based on Hadoop and other big data analytics and object storage systems

                               ____ Up to four independent, hot-pluggable server nodes in one chassis

                               ____ 12 LFF or 24 SFF drive cage options

                               ____ Up to four PCIe slots with flexible performance and IO options

                               ____ The capability to scale in affordable increments in the future

3. HPE Advanced Power Manager replaces rack power distribution units.

    ☐ True

    ☐ False

Learning check answers

1. How can researchers benefit from HPC solutions?

- **HPC solutions increase research agility; lower costs; and allow researchers to process, store, and interpret petabytes of data.**

- **HPC enables simulation and analytical solutions to some of the most vexing problems in areas such as nanotechnology, climate change, renewable energy, neuroscience, bioinformatics, computational biology, and astrophysics.**

2. Match each feature to the correct Apollo system. (Each system matches multiple features.)

Apollo 2000	Apollo 4000
Redundant fans and power infrastructure with up to two 1400 W power supplies	A design based on Hadoop and other big data analytics and object storage systems
Up to four independent, hot-pluggable server nodes in one chassis	Up to four PCIe slots with flexible performance and IO options
12 LFF or 24 SFF drive cage options	The capability to scale in affordable increments in the future

3. HPE Advanced Power Manager replaces rack power distribution units.

   ☐ True

   ☐ **False**

Summary

- HPC customers are focused on how to get the best performance possible with limited resources. Governments, academia, and enterprises use HPC to drive advances in their respective fields. Today, HPC is becoming essential to the continued success of businesses requiring optimal computational performance, unprecedented reliability, memory, and storage scalability.

- HPE Apollo systems are designed to deliver compute, storage, networking, power, and cooling solutions for big data, analytics, object storage, and HPC workloads.

  - The Apollo 2000 offers all the features of traditional enterprise servers and provides twice the amount of density than standard 1U rack servers.

  - Apollo 4000 systems are designed specifically for Hadoop and other big data analytics and object storage systems.

  - Apollo 6000 systems are built to provide power, cooling, and IO infrastructure to support ProLiant XL servers.

  - Apollo 8000 systems are high-density, energy-efficient, HPC solutions that use a groundbreaking warm-water liquid cooling system to deliver a rapid, energy-efficient, sustainable infrastructure.

- In addition to iLO 4, Apollo management options include the HPE Advanced Power Manager, HPE Insight Cluster Management Utility, and HPE Apollo 8000 System Manager.

7 HPE Synergy

WHAT IS IN THIS CHAPTER FOR YOU?

After completing this chapter, you should be able to:

✓ Describe Hewlett Packard Enterprise (HPE) Synergy in the context of current infrastructure challenges

✓ Discuss possible Synergy customer use cases

✓ List the steps to configure a Synergy solution

OPENING CONSIDERATIONS

Before proceeding with this section, answer the following questions to assess your existing knowledge of the topics covered in this chapter. Record your answers in the space provided here.

1. What is your experience with the composable architecture of Synergy?

    _____

    _____

    _____

    _____

    _____

    _____

2. How would you explain the benefits of simplified networking to a customer?

    _____

    _____

    _____

    _____

    _____

    _____

3. How does Synergy address the challenges of integrating with traditional and new IT?

_____

_____

_____

_____

_____

Emergence of bi-modal IT

Many CIOs and data center system administrators feel challenged to deliver traditional business applications while at the same time provisioning new applications such as mobile and cloud-native apps that drive revenue. Ops-driven and cost-focused, traditional IT environments make it difficult to deliver faster value to the business—IT cannot move fast enough for today's application delivery goals.

Traditional business applications are built to run the business. They include applications such as ERP and other large databases that have been pre-packaged and pre-tested and typically go through one or two release cycles a year. IT has been built around these for the last 20 to 30 years.

Apps-driven and agility-focused, new IT environments deliver a new type of apps and services that drive revenue and new customer experiences via mobility, big data, and cloud-native technologies. These new apps challenge IT to maintain a digital enterprise in a digital economy alongside traditional applications. But maintaining two different sets of infrastructure, one designed for traditional apps and another designed for cloud-native apps, increases costs and complexity. This approach is not sustainable.

Gartner calls the strategy of maintaining an existing infrastructure for traditional applications while transitioning to infrastructure and tools for the cloud native applications _bi-modal computing_. The HPE vision is to pull both together with one infrastructure that can deliver the applications while enabling the agility of infrastructure that customers are seeing today in the cloud but on premises.

A composable infrastructure optimizes intelligence and automation via infrastructure as code to seamlessly bridge traditional and new IT environments for huge gains in application speed and operational efficiency. It offers a single infrastructure ready for any physical, virtual, or containerized workload including enterprise applications and cloud-native apps. The HPE composable infrastructure:

- Reduces operational complexity

- Accelerates application deployment

- Is designed for today and architected for the future

Composable infrastructure

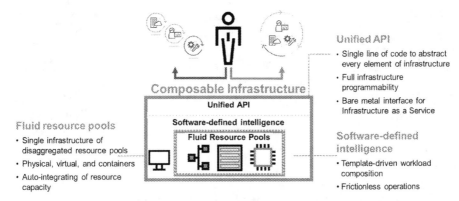

Figure 7-1 Composable infrastructure

Synergy composable infrastructure delivers high performance and composability for the delivery of applications and services. It simplifies network connectivity using disaggregation in a cost-effective, highly available, and scalable architecture. A composable infrastructure creates a pool of flexible fabric capacity that can be configured almost instantly to rapidly provision infrastructure for a broad range of applications.

A composable infrastructure has three key elements as illustrated by Figure 7-1:

- A composable infrastructure starts with fluid resource pools, which is essentially a single structure that boots up, ready for any workload with fluid pools of compute, storage, and network fabric that can be instantly turned on and flexed.

- Software-defined intelligence means embedding intelligence into the infrastructure and using workload templates to tell it how to compose, re-compose, and update quickly, in a very repeatable, frictionless manner.

- The third element uses all of these capabilities and exposes them through a unified API, which allows infrastructure to be programmed like code so it can become infrastructure as a service.

By delivering a highly flexible, high-performance pool of composable resources centered around a unified API to simplify and speed up deployment and management, Synergy offers value and flexibility. This is useful for businesses who just want to write existing apps or code new apps and be able to derive infrastructure directly as code.

The industry's first platform architected for composability

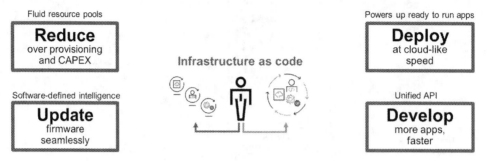

Figure 7-2 HPE Synergy

HPE Synergy, the first platform built specifically for a composable infrastructure, offers an experience that empowers IT to create and deliver new value instantly and continuously. It reduces operational complexity for traditional workloads and increases operational velocity for the new breed of applications and services. Through a single interface, Synergy composes physical and virtual compute, storage, and fabric pools into any configuration for any application. As an extensible platform, it easily enables a broad range of applications and operational models such as virtualization, hybrid cloud, and DevOps.

With Synergy, IT can become not just the internal service provider but the business partner to rapidly launch new applications that become the business. As illustrated in Figure 7-2, with Synergy, IT can continuously:

- **Run anything**—Optimize any application and store all data on a single infrastructure with fluid pools of physical and virtual compute, storage, and fabric.

- **Move faster**—Accelerate application and service delivery through a single interface that precisely composes logical infrastructures at near-instant speeds.

- **Work efficiently**—Reduce operational effort and cost through internal software-defined intelligence with template-driven, frictionless operations.

- **Unlock value**—Increase productivity and control across the data center by integrating and automating infrastructure operations and applications through a unified API.

- **Technically describe Infrastructure as code**—Provision bare metal infrastructure with one line of code—in the same way as virtual machines (VMs) and cloud.

The Synergy platform is unique because it enables customers to create a completely stateless infrastructure. Only HPE allows configuration and provisioning of compute, fabric, storage, and now hypervisor and operating system images as part of a single server profile template. This enables all of the pieces to be configured in one place without having to use multiple tools. It also allows the infrastructure to be provisioned and re-provisioned as needed based on the requirements of the workloads that it is hosting.

Management innovations that enable a composable infrastructure

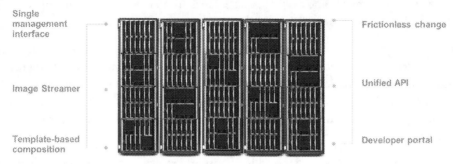

Figure 7-3 Synergy management innovations

Synergy is a single infrastructure of physical and virtual pools of compute, storage, and fabric resources and a single management interface that allows IT to instantly assemble and re-assemble resources in any configuration. Through its single interface, Synergy composes physical and virtual compute, storage, and fabric pools into any configuration for any application.

Synergy eliminates hardware and operational complexity so that IT can deliver infrastructure to applications faster and with greater precision and flexibility. Synergy empowers IT to reduce infrastructure management overhead and focus on optimizing cloud (and traditional) applications for speed, reliability and efficiency, resulting in superior outcomes for business. Synergy eliminates the high costs of overprovisioning and stranding of resources with built-in infrastructure intelligence, dramatically reducing CAPEX. With improved economics, businesses can drive continuous service delivery during workload peak times and significantly reduce risk.

Synergy is ready for any workload—whether physical, virtual, or containerized, including both enterprise applications and cloud-native applications. As illustrated by Figure 7-3, Synergy is built on unique innovations that provide a bare-metal infrastructure that is ready to run any application and delivers infrastructure as code:

- **Single management interface**—Synergy offers a single management interface that is used to discover, compose, update, and troubleshoot the Synergy solution.

- **Image Streamer**—This Synergy component allows you to create and maintain a repository of images for different workloads. Using these images, you can instantly provision operating systems or hypervisors on stateless infrastructure, significantly simplifying and speeding deployments.

- **Template-based composition**—You can configure templates for specific workloads. These templates allow Synergy to seamlessly request the use of the correct resources for a specific workload and release those resources when they are no longer needed, making them available for other workloads.

- **Frictionless change**—You can maintain and update the infrastructure while it is up and running. The updated firmware and driver packages are delivered seamlessly as one pretested package in a manner that minimizes the impact on services and operations.

- **Unified API**—Built into the infrastructure, the unified API abstracts infrastructure complexity so that changes can be automated easily and developers can program the infrastructure as code. The fully programmatic interface integrates into popular management tools such as Microsoft System Center and VMware vCenter. It is also future-proofed by integrating into open source automation and configuration management tools such as HPE Helion, Chef, Python, PowerShell, and OpenStack.

- **Developer portal**—Developers can readily obtain infrastructure as code to rapidly deploy their applications.

Synergy customer use cases

You can configure Synergy in numerous ways to meet a wide variety of customer needs. The following use cases are based on a fictional customer. They are designed to help you understand how Synergy can fit into an existing or planned data center. By generalizing the concepts presented, you can envision Synergy in a real-world scenario. These use cases provide guidelines for some starting points for potential solutions that you might consider when recommending a Synergy solution for a customer.

- **Use Case 1: Database**—The company sells its users on being able to constantly run and return reports to their dashboards that give them real-time data about their applications so they can make informed decisions about their business. The analytics applications run across a variety of applications, which could be handling up to half a million peak concurrent connections. They tried to use Oracle Coherence for this solution, but it was returning three- or four-second response times. They are now migrating to a custom application written for SAP HANA. Because these analytics must be running and reporting all the time, there are multiple IT tiers. The dev, test, and staging tiers are not persistent, but can be spun up and down as needed.

Potential solution

- Synergy Image Streamer is utilized as the appliance that houses and implements the golden templates for the IT tiers. These directly attach to a Synergy D3940 storage module, and this storage can be shared by these tiers.

- The production tier contains two high-performance compute nodes (Synergy 620 Gen9 compute modules) directly attached to a fully populated Synergy D3940 storage module.

- There is a high availability tier that is exactly like the production tier, but is located in another rack within the same data center.

- There are no IO bottlenecks in the fabric, with 20 Gb/s to the compute modules and 40 Gb/s upstream. This is accomplished with Virtual Connect SE 40Gbps F8s for Synergy.

- **Use Case 2: Object Storage**—The company needs about 240 TB of usable capacity to store application media files in addition to support for an object storage-based archive system. They need at least six servers to support object storage hosting and management: two on the front end of the storage, two for identifier/metadata management, and two as proxy hosts. They want a high-capacity storage platform that is optimized around cost per gigabyte. They also need at least two high-performance drives allocated for each of their server compute modules to support that metadata storage specifically, in addition to the object storage itself. They need to ensure that the object storage management traffic is taking place on an isolated network. They also need to ensure that the solution supports an object storage enabled operating system in a storage subsystem, such as Scality, Lustre, or Ceph.

Potential solution:

- Six Synergy 480 Gen 9 compute modules with 1 GB of RAM per TB of disk space and mirrored operating system drives

- Three Synergy D3940 storage modules with dual IO adapters with 36 x 2 TB SATA drives for object storage and 4 x 1.92 TB SSDs for container/metadata storage

- Virtual Connect SE 40Gbps F8s for Synergy

- **Use Case 3: Hybrid Cloud**—In this hybrid cloud, the mobile apps are being developed and will be hosted for user download and consumption. As applications go viral, the solution must be able to expand quickly and use additional compute resources to support massive growth. It must also have maximum uptime.

Revenue is generated by advertising embedded in the applications, so if the applications are not in the hands of users, revenue is lost. This requires the ability to grow or shrink infrastructure used for downloading, depending on the popularity of the application. Use Case 1 talked about the need for real-time data about the applications in order for the customer to make informed decisions about their business. However, it does not do the customer any good to understand the trend if they are unable to respond to it. The company also needs the ability to expand a web farm on the fly, and they are evaluating Docker.

Potential solution

- Compute density provided by 30 Synergy 480 Gen9 compute modules and 256 GB RAM

- Image Streamer with Docker-enabled Linux and Docker Machine plugin

- Virtual Connect SE 40Gbps F8s for Synergy

- Integration with HPE Helion to extend beyond infrastructure during peak user demand

Use Case 4: DevOps—DevOps is the core of this company's business. They work with customers who are developing applications that the company then hosts, so they require a platform that can grow and shrink the infrastructure depending upon the type of development that is taking place at any time. They need infrastructure as code that must be easily programmable, rapidly reconfigurable, and software defined so that the hardware does not need to be manually reconfigured every time a development team expands, each time a new project is created, or if an old project falls off the roadmap. This means continuous testing, monitoring, and deployment. They are using Chef as a DevOps platform for "recipes" (user stories). They also want to provision Docker hosts and flexible clusters.

Potential solution

– Compute density requirement will be met by six Synergy 660 Gen9 compute modules per frame with three frames

– Storage included in the compute modules for boot with mirrored operating system drives

– D3940 Storage for serving up hosting/download files

– Virtual Connect SE 40Gbps F8s for Synergy

– Synergy Composer and Synergy Image Streamer with Docker-enabled Linux for "User stories" for DevOps

Use Case 5: Virtualization—This more traditional use case involves virtualization requirements related to the company's business needs for faster delivery of IT services and solutions. They are focused on streamlining and eliminating stranded and underutilized resources. They want to reduce complexity and improve staff productivity. And they are looking at various automation and orchestration solutions for a roughly 2000–2500 virtual machine environment, most of which is being used for hosting web servers for their user downloads.

Potential solution

– 36 Synergy 480 Gen9 compute modules with 256 GB RAM

– 8 GB MicroSD for hypervisor boot

– Integration with VMware vCenter and Puppet for provisioning and managing the lifecycle of VMs

– Redundant Composers and frame link modules (FLMs)

– Image Streamer with ESXi host support

– Virtualization Performance Viewer for OneView (vPV)

– Virtual Connect SE 40Gbps F8s for Synergy

Use Case 6: Collaboration/Exchange—This use case involves the collaboration aspect of the business, which in this case is Microsoft Exchange. In their operating environment, the company has a certain amount of variability based on mailbox turnover because they host mail for their developing partners. They have about 16,000 mailboxes in three separate Exchange deployments: one for their North America operation, one supporting their Asian operation, and one for global developers and hosted clients. They also do archiving to disk as part of their backup scheme.

Sizing overall is a 1GB mailbox, a 1GB hosted archive, and 30-day recovery. The traffic averages 50 sent and 200 received messages per day. They need flexible capacity built into their solution to account for high mailbox turnover as they add new developers who use the general-purpose mailboxes.

Potential solution

–   18 Synergy 480 Gen 9 compute modules with 512 GB of memory, dual 120Gb uFF boot drives, dual 20Gb Converged NICs and P542D Smart Array controller

–   Nine D3940 storage modules each with dual IO modules

    •   Six modules with 40 x 600 GB SAS HDDs

    •   Three modules for archiving with 40 x 1.2 TB SATA HDDs

–   Dual 12G SAS connection modules

–   Redundant Synergy Composers

–   Virtual Connect SE 40Gbps F8s for Synergy

Configuring a Synergy system

A Synergy solution allows customers to select the right ratios of fabric, storage, and compute to compose the infrastructure necessary for their particular workloads. The process you follow to building a Synergy solution for a customer involves four loosely defined main steps:

1.   First, configure the infrastructure by determining the number of frames the customer requires and the associated Composers, frame link modules, and compute modules.

2.   Determine how the Composers and frame link modules will be connected in a management ring. The Synergy system via the Synergy Composer will automatically discover all of the compute, fabric, and storage resources that have been added to the management ring.

3.   Consider how to configure fabric elements (the data network) as part of a composable infrastructure. Determine whether the Image Streamer is appropriate for the customer's configuration. If you are adding an Image Streamer to the customer's configuration, determine how it fits into the fabric and the management ring.

4.   Lastly, build out storage modules based on the customer's requirements.

Step 1: Configure the infrastructure

Configuring the infrastructure involves making decisions about the frames and the management subsystem.

HPE Synergy 12000 frame

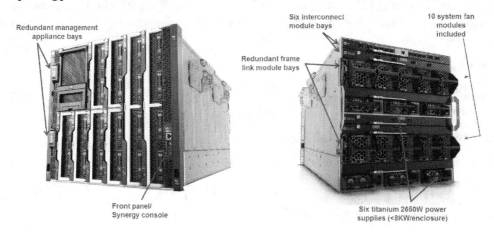

Figure 7-4 Synergy 12000 frame

The Synergy 12000 Frame, as shown in Figure 7-4, is a central element of Synergy that provides the base for an intelligent infrastructure with embedded management and scalable links for expansion as business demand requires. The Frame is the base infrastructure that pools resources of compute, storage, fabric, cooling, power, and scalability. With an embedded management solution combining the Synergy Composer and frame link modules, IT can manage, assemble, and scale resources on demand. The Synergy Frame is designed for today's and future needs with expanded compute and fabric bandwidths and photonics ready.

The Synergy frame reduces complexity through an intelligent autodiscovery infrastructure and delivers performance and to accelerate workload deployment. As the building block for a Synergy infrastructure, a Synergy frame offers substantial expansion and scalability.

Every frame offers dual hot plug integrated appliance bays for redundancy. They have 10 Gb network direct connected to frame link modules for inter- or intra-frame management communications.

A Synergy frame's unique design physically embeds Synergy Composer with HPE OneView management software to compose compute, storage, and fabric resources in any configuration. Synergy frames may be linked into larger groups or domains of frames to form a dedicated management network, increasing resources available to the business and IT efficiency as the size of the infrastructure grows—achieving both CAPEX and OPEX economies of scale.

The Synergy 12000 frame physically embeds management as code into an intelligent infrastructure to offer management and composability of integrated compute, storage, and fabric resources. Whether resources are in a single frame or multiple linked frames, the system offers composability of all resources.

The Synergy 12000 frame eases installation by using the standard power feeds of BladeSystems. Synergy supports up to six 96% Titanium Efficient, 2650 Watt power supplies that offer redundant

N + N, N + 1 power setup. The Synergy frame provides an efficient cooling system and has 10 built-in fans in every frame. It delivers the frame link topology (the ring architecture) through 10Gbase-T RJ-45 jacks and CAT6 cables, providing resource discovery and status, management commands, and inventory reporting.

The Synergy 12000 frame provides walk up diagnostic and configuration link through display port and USB connections either at the rear or at the front panel of the frame.

Synergy management subsystem

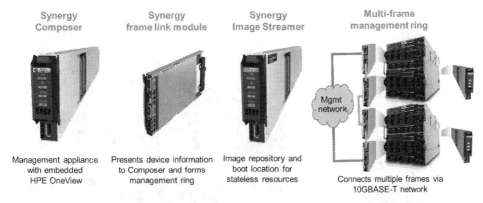

Figure 7-5 Synergy management subsystem

As shown in Figure 7-5, the Synergy management subsystem comprises the following components:

- **Composer**—A management appliance that directly integrates into the frame of the system and provides a single interface for assembling and re-assembling flexible compute, storage, and fabric resources in any configuration. Its infrastructure-as-code capability accelerates transformation to a hybrid infrastructure and provides on-demand delivery and support of applications and services with consistent governance, compliance, and integration.

- **Frame link modules**—The integrated resource information control point. Frame link modules report asset and inventory information about all the devices in the frame. As resource controllers, they provide functions such as inventory and configuration checking. They also provide the management uplinks to the customer's network.

 Note

An uplink port (or link aggregation group) is used to expand a network by connecting to another network or a device such as a router, switch, or server. A downlink port is used to receive data from another (often larger) network.

A single Synergy Composer manages one frame or multiple racks of frames linked through the frame link modules. The Synergy Composer option that you select determines the number of frames linked and managed. HPE recommends using two Synergy Composer modules for redundancy and high availability.

- **Image Streamer**—A new approach to deployment and updates for a composable infrastructure. This management appliance works with Composer for fast software-defined control over physical compute modules with operating system provisioning. Image Streamer enables true stateless computing combined with instant-on capability for deployment and updates. This management appliance deploys and updates infrastructure at extreme speed.

HPE Synergy Composer

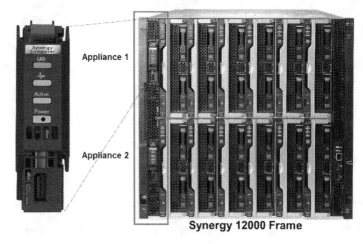

Figure 7-6 HPE Synergy Composer

As shown on the left side of Figure 7-6, Composer is a physical appliance integrated within the Synergy frame. It plugs in to an appliance bay in the left side of the Synergy frame as shown on the right of Figure 7-6.

 Note

Composer does not use a compute module slot.

Composer provides a single interface for assembling and re-assembling flexible compute, storage, and fabric resources in any configuration. Its infrastructure as code capability accelerates transformation to a hybrid infrastructure and provides on-demand delivery and support of applications and services with consistent governance, compliance, and integration.

Composer embeds the HPE OneView management solution to manage compute modules, fabrics, and storage, which is the essence of software-defined intelligence in Synergy.

Composer deploys, monitors, and updates the infrastructure from one interface and one unified API. It allows IT departments to deploy infrastructure for traditional, virtualized, and cloud environments in a single step, in just a few minutes. Resources can be updated, flexed, and redeployed without service interruptions. This allows infrastructure to be deployed and consistently updated with the right configuration parameters and firmware versions, streamlining the delivery of IT services and the transition to a hybrid cloud. Its reduced complexity and faster service delivery times ultimately enable IT to better respond to changing business needs.

Server templates are a powerful new way to quickly and reliably update and maintain an existing infrastructure. Composer uses templates to simplify one-to-many updates and manage compute module profiles. This feature adds inheritance to the process, meaning updates can be made once, in the template, and then propagated out to the profiles created from that template. Elements that can be updated via a template include firmware, BIOS settings, local RAID settings, boot order, network configuration, shared storage configuration, and many others.

Composer templates also provide *monitor and flag* capabilities with remediation. Profiles created from the template are monitored for configuration compliance. When inconsistencies are detected, an alert is generated indicating that the profile is out of compliance with its template. When a new update is made at the template level, all profiles derived from that template will be flagged as inconsistent. From there, the user has complete control over the remediation process for bringing individual modules or multiple systems back into compliance.

HPE Synergy frame link modules

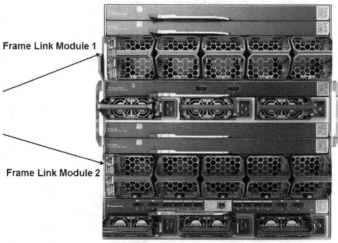

Rear of Synergy 12000 frame

Figure 7-7 Synergy frame link modules

Figure 7-7 shows a rear view of a Synergy 12000 frame and the frame link modules. The frame link module is the intelligence behind the frame and the management architecture. It provides shared frame services such as power, cooling, auto-discovery, and inventory of all installed components on the management interface. For example, they report, in real time, the power each module uses and the total power used per frame.

A frame link module enables you to take one composable element or frame and connect several of them together to allow them to automatically scale together. You can start off with one element and add more infrastructure as needed. As you add elements, they auto-assemble together into one larger infrastructure that can still be managed as a single infrastructure. You can scale up to a rack or even a row scale deployment and have it all managed as a single element. Appliance bays link directly to the frame link modules in the rear of the frame to provide detailed information of compute, storage, and fabric resources for management of the Composer with HPE OneView.

Frame link modules have a plug-and-play system assembly, which means the installation technician can cable the modules and then automatically assemble the system. Two frame link modules per frame are used to provide fault tolerance.

Frame link modules link to the management appliances and provide control points for providing resource and health information of the frame to the management appliances. A management port on each frame link module provides access to the management appliance and link ports for linking modules for multi-frame linking and setup.

Synergy Image Streamer

Figure 7-8 Synergy Image Streamer

To accelerate workload deployment, Synergy uses the Image Streamer, as shown in Figure 7-8, a physical appliance repository of bootable (golden) images that can be streamed across multiple compute modules in a matter of seconds. This unique capability enables Synergy to set up and update infrastructure quickly and consistently. This is significantly faster than the traditional, sequential process of building compute modules—physical provisioning followed by operating system or hypervisor installation. It is ideal for situations such as web-scale deployments where IT needs to provision an operating environment across a large number infrastructure blocks.

Traditional server deployment is a sequential process of provisioning the physical hardware, followed by provisioning an operating system, and then by provisioning a hypervisor installation. Traditional memory-based server deployments use general deployment/provisioning tools for service operating system deployment, which uses a RAM-based operating system, and is also known as a *pre-boot (pre-install) environment.*

Image Streamer enables true stateless operation by integrating server profiles with golden images (operating environment and IO driver) and personalities (operating system and application) for rapid implementation onto available hardware. The fast deployment and compliance management capabilities leverage software-defined intelligence and are accessible via the unified API. These capabilities set Image Streamer apart from traditional approaches.

 Note

Image Streamer provisions boot volumes as Thin Provisioned Smart Clones of a read-only "golden volume" so that the boot volumes only contain delta changes relative to the "golden volume."

Updates to highly replicated physical compute nodes with their operating environments at extreme speeds enables Image Streamer to deliver fast virtualized image changeovers (for use in Test and Dev, DevOps, multiple PaaS) or secure boot and image compliance (for use in defense, government, or financial services institutions). These capabilities are ideal for web-scale deployments where IT needs to provision an operating environment across a large number infrastructure blocks.

Image Streamer ensures high availability by providing redundant repositories of bootable images, which are used in a secure manner. These golden images can be rapidly cloned to create unique bootable images for compute nodes. It enables Synergy to quickly deploy a new compute module or update an existing one. This is far faster than the traditional, sequential process of building servers—physical provisioning followed by operating system, hypervisor installation, IO drivers, and application stacks.

Administrators using Image Streamer can design bootable images for compute nodes, with the operating system and application stacks included, for ready-to-run environments.

 Important

Image Streamer requires a minimum of three Synergy frames with redundant Composers for operation and must be implemented as redundant pairs. This minimal system requires four cables, two transceivers, and one interconnect module (ICM) for complete operation.

Synergy Gen9 compute module portfolio

Table 7-1 Synergy Gen9 compute module portfolio

	Synergy 480 Gen9	Synergy 660 Gen9	Synergy 620 Gen9	Synergy 680 Gen9
CPU	Xeon E5-2600 v4 2-socket	Xeon E5-4600 v4 4-socket	Xeon E7-4800 v4 or Xeon E7-8800 v4 2-socket	Xeon E7-4800 v4 or Xeon E7-8800 v4 4-socket
Form Factor	half-height	full-height	full-height	full-height, double-wide
Number per frame/ rack	12/48	6/24	6/24	3/12
DIMM slots	24	48	48	96
Max Memory Size	1.5 TB	3 TB	3 TB	6 TB
Local Storage Opt	Diskless, 2 SFF, 4 uFF, USB, MicroSD	Diskless, 4 SFF, 8 uFF, USB, MicroSD	Diskless, 2 SFF, 4 uFF, USB, MicroSD	Diskless, 4 SFF, 8 uFF, USB, MicroSD
Graphics Adapter (optional)	NVIDIA Tesla M6	N/A	N/A	N/A
Mezz Connectors	3 x16 PCIe 3.0	6 x16 PCIe 3.0	2 x16 and 3 x8 PCIe 3.0	4 x16 and 6 x8 PCIe 3.0
Fabric	Broad choice of storage controllers, CNAs, and Fibre Channel HBAs			

The Synergy Gen9 compute modules deliver general-purpose to mission-critical x86 levels of availability with real-time performance enabled by Intel Xeon processors and a variety of memory, storage, and fabric choices. Synergy supports both two-socket and four-socket x86 compute modules, which provide the performance, scalability, density optimization, and storage simplicity.

As shown in Table 7-1, available Synergy compute modules are:

- **Synergy 480 Gen9 compute module**—Has memory capacity of up to 1.5 TB and 24 DIMM slots. It supports the entire Intel Xeon E5-2600 v4 processor family without any DIMM slot restrictions. Greater consolidation and efficiency are achieved through an increase in VM density per compute module.

- **Synergy 660 Gen9 compute module**—Handles data-intensive workloads with uncompromised performance and exceptional value. The Synergy 660 Gen9 compute module is a full-height, high-performance module with Xeon E5-4600 v4 processors, 48 DIMM slots providing up to 3 TB of available memory, flexible IO fabric connectivity, and right-sized storage options.

- **Synergy 620 Gen9 compute module**—Delivers mission-critical availability and performance, as well as flexible memory, storage, and fabric options in a full-height, two-socket form factor. The Synergy 620 Gen9 compute module is a full-height, high-performance module with Xeon E7-4800 v4 or Xeon E7-8800 v4 processors. With 48 DIMM slots, it supports up to 3 TB of HPE DDR4 memory and has five IO PCIe 3.0 connectors (two x16 and three x8).

- **Synergy 680 Gen9 compute module**—Is a four-socket, full-height, double-wide compute module with 96 DIMM slots for up to 6 TB of HPE DDR4 memory. The Synergy 680 Gen9 compute module is a high-performance module with Xeon E7-4800 v4 or Xeon E7-8800 v4 processors. It features 10 PCIe 3.0 IO connectors (four x16 and six x8)

Ideal workloads for Synergy composable computing

Table 7-2 Ideal Synergy workloads

	Synergy 480 Gen9	Synergy 660 Gen9	Synergy 620 Gen9	Synergy 680 Gen9

Types of workloads	• Mainstream workloads and business applications:
	  – Collaborative
	  – Content management
	  – IT/Web Infrastructure
	• Workload consolidation
	• Composable compute
	• Optimum performance for mainstream applications with headroom to grow
		• Mainstream compute with greater CPU and memory requirements:
		  – Enterprise IT consolidation
		  – Virtualization
		  – Business processing
		  – Decision support
			• Enterprise class reliability, availability, and serviceability
			  – Enterprise IT consolidation
			  – Virtualization
			  – Structured database
			  – High memory-intensive workloads
				• Enterprise class performance, scalability, and RAS
				• Maximum memory-intensive workloads
				  – Scale up applications
				  – Enterprise IT consolidation
				  – Large in-memory database

Synergy compute modules have the configuration flexibility to power a variety of workloads, including business processing, IT infrastructure, web infrastructure, collaborative, and high-performance computing. As shown in Table 7-2, ideal workloads include:

- **Synergy 480 Gen9 compute module**—Delivers superior capacity, efficiency, and flexibility to power more demanding workloads and increase VM density by providing a full range of processor choices, right-sized storage options, and a simplified IO architecture. It is designed to optimize general-purpose enterprise workload performance including business processing, IT and web infrastructures, collaborative applications, and high performance computing (HPC) in physical and virtualized environments while lowering costs within a composable infrastructure.

- **Synergy 660 Gen9 compute module**—Supports demanding workloads such as in-memory and structured databases. It is also ideal for enterprise IT consolidation and virtualization.

- **Synergy 620 and Synergy 680 Gen9 compute modules**—Are designed to meet the needs of almost any enterprise IT tier and workload. These two-socket and four-socket x86 compute modules are ideal for financial, insurance, healthcare, manufacturing, and retail enterprises that require mission-critical levels of availability, extended versatility, and real-time performance. They are also designed for HPC applications with large memory demands.

 Note

Scan this QR code or enter the URL into your browser for more information on Synergy compute modules.

http://www8.hp.com/us/en/products/synergy-compute/index.html

Step 2: Configure the management ring

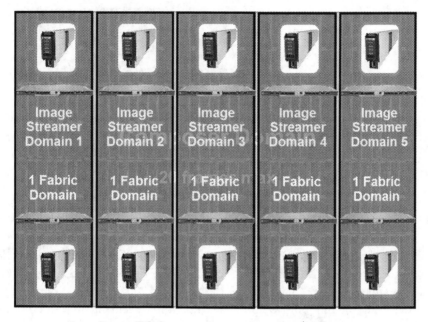

Figure 7-9 Synergy management topologies

A Synergy management ring is formed by using frame link modules across multiple frames. Figure 7-9 shows 20 frames as an air-gapped secure management network that has been automatically discovered and can automatically detect changes (such as frame removals or additions).

Adding two Composer appliances to manage the network makes this a high-availability configuration.

This management architecture prevents over-subscription issues on the production network because the management (control) plane is separated from the fabric (data) plane. This also helps prevent malicious takeover of control from denial of service (DoS) attacks.

For simplicity, Figure 7-9 shows all the fabric domains to be the same size (four frames each). Technically, better terms for these fabric domains are *logical enclosures* or *frame link domains*. It is important to note that for high availability, each fabric domain has:

- Two Synergy Virtual Connect master interconnect modules
- Two Image Streamer appliances (only the Image Streamer appliances are shown in Figure 7-9)

Although this is a simplified example, it shows how the different management elements in Synergy work together to produce a composable infrastructure with:

- Fluid pools of resources
- Software-defined intelligence
- Unified API access

Synergy management ring—Single frame

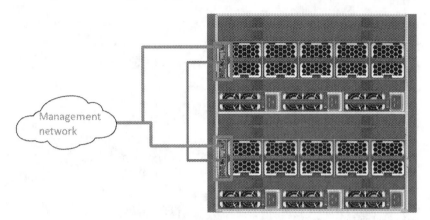

Figure 7-10 Synergy single frame

Even in a single-frame configuration, as a best practice, you should design a redundant solution with two Composers and two frame link modules. Although there is a 10 GbE connection between the two frame link modules across the midplane of the Synergy 12000 frame, you must cable the two modules together using the link port to complete the full management ring. In addition to being connected to each other, the frame link modules are also connected to the management network using the uplink port as shown in Figure 7-10.

This redundant design ensures that you maintain operational use of the Synergy management network, eliminating a single point of failure. This design also allows you to update the frame link module firmware without disrupting the entire system.

Connection locations in a single-frame management ring are:

- Frame link connection location

  - Frame 1 FLM 1 Link Port to Frame 1 FLM 2 Link Port

- Management network connection locations

  - Frame 1 FLM 1 management port to management network

  - Frame 1 FLM 2 management port to management network

Synergy management ring—Two frames

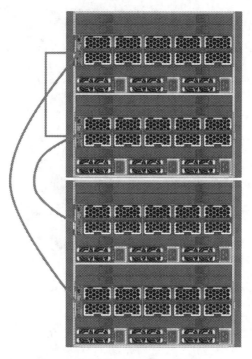

Figure 7-11 Two Synergy frames

As shown in Figure 7-11, to scale this one-frame configuration to a two frame configuration, you must disconnect the existing management ring and then cable the frame link modules in the first frame to the frame link modules in the second frame, creating a management ring across frames. The frame link modules then auto-discover all the devices within the additional frame.

As a best practice, you should move one of the Composers to the second frame. You should also move one of uplinks to a frame link module in the second frame. This creates a highly redundant management ring, which spans the physical architecture and creates one logical unit, from which the IT staff can compose resources.

A benefit to the newly configured ring is that new frames are automatically discovered and inventoried.

Synergy management ring—Three frames

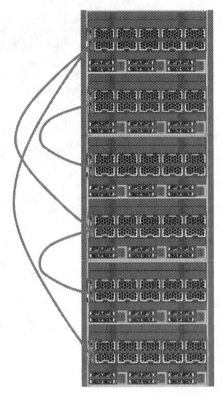

Figure 7-12 Three Synergy frames

To move to a three-frame configuration as shown in Figure 7-12, you need to change the cabling that connects the frame link modules so that you can add the additional frame. The frame link modules are connected as follows.

- Frame 1 FLM 2 Link Port to Frame 2 FLM 1 Link Port

- Frame 2 FLM 2 Link Port to Frame 3 FLM 1 Link Port

- Frame 3 FLM 2 Link Port to Frame 1 FLM 1 Link Port

The frame link modules auto-discover components in the third frame, dynamically adding resources into the same pool. This configuration is highly available because Composers are spread across the composable infrastructure and multiple uplinks are connected to the management network.

Whether the environment is configured with two frames per rack or three, customers can expand their infrastructure without adding management complexity. They can simply move frame link cables to provide a single, redundant management ring across enclosures and then manage the solution through a single pair of Composers.

Step 3: Configure the composable infrastructure fabric

Scaling is fast and simple with a Synergy composable fabric. When you add a new frame by using the fabric interconnect link module (ICLM), the new frame becomes an extension of the existing fabric, and the east/west design scales so the performance of the existing workload is not negatively impacted.

Rack-scale fabric architecture

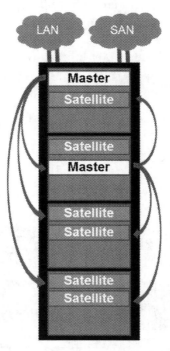

Figure 7-13 Rack-scale fabric architecture

As shown in Figure 7-13 the disaggregated, rack-scale Synergy design uses a master/satellite architecture to consolidate data center network connections, reduce hardware and management complexity, and scale network bandwidth across multiple frames. This architecture enables several Synergy frames to establish a single logical fabric, interconnecting compute modules at high speeds and low latency. Satellite modules—either Synergy 10Gb ICLMs or Synergy 20Gb ICLMs—connect to one of two master modules, which also connect to each other over ICM cluster links. The combined modules form a logical fabric. The modules in the logical fabric connect over links with 1:1 subscription (no oversubscription) and ultra-low latency. This one-hop design for traffic across the aggregated frames maximizes data throughput and minimizes latency for large domains of VMs or compute modules.

The master modules provide the uplinks for the complete logical fabric, consolidating the connections to the data center network. They contain intelligent networking capabilities that extend connectivity to satellite frames through the ILM, eliminating the need for a ToR switch and substantially reducing cost. The reduction in components simplifies fabric management at scale while consuming fewer ports at the data center aggregation layer.

A satellite module extends the composable fabric to additional satellite frames and replaces fixed ratios of interconnects in each frame by extending the fluid pool of network resources from the master module. Satellite frames inherit all the benefits of master module. Up to four additional satellite frames can be connected to the master module with one Synergy 20Gb ICLM.

Advantages of a master/satellite architecture include:

- 40% or more lower fabric hardware costs
- 10 Gb and 20 Gb bandwidth with a future path to 40 Gb/100 Gb to compute modules
- Flexible bandwidth allocation in 100 Mb increments
- Ethernet, Fibre Channel, Fibre Channel over Ethernet (FCoE), and iSCSI
- Zero touch change management
- Upgrade with minimum downtime

Composing a rack-scale fabric

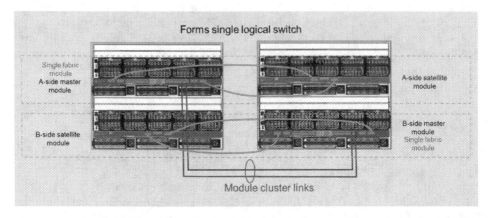

Figure 7-14 Composing a rack-scale fabric

The Synergy composable fabric has extreme scale. It can address workloads from a traditional data center to a cloud environment.

You can deploy an Ethernet or converged fabric using a single-frame, two-frame, or three-frame configuration. Four-and five-frame configurations are also supported if you are using 10Gb ICLMs.

You can include an additional 10 Gb satellite frame or frames containing a Synergy 20Gb ILM, for a total not exceeding five frames within a master/satellite stacking domain—that is, up to 60 compute modules. The same rule applies to 20 Gb satellite frames, frames containing Synergy 10Gb ILM, except that the total number of the frames within master/satellite domain cannot exceed three—that is, up to 36 compute modules.

The configurations depend on the use case. The size of the fabric is determined by three customer requirements:

- **Bandwidth to compute node**—Does the customer want 10 Gb or 20 Gb per port on the compute node?

- **Level of fabric redundancy**—Does the customer need a redundant fabric or highly available fabric?

- **Is Image Streamer in play?**—Does the customer need the fabric ring to be redundant or highly available?

For a proof of concept configuration, a single frame in a nonredundant configuration is a good place to start. For the example configuration shown in Figure 7-14, populate the fabric modules on the A Side (for example, start by placing a master module in Bay 3). If you want a single-frame configuration for a production environment, a redundant configuration is preferred. In this case, populate both the A and B Sides and use the stacking links to tie the master modules together.

Composing a two-frame rack-scale fabric

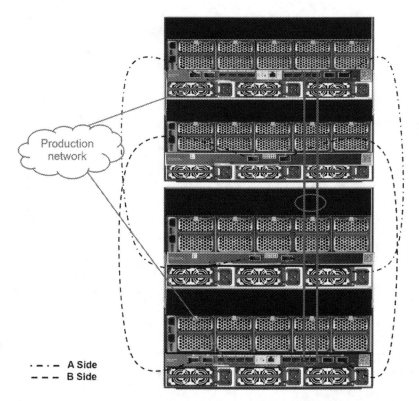

Figure 7-15 Two-frame rack-scale fabric

A multi-frame composable fabric is a cost-effective scaling of frames and compute modules without always needing additional pairs of master interconnects. The multi-frame composable fabric is orchestrated through the Synergy Composer and the logical interconnect group is managed by HPE OneView. The ICLMs create a flat, rack-scale fabric.

A two-frame rack-scale fabric, as shown in Figure 7-15, is a master/master configuration with an associated satellite. Each frame has a master and each A Side or B Side has a satellite.

The 2.56 Tb/s bandwidth of the master module is spread across 10 Gb and 20 Gb compute modules connected to both master and satellite frames in the stacking fashion using interconnect link ports. Adding additional frames to the master frame does not impact the performance of existing compute modules traffic. Because the subscription ratio between the master frame and the satellite frame is 1:1, performance of east/west traffic is not impacted within the stacking domain of the master and satellite modules when more satellite frames are added to stacking domain.

Only one type of fabric is allowed per master module or across a pair of master modules. A master can operate at 10 Gb connectivity to compute modules or 20 Gb connectivity to compute modules but not both at the same time.

A two-frame rack-scale fabric is configured as follows:

- Location of the module cluster links

  - Frame 1 Bay 3 Port Q7 to Frame 2 Bay 6 Port Q7

  - Frame 1 Bay 6 Port Q8 to Frame 2 Bay 6 Port Q8

- Location of data center uplinks

  - Frame 1 Bay 3 Ports Q1 through Q6 to network

  - Frame 2 Bay 6 Ports Q1 through Q6 to network

Composing a three-frame rack-scale fabric

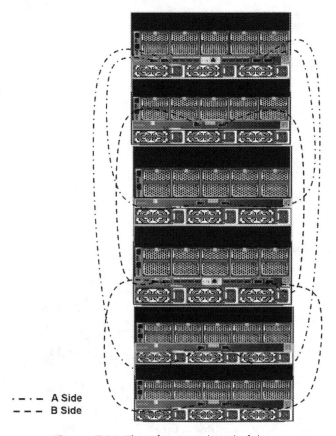

Figure 7-16 Three-frame rack-scale fabric

Adding a third frame to a fabric results in having one more satellite, which is cabled to both fabrics as shown in Figure 7-16. The third frame does not contain any masters.

A three-frame rack-scale fabric is configured as follows:

- Foundation for setup

  - Two VC SE 40Gb F8 ICMs

  - Four 20Gb ICLMs

- Location of the ICMs—Frame 1 Bay 3 and Frame 2 Bay 6

- Location of the ICLMs—Frame 1 Bay 6, Frame 2 Bay 3, and Frame 3 Bay 3 and Bay 6

- Cabling configuration

  – Frame 1 Bay 6 Port 1 to Frame 2 Bay 6 Port L1

  – Frame 1 Bay 6 Port 2 to Frame 2 Bay 6 Port L4

  – Frame 2 Bay 3 Port 1 to Frame 1 Bay 3 Port L1

  – Frame 2 Bay 3 Port 2 to Frame 1 Bay 3 Port L4

  – Frame 3 Bay 3 Port 1 to Frame 1 Bay 3 Port L2

  – Frame 3 Bay 3 Port 2 to Frame 1 Bay 3 Port L3

  – Frame 3 Bay 6 Port 1 to Frame 1 Bay 6 Port L2

  – Frame 3 Bay 6 Port 2 to Frame 1 Bay 6 Port L3

Synergy fabric layout

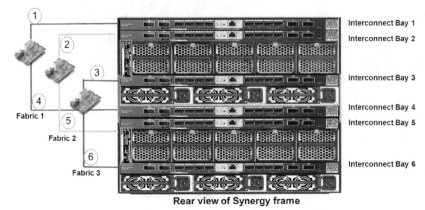

Figure 7-17 Rear view of Synergy frame

The rear of a Synergy frame, shown in Figure 7-17, provides six interconnect bays to house the fabric modules. The bays divide into an A Side—the top three bays (1, 2, and 3)—and a B Side—the bottom three bays (4, 5, and 6). Each A Side bay is paired with a bay on the B Side. Together, the two modules installed in these bays provide a redundant fabric.

Each Synergy frame supports up to three redundant fabrics when the interconnect module bays are fully populated with Synergy ICMs. These redundant fabrics are supported by SAS modules, Ethernet modules (including composable and traditional options), and Fibre Channel modules (including composable and traditional options), or a combination.

- Fabric 1 consists of Fabric 1-A in Bay 1 and Fabric 1-B in Bay 4.

- Fabric 2 consists of Fabric 2-A in Bay 2 and Fabric 2-B in Bay 5.

- Fabric 3 consists of Fabric 3-A in Bay 3 and Fabric 3-B in Bay 6.

Note

It is a best practice to populate Fabric 3 and work upward because Fabric 1 is always reserved for storage. If you start by populating Fabric 1 with a mezzanine card and want to add a storage fabric later, you will need to reconfigure the entire system (including, for example, interconnect modules).

The module's QSFP+ unified ports can be configured as converged ports, which support Ethernet, iSCSI, and FCoE traffic, or as Fibre Channel ports. A converged fabric enables compute modules to consolidate traditional Ethernet and block storage traffic on the same adapters, lowering TCO and reducing complexity. Both the Virtual Connect SE 40Gb F8 module and 40Gb F8 switch module (and their connected satellites) provide convergence for iSCSI, FCoE, and traditional Ethernet traffic. With a Fibre Channel license on the module, you can also enable native Fibre Channel on uplinks, enabling the fabric to connect to an SAN fabric or an HPE 3PAR StoreServ system.

Native Fibre Channel options can provide high-speed Fibre Channel connectivity all the way to the server (compute module) ports. However, they do not support a disaggregated, master/satellite architecture. If you want native Fibre Channel, you have a choice between composable and traditional fabrics. A composable fabric is managed by HPE OneView, creates a flexible pool of fabric resources, and provides frictionless updates and scaling. If you want these benefits, choose the Virtual Connect SE 16Gb FC Module for Synergy. This module requires no licensing for using the Fibre Channel ports. If you want a traditional Fibre Channel fabric, select one of these SAN switches, which can be managed with Brocade management tools, including Brocade 16Gb/24 FC Switch Module, Brocade 16Gb/24 FC Switch Module Pwr Pk, and Brocade 16Gb/12 Fibre Channel SAN Switch Module.

Synergy fabric portfolio

Table 7-3 Synergy fabric portfolio (1 of 3)

	Virtual Connect SE 40Gb F8	20Gb Interconnect Link Module	10Gb Interconnect Link Module	10/40Gb Pass Thru Module	40Gb F8 Switch Module
Management and config	OneView managed	OneView managed	OneView managed	Zero Config	CLI configured

	Virtual Connect SE 40Gb F8	20Gb Interconnect Link Module	10Gb Interconnect Link Module	10/40Gb Pass Thru Module	40Gb F8 Switch Module
Downlink ports (to compute)	12 x 10/20 Gb Enet/FCoE	12 x 20 Gb Ethernet/ FCoE	12 x 10 Gb Ethernet/ FCoE	12 x 10/40 Gb Ethernet	12 x 20 Gb Ethernet/ FCoE
Uplink ports (to data net)	8 x 40 Gb Ethernet	N/A	N/A	12 x 10/40 Gb Ethernet	8 x 40 Gb Ethernet
Interconnect link ports	4 x 120 Gb ports to satellite	2 x 120 Gb ports to master	1 x 120 Gb port to master	N/A	4 x 120 Gb ports to satellite

Table 7-3 shows the capabilities of Synergy Virtual Connect SE 40Gb F8, 20Gb and 10Gb ILM, 10/40Gb pass thru, and 40Gb F8 switch modules.

Table 7-4 Synergy fabric portfolio (2 of 3)

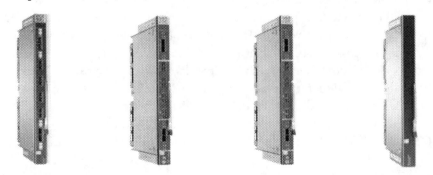

	Virtual Connect SE 16Gb FC	Brocade 16Gb/24 FC Switch	Brocade 16Gb/12 FC Switch	12Gb SAS Connection Module
Management and config	OneView managed	CLI configured	CLI configured	OneView managed
Downlink ports (to compute)	12 x 16 Gb Fibre Channel	12 x 16 Gb Fibre Channel	12 x 16 Gb Fibre Channel	12 Gb SAS
Uplink ports (to data net)	12 x 4/8/16 Gb Fibre Channel	24x 4/8/16 Gb Fibre Channel	12 x 4/8/16 Gb Fibre Channel	N/A
Interconnect link ports	N/A	N/A	N/A	N/A

Table 7-4 shows the capabilities of Synergy Virtual Connect SE 16Gb Fibre Channel, Brocade 16Gb/24 and 16Gb/12, and 12Gb SAS modules.

Table 7-5 Synergy fabric portfolio (3 of 3)

	Virtual Connect	**Ethernet/SAN switches**	**Pass Thru modules**
When to use	Server admin supports a Virtual Connect solution Enterprise infrastructure management desired HPE BladeSystem customer Cisco networking both LAN and SAN	Networking team influences networking choice at the edge Rich networking features for instance L2/L3 for LAN and fabric servers for SAN at edge Customer has HPN or Brocade networking gear	Server Admin wants 10 Gb connectivity for a single frame Clear demarcation between server and network admin Networking team controls all networking aspects
Advantages	Software defined infrastructure template-based management Wire-Once—easiest server moves, additions, and changes Simplest, most flexible infrastructure to manage Flat SAN with 3PAR storage Minimal impact on existing SAN/LAN infrastructure	CLI based interface can be used with home-grown scripting tools A high degree of flexibility	Simple to install and manage Cost-effective with single frame

Table 7-5 compares the advantages of Synergy Virtual Connect, Ethernet/SAN switches, and pass-thru modules.

 Note

Scan this QR code or enter the URL into your browser for details on the Synergy fabric portfolio.

http://www8.hp.com/us/en/products/synergy-fabric/index.html?#!view=grid&page=1

Synergy 10Gb Interconnect Link Module

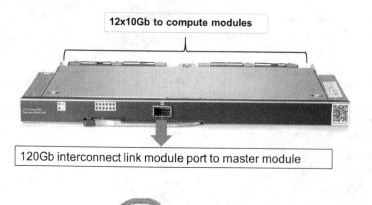

Figure 7-18 Synergy 10Gb Interconnect Link Module

The Synergy 10Gb ICLM, as shown in Figure 7-18, is a 10 Gb satellite module. It connects only to the master module, which in this case is a Synergy Virtual Connect module. It has one 120 Gb uplink and 12 x 10 Gb downlinks to the compute modules. The uplink is connected to a master module with zero latency AOC or DAC interconnect link cables. After a satellite module is connected to a master module, it automatically extends the respective compute modules in that satellite frame to the master module. In essence, all the compute modules in the satellite frame become extended ports of the master module.

The Synergy 10Gb ICLM has no intelligence other than circuitry to amplify the signal. Because there is no processing of any signal with any silicon logic, the latency of satellite modules is almost negligible. The satellite module can be thought of as a link extender from compute modules to the master module.

Synergy 20Gb Interconnect Link Module

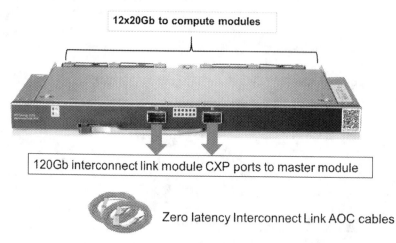

Figure 7-19 Synergy 20Gb Interconnect Link Module

As shown in Figure 7-19, the Synergy 20Gb ICLM is a 20 Gb satellite module. It connects only to the master module, which in this case is Synergy Virtual Connect. It has two 120 Gb uplinks and 12 x 20 Gb downlinks to the compute modules. The uplink is connected to a master module with zero latency AOC or DAC interconnect link cables. After a satellite module is connected to a master module, it automatically extends the respective compute modules in that satellite frame to the master module. In essence, all the compute modules in the satellite frame become extended ports of the master module.

Like the 10Gb ICLM, the Synergy 20Gb ICLM has no intelligence other a silicon timer to amplify the signal. Because there is no processing of any signal with any silicon logic, the latency of 20 Gb satellite modules is almost negligible.

Virtual Connect SE 40Gb F8 module

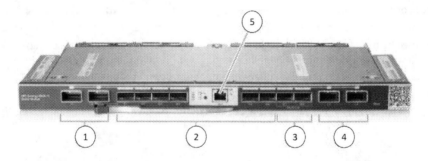

Figure 7-20 Virtual Connect SE 40Gb F8 module

The Virtual Connect SE 40Gb F8 Module operates as the master module. The ports shown in Figure 7-20 are as follows:

1. Interconnect link ports
2. 40Gb QSFP+ uplink ports
3. ICM cluster ports
4. Interconnect link ports
5. Reserved for future use

It has Eight QSFP+ uplink ports; six are unified Fibre Channel and Ethernet (labeled **2** in Figure 7-20) and dedicated for the upstream switches. The last two (labeled **3** in Figure 7-20) are reserved exclusively for ICM cluster ports that enable multi-chassis link aggregation groups (MLAG) between two VC modules and cannot be used as Ethernet uplink ports.

> **Note**
>
> A Fibre Channel license is needed to leverage the Fibre Channel interface on the uplinks. After the Fibre Channel uplinks are activated, they can be used for either NPIV or Flat SAN.

The four interconnect link ports (labeled **1** and **4** in Figure 7-20), each with 120 Gb bandwidth are reserved for connecting to ICLMs. In the case of 10 Gb satellite modules, up to four ICLMs can be connected to a single Virtual Connect master module. In case of 20 Gb ICLMs, only two modules can be connected.

The Virtual Connect SE 40Gb F8 Module has 12 downlinks ports (not shown in Figure 7-20). Each downlink port can operate at 10/20 Gb with support planned for 40 Gb.

Because this module is a Virtual Connect module, it is compatible with Virtual Connect features such as edge-safe, profiles, support for Flex-10/20, Flat SAN, and so forth. In addition, it supports M-LAG on uplinks and supports firmware upgrades with minimal traffic disruption. When combined with satellite modules, it offers itself as composable fabric for Synergy. Virtual Connect capabilities can be extended to satellite frames.

Fabric considerations with Image Streamer

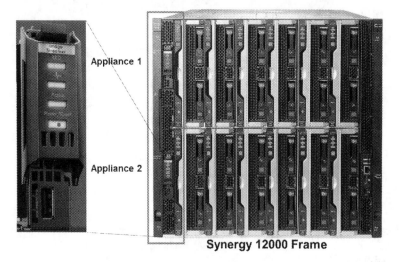

Figure 7-21 Synergy and Image Streamer

Each Synergy Frame has two dedicated management appliance bays, as shown in Figure 7-21, which are available for use by either Composer or Image Streamer

To benefit from the advantages that Image Stream offers, you can add it to the configuration. As shown in Figure 7-21, Image Streamer has the same form factor as the Synergy Composer, and you can install Image Streamer in a spare frame management slot in a Synergy frame. Image Streamer can be inserted into any empty slot number. For high availability, use two Image Streamers and install each in a different frame.

To understand how Image Streamer changes the configuration of the Synergy management ring, consider a management ring that does not include Image Streamer. All of the frame link modules are connected through their link ports. The management ring is then connected to the management network through the management uplinks on two frame link modules. In a multi-frame configuration, the frame link modules connecting to the customer's management network are located in different frames.

When you install Image Streamer into a frame, however, the infrastructure changes. The management uplink port on the frame link modules in that frame now becomes the Streamer port. You must then re-cable the frame link modules. The Streamer port (formerly, the management port) on each frame link module must be connected to a fabric module via the production network, not the management network.

The management network and the production network are two separate networks, which prevents the one from interfering with the other. The data network communicates via the ports on the interconnect modules, including the master modules and satellite modules. It connects compute modules to each other and to the data center network through uplinks. The data network is used primarily for production traffic, although Image Streamer can leverage it. Each interconnect module also offers a management port, which enables administrators to reach the module's CLI remotely.

Differences between the two networks are summarized as follows:

- Management ring

    - Link port to cascade frame link modules

    - Connection to Composer and Image Streamer

- Data network

    - Interconnects including master and satellite modules

    - Uplinks to network and downlinks to compute modules

Step 4: Configure storage

Database	VM, VDI	Collaboration, analytics	Maximum capacity
40 SFF drives	80 SFF drives	120 SFF drives	200 SFF drives

Figure 7-22 Flexible capacity for a wide range of uses and workloads

When configuring storage for a Synergy system, consider the number of drives that can be included in each frame. Figure 7-22 shows sample storage configurations. There is no fixed ratio of compute to storage, so the numbers are average drive calculations per compute.

- The first example shows 40 small form factor (SFF) drives, which would require up to 10 half-height compute modules or four full-height compute modules.

- The next example includes 80 SFF drives. This configuration would use up to eight half-height compute modules and four full-height compute modules.

- For 120 SFF drives, you would need six half-height compute modules or two full-height compute modules.

- For 200 SFF drives or 5 D3940 storage modules, there would be no room for full-height compute modules in the frame. Because the frame needs compute modules, you could install up to two half-height compute modules.

Synergy storage portfolio

Figure 7-23 Synergy storage portfolio

The Synergy storage portfolio, as shown in Figure 7-23, includes Synergy-specific options such as storage local to the compute modules, the D3940 direct-attached composable storage module, as well as traditional NAS and SAN storage such as HPE 3PAR solutions.

The D3940 storage module can provide DAS for compute modules in the same frame. It provides up to 40 SFF hard disk drives (HDDs) or SSDs. You should choose SAS HDDs, which provide a maximum capacity of 2 TB each, if the customer values a lower TCO and can tolerate the tradeoff in performance. Choose SAS SSDs, which support capacities up to 3.86 TB each, if the customer is willing to pay more to receive the best performance.

All D3940 storage modules are fully connected to all of the compute blades through SAS interconnect modules. You can map any number of the drives—from one to 200—to a compute module. For example, you install one storage module and 10 compute blades. You can evenly distribute the 40 drives across the compute blades, with four drives assigned to each. Alternatively, you could map all 40 drives to a single server. Any combination in between is valid as well. The D3940 storage module uses a one-to-one initiator to target model; compute modules cannot share drives. Each server that attaches to the drives directly controls the drives via an integrated Smart Array controller.

 Note

Scan this QR code or enter the URL into your browser for more information on the Synergy storage module.

http://www8.hp.com/us/en/products/synergy-storage/index.html?#!view=grid&page=1

Synergy required modules for storage

Figure 7-24 Required modules for storage

As shown in Figure 7-24, the Synergy D3940 storage solution requires the following three components:

- **Synergy D3940 storage module**—A 40 SFF drive bay that enables you to create logical drives for any compute module in the Synergy frame. Each D3940 module supports between 8 and 40 hot-plug SFF SSDs via SAS and SATA. Each Synergy frame supports up to five such modules, thus enabling a single frame to deliver up to 200 disk drives, total. D3940 Storage Modules feature simple configuration and setup via HPE Synergy Composer. They are easy to maintain and troubleshoot with industry-standard tools. These modules use a high-performance SAS connection with 16 x 12 Gbps SAS lanes. With this design, Synergy storage can deliver up to eight times the bandwidth of other JBOD options and provide up to two million IO operations per second (IOPS). It supports single or dual 12 Gb IO modules, with dual modules providing nondisruptive updates.

- **Synergy 12Gb SAS connection module (single or dual)**—HPE Synergy D3940 storage modules connect to compute modules within the Synergy frame via Synergy 12Gb SAS connection modules. These modules reside in interconnect module Bays 1 and 4, where they create a non-blocking fabric for storage traffic routed from storage controllers inside compute modules through 12 Gbps SAS ports, each of which has four 12 Gbps second channels for an aggregated total of 48 Gbps per port. With the Synergy D3940 storage module, the Synergy 12Gb SAS connection module connects composable direct attached storage for up to 10 compute modules in a single frame.

- **Smart Array P542D controller module**—A PCIe 3.0 mezzanine, four-port, 12 Gbps SAS RAID controller that connects Synergy compute modules to local and zoned DAS. Its advanced RAID capabilities provide enterprise-level reliability. The latest SCSI technology delivers enterprise-level connectivity and performance. The controller supports 2 GB flash-backed write cache (FBWC) to maximize data retention in case of power failure. It provides a consistent set of tools and works across multiple applications to deliver a lower TCO. In addition to RAID mode, the controller supports HBA mode, which allows the controller to present drives to operating systems and applications as JBOD—a requirement for some solutions.

Together, these three components enable you to assign storage module drives to any Synergy compute module as JBODs, which can be configured as RAID volumes by the module's P542D controller or not, as you choose.

Synergy storage configurations

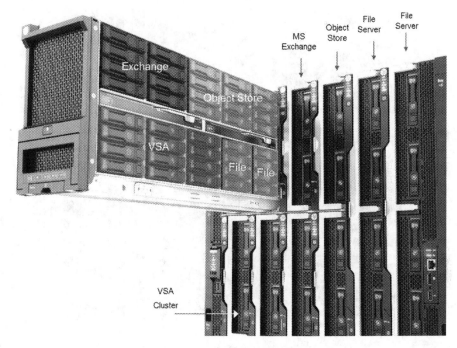

Figure 7-25 Synergy storage configurations

Currently, the storage modules cannot span multiple frames; they provide DAS only for compute modules in the same frame. However, as shown in Figure 7-25, customers can set up multi-frame clusters of compute modules that act as block, file, or object servers by connecting the servers on a backend network, as required by the particular application. For example, object storage servers can connect over an Ethernet network.

DAS environments can be built up dynamically with the optimal number of drives assigned per server as needed. The graphic illustrates an example. A Synergy frame includes compute modules that host a Microsoft Exchange server, an object storage server, two file servers, and a cluster of six VSA servers. The architect has created a plan for assigning the appropriate number of drives in the storage module to each of these compute modules.

Many newer applications, including Hadoop, object stores, file servers, and data analytics applications such as Vertica, expect compute resources to operate on data in DAS. This architecture enables all of these applications. Additionally, HPE VSA software can be loaded on two or more compute blades and re-present the DAS as SAN-based block storage. This allows any compute node or server that can reach the VSA cluster over Ethernet to connect to the same storage.

StoreVirtual VSA in Synergy

Figure 7-26 StoreVirtual VSA

Figure 7-26 illustrates a sample configuration for a VSA cluster. This cluster has eight 10 TB VSA licenses to support up 80 TB total. This compute-dense configuration includes eight two-socket compute modules and two storage modules. The compute modules have P542D controllers and use their local flash drives for booting. Each D3940 storage module provides a mix of HDDs and SSDs, specifically 28 2TB HDDs and 12 1.92TB SSDs, creating a tiered design. Auto-tiering will dynamically move data to the correct tier for optimizing performance.

A redundant solution using dual IO adapters in each D3940 storage module and dual 12Gb SAS connection modules is recommended for optimal availability.

Taking into account data that is replicated, the solution provides about 65 TB of usable capacity. HPE estimates its performance at 224,000 IOPS.

SDS solution general requirements

Customers who require software-defined storage (SDS) need the following hardware components, software licensing, and management solutions:

- **Hardware components**—Each Synergy 12000 frame requires the following storage components:

  - One or more Synergy D3940 storage modules

  - Smart Array P542D controller modules installed on compute notes that access D3940 modules

  - Up to 40 (total) SAS HDD, or SAS SDD, or both in each D3940 module

  - One or more Synergy 12Gb SAS connection modules

  - Physical components for backend network connections

- **SDS options and licensing**—You can opt for a solution that includes either HPE StoreOnce VSA or VMware VSAN. If you choose a StoreVirtual solution running on VMware or Microsoft Hyper-V, you will need licenses for each VSA. HPE licenses StoreVirtual appliances per TB; in increments of 4 TB, 10 TB, and 50 TB; starting with 4 TB, three-pack, three-year licenses to use (LTUs) and going up to 50 TB, 300-pack, five-year Stock LTUs. A variety of 4 TB–10 TB and 4 TB–50 TB upgrade license options are also available.

  If you opt for SDS via VMware VSAN, you will license your SDS solution per CPU and need vSphere 6.x Enterprise Plus licenses, which are available in a variety of standard packages for hybrid (SSD and HDD) and advanced (all flash) packages.

- **Management options**—You can manage Synergy frames via Synergy Composer, which is powered by HPE OneView (which is embedded) and can manage compute, storage, and fabric resources on one or across multiple Synergy frames. Composer automatically discovers resources, instantly composes them into logical infrastructures that meet applications' needs, and in general, provides software-defined intelligence for template-driven, frictionless operations. Composer manages other Synergy management interfaces, which include the frame link module and Image Streamer.

You can use the HPE Central Management Console (CMC) to manage StoreVirtual SDS implementations, or HPE OneView for VMware vCenter plug-in for managing servers, storage, and networking from the native vSphere management console.

If you have implemented SDS on Synergy using VSAN Cluster, you can manage the solution's storage via the vCenter web client interface.

Learning check

1. What are the four main steps of Synergy configuration?

_____

_____

_____

2. Match the Synergy component with its function.

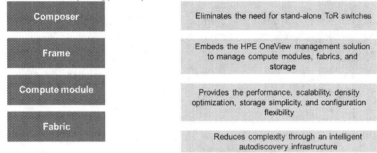

3. Match the Synergy management subsystem component with its function.

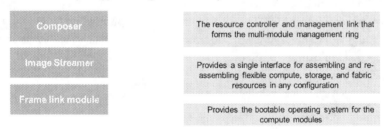

Learning check answers

1. What are the four main steps of Synergy configuration?

    - **Configure infrastructure**
    - **Configure management ring**
    - **Configure fabric**
    - **Configure storage**

2. Match the Synergy component with its function.

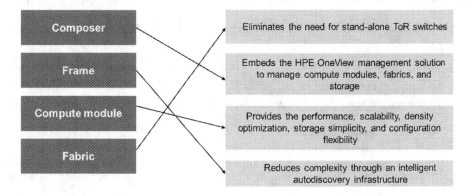

3. Match the Synergy management subsystem component with its function.

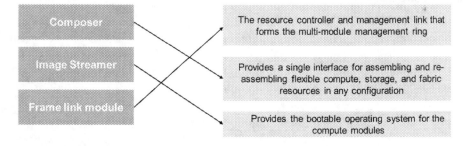

Summary

- Synergy is a single infrastructure of physical and virtual pools of compute, storage, and fabric resources and a single management interface that allows IT to instantly assemble and re-assemble resources in any configuration.

- Synergy eliminates hardware and operational complexity so IT can deliver infrastructure to applications faster and with greater precision and flexibility.

- The steps to build a Synergy solution are:

    1. Configure the infrastructure by determining the number of frames the customer requires and the associated Composers, frame link modules, and compute modules.

    2. Determine how the Composers and frame link modules will be connected in a management ring.

    3. Consider how to configure fabric elements (the data network) as part of a composable infrastructure. Determine whether the Image Streamer is appropriate for the customer's configuration. If you are adding an Image Streamer to the customer's configuration, determine how it fits into the fabric and the management ring.

    4. Build out storage modules based on requirements.

- The Synergy management subsystem comprises Synergy Composer, Synergy Image Streamer, and the frame link modules.

- Synergy Composer uses software-defined intelligence with embedded HPE OneView to aggregate compute, storage, and fabric resources.

8 Planning and Designing HPE Server Solutions

WHAT IS IN THIS CHAPTER FOR YOU?

After completing this chapter, you should be able to:

✓ Describe how to assess each customer's requirements and environment in order to develop a Hewlett Packard Enterprise (HPE) server solution, including how to perform:

  – Needs analyses

  – Requirements, segment, and workloads analyses

  – Site surveys

✓ Name the design considerations that should be taken into account when planning server solutions

✓ Identify the HPE tools that can be used to select solution components when designing a solution

✓ Explain how to use the return on investment (ROI) and total cost of ownership (TCO) tools available from HPE

✓ Describe the process of developing solution proposals

OPENING CONSIDERATIONS

Before proceeding with this section, answer the following questions to assess your existing knowledge of the topics covered in this chapter. Record your answers in the space provided here.

1. What experiences have you had with conducting a needs analysis for a customer? What type of questions do you think would help evaluate a customer's business requirements?

   _____

   _____

   _____

_____

_____

_____

_____

2. Name some of the challenges you have encountered when working with HPE solution design and configuration tools. Which functionalities have worked best, and which features need improvement?

_____

_____

_____

_____

_____

_____

_____

3. What do you know about HPE TCO and ROI tools? How can these tools benefit customers?

_____

_____

_____

_____

_____

_____

_____

Assessing the customer's requirements and environment

When you are planning an HPE server solution, one of the first tasks to perform is an analysis of the customer's current infrastructure and the applications the customer plans to run on the servers. Several factors should be considered, including:

- **The number of users for the application**—The number of users has a direct impact on the amount of CPU, memory, networking, and storage resources that will be required.

- **IT resources required by the application**—The application installation or user guide should offer recommendations regarding the amount of CPU, memory, networking, and storage resources that will be required for the application.

- **Applications that can be consolidated**—Are there any applications that no longer provide business value? If so, could they be retired, or could a different application deliver results?

- **Service-level agreements (SLAs) in place for the various business organizations**—SLAs have a direct impact on the hardware and the software that will be required. For example, a solution capable of achieving 99.999% availability will need to be designed with no single points of failure and will require redundant components, duplicated systems, and clustering software. If the solution is not considered business critical and will only be expected to achieve 99% availability, redundant components, duplicated systems, and clustering software will not normally be required.

- **Customer's current methodology**—Assess the customer's current infrastructure and processes to ensure that any new solution will integrate with their existing framework.

Conducting a needs analysis

When planning and developing an IT solution, the business and IT needs of the customer should be assessed. These assessment results should be used to guide the implementation planning process. When all customer information has been gathered, your experience and knowledge can be used to recommend the best possible solution. The IT recommendation is ultimately described in the statement of work. However, before a solid statement of work can be developed, the specific needs, expectations, and environment for the server solution must be understood. Thorough planning helps to avoid potential costly mistakes and prepares an upgrade path for the future.

A needs analysis starts with a customer interview. Ask questions to determine current challenges, ways to address those challenges, and business goals. Understanding a customer's needs is crucial to developing a positive, long-term relationship.

If a customer has experience with server technology, ask questions about intended solutions. Consider asking questions in the following categories:

- Future plans

  - What are the business goals?

  - What is the projected role of the servers?

  - What are the projected operating systems?

  - To which kind of network will the servers be connected?

- Business requirements

  - What is the capital budget?

  - Are there any TCO requirements?

- Are there any ROI requirements?

- How does a solution impact the users?

- How does a solution help business analytics?

● Current environment

- How much storage currently is used?

- Have storage needs grown over the last 12 months?

● Technical requirements

- What is the expected availability of the servers?

- Is server price more important than functionality?

- Is a rack or tower configuration preferred?

- How frequently will backups be performed?

- Is power protection needed?

- Which kinds of system management tools are needed?

- Does the server need to be configured from the component level, or does it need to be ready to install out of the box?

- What level of maintenance and support is desired?

● Obstacles

- What is the biggest IT problem facing the business today?

- What does the customer believe are possible solutions?

- What are the barriers to the solution?

● Resources

- Is the customer willing to commit resources to achieving these goals?

- Is the customer willing to let technical professionals help guide the way?

● Nontechnical considerations

- Are there any open service calls or other customer sensitivities?

- Who is the contact person for IT solution implementation within the customer's organization?

- Are the customer's applications standardized?

- Is there a long-term IT strategy in place?

- Are there any rules in place for hardware isolation among business units?

Based on the answers to these questions, recommendations can be made about which server components are required and which are optional. For example, if file and print is the projected role for the server, storage capacity and transfer rate are important selection factors. Alternatively, if the server will be a database server, processor speed and memory are the primary considerations.

Matching customers to the best platforms

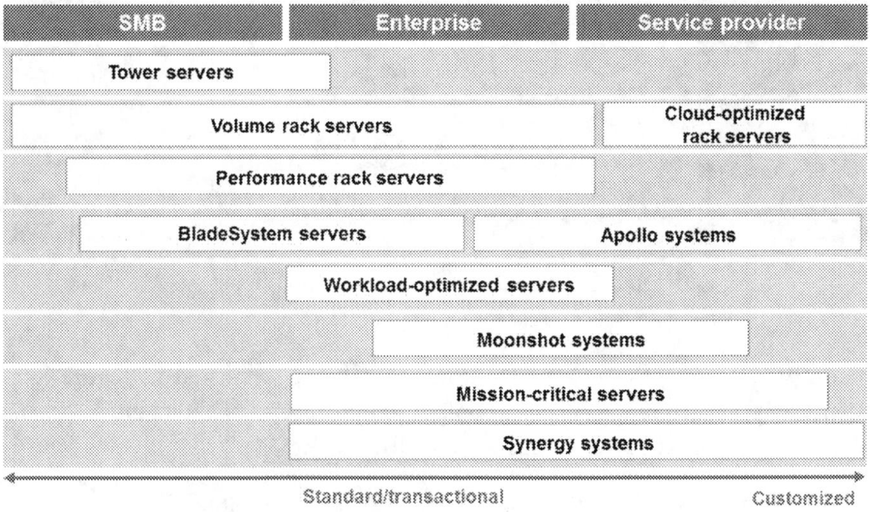

Figure 8-1 Matching customers to the best platforms

All customers have different needs, and there is no substitute for completing a full needs analysis. However, Figure 8-1 provides useful high-level guidance regarding which HPE servers might be suitable for small- to medium-sized businesses (SMBs), enterprises, and service providers.

There are three tiers of service providers, including:

- **Tier 1** service providers are considered the highest ISP class. A Tier 1 ISP has its own IP network in a particular region connected with the primary Internet backbone or other Tier 1 ISP. Typically, a Tier 1 ISP sells bandwidth to Tier 2 and Tier 3 ISPs, which provide Internet connectivity to businesses and individual customers.

- **Tier 2** ISPs purchase their Internet service from a Tier 1 ISP and tend to cover a specific region and focus on business customers.

- **Tier 3** ISPs also purchase their Internet service from Tier 1 ISPs. Tier 3 ISPs tend to cover a specific region and focus on the retail market.

Evaluating the business requirements

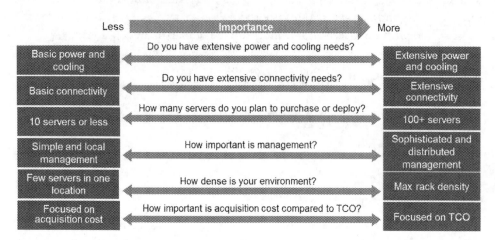

Figure 8-2 Evaluating business requirements

When evaluating the customer's business environment, the factors listed in Figure 8-2 should be assessed and the importance of power and cooling, connectivity, deployment scale, management tools, server density, and TCO focus should be recorded.

Examples of questions that can be asked are:

- **Infrastructure**—Are there extensive power and cooling and interconnect needs?

- **Node count**—How many servers is the customer buying?

- **Management tools**—How important are the management tools? Does the customer provide in-house training on these tools?

- **Deployment density**—How dense is the environment?

- **Price**—How important is acquisition cost compared to TCO and ROI?

Three analyses should be performed in order to thoroughly assess each customer:

- **Requirements analysis**—Using the answers to the business environment evaluation can guide the selection of a particular HPE server portfolio. For example, if the customer has a need for a deployment on a massive scale, HPE Synergy and HPE Apollo solutions should be explored. If the customer has a need for very high server density, HPE BladeSystem, Apollo, and Synergy products should be considered. Specific factors to evaluate include power/cooling requirements and interconnectivity.

- **Segment analysis**—Customer segment analysis can also be used to guide the selection of a particular server family. For example, customers in need of a high-performance computing (HPC) solution might benefit from Apollo, BladeSystem, and rack servers. For Tier 3 service providers, rack and BladeSystem solutions should be considered.

- **Workload analysis**—Analysis of the types of workloads the customer is running can also provide guidance in selecting a server family. It is important to find out if the customer needs support for virtualization, cloud, web infrastructure, database, app development, and so on. For example, SMB customers requiring servers for a small IT infrastructure should consider the HPE ProLiant ML family. Enterprise customers who need large deployments of servers in an application development environment should consider BladeSystem or Synergy solutions.

Conducting a site survey

After the needs analysis interview, it is important to conduct a site survey to assess the facility and evaluate its suitability for the proposed IT solution. Factors to consider include:

- **Site/facility suitability**—Server room size, layout, limitations, and interference
- **Site services/utilities**—Power delivery, fire suppression, and environmental controls
- **Physical security**—Key locks and card, code, or fingerprint access
- **IT integration**—Existing computing infrastructure
- **Applications/software**—Loads and availability
- **Human resources**—Ownership and internal or external support
- **Projected growth**—Computer, employee, and business expansion

To evaluate these factors, survey questions can be used to gather data. Example questions include:

- How large is the facility?
- Does the facility currently have any radio frequency interference (RFI) problems?
- Is there any extra space available?
- Will an existing space need to be modified?
- Are adequate utility outlets available in the proposed space?
- Are the electrical circuits of sufficient capacity?
- Are the electrical circuits shared or isolated? Are they properly grounded?
- Are there additional electrical circuits available in the facility?
- What type of fire-suppression system is in place, if any?
- If overhead sprinklers are installed, are they water- or halon-based?
- What type of floors or floor coverings exist?

- Does the ceiling allow cabling to be run easily?

- Is there adequate ventilation for the space?

- Is extra cooling capacity available?

- Is the proposed space in the interior of the building or does it have an outside window?

- Does the facility already use keypads, card readers, or other physical security devices to control access?

- How many workstations does the company have now?

- How many servers are in use?

- Is a recent inventory of company IT assets available?

- If there is an existing network, what is its topology?

- What kinds of IT equipment purchases are planned for the next 12 months?

- What kinds of software are used regularly?

- Is the software workstation- or server-based?

- What software purchases are planned for the next 12 months?

- How are IT support issues currently handled?

- Who is responsible for IT issues at the company?

- Will any IT staff be added?

- How many other employees will be added in the next 12 months?

- Does the company plan to open other offices in the next 12 months?

This list is not complete, but it should serve as a guide to effectively define a specific customer's particular environment. It is important to identify existing resources, such as network capacity or IT assets, which might be required during the transition to the new environment. The availability of resources also affects the implementation timeline. For instance, if no network cabling exists in the facility, the plan must include time to install the wires.

Although the plan can recommend simultaneous tasks, it is inefficient to have servers and workstations in place waiting for interconnectivity to complete their configuration. The amount of time needed to set up the server and address dependencies must be determined.

HPE Converged Infrastructure Capability Model

The HPE Converged Infrastructure Capability Model (CI-CM) is delivered by HPE Enterprise Group presales architects or HPE Technology Services Consulting resources. It provides a high-level assessment of the current state of a customer's IT infrastructure, including suggested next steps and recommended projects to help drive toward a converged data center. It also helps define an action-oriented, customer-specific road map. A CI-CM evaluation also compares a customer's IT environment to others in their industry and region. CI-CM should be used when customers are interested in moving toward a converged infrastructure and want an external assessment of where they are in the journey.

The next-generation data center moves away from traditional silos and uses a converged infrastructure across servers, storage, and networking. Leading analysts agree that this new data center improves IT across three characteristics:

- Operational efficiency

- Quality of Service (QoS)

- Speed of IT project implementation

IT domain areas that drive capability

A successful strategy for developing a next-generation data center requires both tactical and strategic dimensions. Tactically, companies need to break down the IT silos by standardizing, modularizing, and virtualizing their IT environment. They should also focus on automating error-prone manual processes, along with simplifying and tightening management control. Strategically, they need to change how IT interacts with the business and realign people, processes, and technology to the new vision.

First, CI-CM is used to establish the current state of the IT infrastructure for each business. Then, the model is used to define the target state for each metric.

CI-CM domains

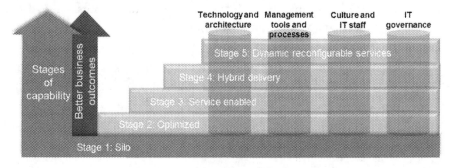

Figure 8-3 CI-CM

CI-CM measures a company's maturity in four domains as illustrated by Figure 8-3:

- **Technology and architecture**—This focuses on data center hardware and facilities, including how the company's infrastructure is used and how well IT services can withstand failures. The goal here is for hardware, software, and network resources to be shared, virtualized, and automated. This means moving from dedicated, high-cost IT silos to low-cost, highly available, pooled IT assets.

- **Management tools and processes**—This domain focuses on deploying resources to plan, manage, and deliver IT services more efficiently. The aim is to move toward more standardized, integrated processes using tools that align IT more closely with business goals.

- **Culture and IT staff**—Moving toward a business-centric IT environment calls for the IT organization structure, roles, and functions to be redefined. This area, though challenging, is crucial to achieve and sustain a more adaptive IT environment.

- **Demand, supply, and IT governance**—This involves outlining the organizational structures and processes in order to better align the enterprise IT environment with overall business strategies and objectives. Here, roles in the business strategy development process are defined, and the mechanisms for executing the business strategy with adherence to the IT strategy are established. In essence, this aligns IT supply with business demands.

CI-CM stages

The infrastructure's current state and targeted state are rated against five stages of capability in each of the domains:

- **Stage 1: Silo**—This domain is primarily a dedicated, project-based technology. A business unit specifies an application, which specifies the infrastructure. Then, the application and infrastructure are moved into the data center, and IT managers run the operations. In this stage of evolution, the technology is built around the project itself. This, in turn, limits the sharing assets and the consistency of infrastructure, operating systems, and processes.

  The results of this type of project-based architecture is a one-application-to-one-server environment—in essence, a compartmentalized approach. This approach tends to lead to old legacy systems that are not upgraded because systems "belong" to the business unit and after depreciation, they are perceived to be "free." As a result, data centers can slowly collect old equipment and applications that are high maintenance, often with a very inefficient cooling architecture.

- **Stage 2: Optimized**—This stage consists of rigid standardization of all IT architecture elements. It also requires standardization of management tools and processes. From an operational governance perspective, there is a significant shift in management within a standardized IT environment. For example, in Stage 1, a business unit decides what a new system does and how it does it. In Stage 2, a business unit stills decides what a new system does, but IT decides how it will be done— application, infrastructure, and so on. This approach enables IT to have:

  - Consistency of technology and operating system images

  - Standardized applications and databases

  - Standardized operational practices often independent of technology type

  - Reduced maintenance labor

  - Higher reliability

- **Stage 3: Service enabled**—In this stage, IT moves to a consolidated, virtualized, and shared infrastructure. All aspects of IT are rationalized including architectures, processes, and tools. All technology aspects are managed by IT and are not under business unit control. This enables IT to share the assets across applications without business unit intervention. In this stage, there tends to be pervasive use of virtualization; a standards-based, shared infrastructure; disaster tolerance; and consolidated, environmentally efficient data centers. The objectives include increased asset utilization, greater ability to handle periods of peak business demand, faster deployment of new applications, improved service levels, maximized reliability, and operational efficiency.

- **Stage 4: Hybrid delivery**—Infrastructure and other enterprise architecture artifacts are offered as a service on demand with tiered service levels. These offerings are supported by service-centric, integrated IT processes. The service levels are not measured on the technology, but on the service as whole. This means that each layer of the IT environment is offered as a service, starting with the infrastructure services that provide server and storage. Tangible benefits include faster time to market for practically any IT service.

- **Stage 5: Dynamic reconfigurable services**—IT provides all services from a pooled, shared, automated source of supply, which can be inside or outside the enterprise IT environment (for instance, cloud-based). At this stage, automated management is enabled and infrastructure is reallocated based on business process needs. Manual labor is eliminated from most provisioning and resource utilization improves. The IT infrastructure is virtualized, resilient, orchestrated, optimized, modular, and fully converged.

Designing a solution
Service-level agreements (SLAs) drive the solution

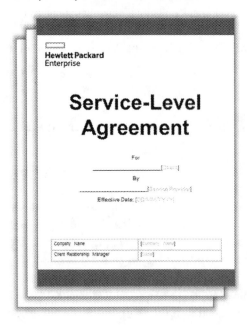

Figure 8-4 Service-level agreement

SLAs, like the one shown in Figure 8-4, provide a common understanding about services, priorities, responsibilities, guarantees, warranties, and penalties. When necessary, the SLA can be modified to enable the deployment of less equipment and, therefore, reduce costs.

Several factors need to be determined and agreed upon with the customer:

- What performance levels are expected? Consideration needs to be given to CPU, memory, networking, and storage performance.

- What availability is expected? How will it be calculated?

- Are there any legal compliance requirements?

- How will the systems and data be secured?

- What are the data-retention policies?

    - Where is the data located?

    - What are the legal implications of storing data in a different jurisdiction?

    - When data is deleted (either accidentally or on purpose), can it be recovered?

Solution design considerations

The information gathered during the needs analysis and site survey should narrow the choices for the server solution recommended to the customer. When applying the data, the following areas should also be considered:

- **Servers**—Will the solution consist solely of physical servers or a mix of physical and virtual servers? If the latter, what virtualization technology will be used? For physical servers, memory and processor technology components are important. Fault-tolerant memory or redundant processors are less crucial to a file and print server than they are to a database server, which primarily performs computations and requires temporary storage. Determining the relative importance of server technologies, together with the server's projected role, narrows the focus to a server with the required capabilities.

- **Storage**—Every server has storage, but deciding on a RAID or non-RAID configuration depends on factors such as cost, storage availability, and fault tolerance. For a file and print server, a non-RAID configuration leaves data vulnerable to disk failure or data corruption. This would be less important to a network-centric firewall server.

- **Networking**—Depending on the existing network topology or the decision for a new topology, server networking capabilities must also be determined. Current corporate networks are at least 100 Mb/s (Fast Ethernet), with many customers already using 1 Gb/s Ethernet. Corporate backbones can exceed 10 Gb/s (Gigabit Ethernet).

- **Operating system**—The choice of operating system directly affects the server components. As a general rule, the more recent (and thus more advanced) the operating system, the greater its demands on system hardware. Certain operating system features can also steer the decision. Potential server purchases should be made based on careful consideration for meeting or exceeding the highest minimum system requirements.

- **Applications**—Often referred to as *workloads*, the applications that the customer needs to use will have a major impact on several design considerations. For example, if the main application is a mission-critical, multi-petabyte database application, it will be necessary to make sure that the server solution is designed with minimal single points of failure and with sufficient storage, networking, memory, and processor resources to ensure the smooth operation of the database application. Availability clustering should also be considered to make sure that in the event of a catastrophic failure, the application can continue to run (by failing over to a standby or a secondary system).

- **Availability**—Workloads should be assessed for their level of business importance and housed on an appropriately available server solution. There might be applications that are not considered mission-critical and can, therefore, be unavailable without significant business impact. These might be located on nonclustered virtual or physical servers.

- **Security**—When planning where to place the server and how it should be configured, it is important to consider security. Be alert for physical and virtual security holes. When an employee leaves an organization, it is important to recover any keys and access cards. It might be necessary to change locks and codes. Disabling the user account and changing high-level passwords to which an employee had access are good practices.

  - **Physical**—Security measures also involve locks, codes, and location. Deciding to place a server in an interior room with a locked door sufficiently addresses most physical security needs. Because the temperature and humidity in a windowless interior room remain relatively constant, there should be no need to keep the door open to enhance airflow. A closed and locked door ensures that only individuals with authority and access can enter.

  - **Virtual**—Passwords, permissions, and access control lists should also be secured. Setting up users, groups, and permissions addresses virtual security needs. Each user needs a password to access project files stored on the server, and being a member of a particular group allows or denies access to other network resources. Grant each user only as much access as he or she typically needs using access control lists. All passwords should be changed regularly; meet a minimum length (as defined by the operating system); and include letters, numbers, and special characters.

BTO and CTO product SKUs

Figure 8-5 BTO and CTO

When designing a solution, it is important to understand two different approaches:

- **Build to order (BTO)**—BTO products are prepackaged bundles kept in stock by HPE partners. They provide competitive pricing, off-the-shelf fast delivery, and worldwide availability. However, they only offer field integration of components. They are typically available from HPE within five to eight days or immediately if in stock at distributor.

- **Configure to order (CTO)**—HPE partners or customers can request CTO products directly from HPE. This approach offers maximum flexibility in product customization and factory integration of components. There are two disadvantages to CTO products: re-orderable configurations are not possible across regions, and they carry longer quote and order cycle times. Availability ranges from 10 to 15 days.

Physical or virtual servers?

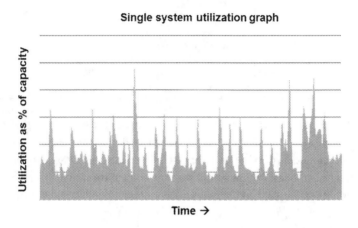

Figure 8-6 System utilization

Traditionally, servers have been acquired and deployed on an application basis. Each environment was configured to handle peak—not average—workload requirements. As a result, utilization rates could be as low as 30%. Although not an issue from a technical perspective, business leaders view this situation as a waste of valuable resources.

Many workloads exhibit utilization patterns similar to the workload shown in Figure 8-6. The peaks vary by time of day, and they tend to be for short durations. This means that most of the time resources are idle, but when they are required by a workload, they are required instantly.

Server virtualization enables customers to multiply their resource utilization by consolidating multiple applications onto a single server and dynamically balancing processing resources among applications. Server virtualization can optimize the use of available and often limited space. A recent technology survey from Tech Pro Research found that 74% of businesses either currently use server virtualization or plan to implement it in the near future, with respondents citing benefits such as a significant reduction in the time needed to deploy new applications, system downtime, and overall IT spending.

 Note

To read the full report on the Tech Pro Research survey, scan this QR code or enter the URL into your browser.

http://www.techproresearch.com/downloads/
research-smbs-discuss-current-status-and-future-adoption-plans-of-new-technologies/

IT departments are also frequently asked to provide more storage using budgets that are not increasing. Virtualization software residing between server hardware and software allows more applications to run on the server, and related hypervisor software enables the operating system and applications to move between servers. This allocates more resources where they are needed during periods of high demand. These server virtualization features also provide important redundancy capabilities and might be deployed as part of a comprehensive IT resiliency strategy.

VM design goals

Virtual machines (VMs) should be designed without bottlenecks so that they can adequately meet the needs of users. To assist in the sizing process, it is possible to use an equivalent of the Amazon Compute Cloud (EC2) Compute Unit (ECU) VM instance. All Amazon EC2 instances are priced based on hourly usage and instance type. Each instance type consists of a certain number of ECUs and a set RAM size. The EC2 general purpose instances, which are well suited for many applications, include:

- **t2.small**—Variable ECU (one virtual CPU [vCPU]), 2 GB memory

- **t2.large**—Variable ECU (two vCPU), 8 GB memory

- **m4.large**—6.5 ECUs (two vCPUs), 8 GB memory

- **m4.xlarge**—13 ECUs (four vCPU), 16 GB memory

One ECU provides the equivalent CPU capacity of a 1.0 GHz–1.2 GHz 2007 AMD Opteron or 2007 Intel Xeon processor (roughly equivalent to a PassMark CPU score of 400). The ECU of a particular processor can be calculated by taking the PassMark CPU score and dividing this by 400. As an example, the Xeon E5-2620 v4 processor @ 2.1 GHz with a PassMark score of 11887 has an ECU of approximately 30 (11887/400), meaning that it should be able to host roughly four m4.large (30/6.5) or two m4.xlarge (30/13) EC2-equivalent VMs.

 Note

For more information on PassMark CPU scores, scan this QR code or enter the URL into your browser.

http://www.cpubenchmark.net/cpu_list.php

According to Gartner, the recommended network bandwidth for each VM should be at least 100 Mb/s for IP traffic and 200 Mb/s for SAN traffic. iSCSI handles only IP traffic, so the bandwidth becomes 300 Mb/s for IP traffic.

Although 100 Mb/s for IP traffic is not always considered ideal, it is generally sufficient as a starting point. In a highly virtualized environment, there is a large amount of IP traffic coming from the system, and this can result in many cables emerging from a single enclosure. In this scenario, it might be necessary to specify a switch such as the HPE 59xx or HPE 125xx to provide ToR functions.

VM design guidelines

When designing a VM oversubscription ratio, there are four aspects to consider:

- Processing capacity per VM
- Memory per VM
- IO per VM
- Licensing cost

Together, these elements create a balanced virtualization design for the data center.

Processing capacity per VM

To enable customers to make a fair comparison of VM solutions, HPE has made use of an absolute VM processing capability. This is equal to an ECU m1.small (one ECU, one CPU core, and 1.7 GB RAM) and close to the ratio that customers often use. A ratio of 36:1 works well with loads that typically use 2%–3% of a dedicated server. If there are loads with lower typical use and no correlated peaks in traffic, then customers sometimes use in excess of 100:1. However, this can cause an imbalance with other design constraints. Also, using very high virtualization oversubscription ratios requires consideration of the number of threads supported. This depends on the CPU choice and knowledge of the workloads.

Memory per VM

Initially, HPE recommends using guidance from the virtualization platform provider (VMware, Microsoft, Citrix, and others) to size memory. Additional memory might be required to satisfy the needs of the applications being hosted on the VM.

 Note

Dropping under the VMware-recommended memory per VM might cause instability with some workloads.

Use caution when specifying memory, because increasing server memory to support more VMs has two significant downsides:

- The use of high-density, high-power devices to achieve very large amounts of memory can significantly increase the cost of memory. High-density DIMMs such as 32 GB and 64 GB are relatively expensive when compared with lower-density DIMMs such as 8 GB and 16 GB. The current and future requirements for RAM need to be assessed and the most cost-effective DIMMs should be selected.

- The cost of VM licenses can increase significantly with very large memory configurations. Microsoft recommends accounting for overhead required by Hyper-V by configuring an additional 10%–25% of hardware resources.

 Note

Scan this QR code or enter the URL into your browser for more information on optimizing performance on Microsoft Hyper-V.

https://msdn.microsoft.com/en-us/library/cc768529(BTS.10).aspx

Minimum memory size for VMware vSphere is 4 GB; the maximum size depends on the host. The maximum size for best performance represents the threshold beyond which the host's physical memory is insufficient to run the VMs at full speed. This value changes in alignment with host conditions (as VMs are powered on or off, for example). In addition to a complete portfolio of VMware licenses, HPE offers HPE OneView for VMware vCenter, which seamlessly integrates the manageability features of ProLiant and BladeSystem servers, Virtual Connect, and storage solutions with VMware products.

 Note

For more information on minimum hardware requirements for vCenter Server, scan this QR code or enter the URL into your browser.

https://pubs.vmware.com/vsphere-50/index.jsp?topic=%2Fcom.vmware.vsphere. install.doc_50%2FGUID-67C4D2A0-10F7-4158-A249-D1B7D7B3BC99.html

---

With Citrix XenServer, three factors determine the memory size of a XenServer host:

- Memory consumed by the Xen hypervisor itself
- Memory consumed by the control domain on the host
- Memory consumed by the XenServer crash kernel

 Note

For more information, refer to the *XenServer 6.5.0 Administrator's Guide*. To download the guide, scan this QR code or enter the URL into your browser.

http://support.citrix.com/servlet/KbServlet/download/38321-102-714737/ XenServer-6.5.0_Administrators%20Guide.pdf

---

IO per VM

The final technical constraint on oversubscription of VMs is on the IO bandwidth from the server to the core switch. HPE recommends 200 Mb/s storage and 100 Mb/s IP for each VM. With converged network adapters (CNAs) in a dual-resilient high-availability configuration, each 10 Gb/s port pair can support 33 VMs. HPE uses dual CNAs on server blades; therefore, each blade can support 66 VMs in high-availability configurations. Additional CNAs can be added if required.

Licensing cost

The cost constraint on VM oversubscription is based on the cost of the hypervisor licenses. The licenses can dominate the cost of the raw compute elements and might be more expensive than the servers they run on. In this case, a joint calculation on hypervisor and server cost will reveal the best oversubscription ratio.

Highly available designs

There are several approaches to designing a highly available IT environment. Most highly available environments use some form of hardware redundancy. The most basic form of hardware redundancy in a rack and tower server solution revolves around having redundant power supplies, fans, NICs, and more, so that if one component fails, the server continues running applications without interruption. In an HPE BladeSystem environment, redundant power supplies, fans, NICs, and other elements should be complemented by redundant Onboard Administrator and Virtual Connect modules.

Beyond protecting hardware resources with redundant components, software-based clustering solutions such as VMware high availability and Hyper-V clustering are available. These options are capable of providing greater levels of availability instead of solely using the hardware redundancy method.

With software-defined data centers (SDDCs), high availability can be implemented using programmatic software control that can move server personalities (profiles) between servers automatically in order to maintain the required level of performance and availability.

Activity: Designing customer solutions

Read the following case study that represents challenges faced by many large enterprises and develop a potential IT solution to fit the customer's environment, workloads, and so on. You can use any materials from this book, along with documentation from the HPE website.

Case study

AB Technologies is an American multinational technology company headquartered in Palo Alto, California. It is in the top ten of the world's largest semiconductor makers, based on revenue. AB

Technologies supplies components for computer system manufacturers such as Apple, Samsung, HPE, and Dell. AB Technologies also makes NICs and integrated circuits, flash memory, graphics chips, embedded processors, and other devices related to communications and computing.

AB Technologies has 1500 hardware engineers that need access to development applications such as Cadence, Mentor Graphics, and other third-party tools when designing new devices. The different stages of the development cycle require separate IT resources that must be made available in a rapid, seamless way. The IT department needs to allow the design engineers to choose their own resources and tools. This significantly reduces the amount of time needed for each phase of product development, speeds product readiness, and increases potential revenue.

Business needs

AB Technologies' CIO Alice Turnbull is under enormous pressure to deliver innovative applications and services. She is being asked to transform the IT organization from a cost center to a value creator. Alice believes that IT can become a value creator by taking the lead in driving innovation, speeding up service delivery, and supporting revenue growth.

Alice's list of IT business goals include:

- Build an infrastructure that can be highly optimized for virtualized workloads now and in the future.

- Provide operational simplicity and maximum utilization of data center resources.

- Combine hardware infrastructure, software, and services to deliver a single platform that positions the business for future growth opportunities.

- Enhance workplace productivity by intelligently managing the provisioning and deprovisioning of compute, storage, and network resources for the design team

- Allow the data center staff to spend 75% of their time delivering business applications and value-added services.

- Reduce costs, decrease time to value, and provide 99.999% availability for business-critical applications.

- Transform operations to function more like a cloud provider to the lines of business (LOBs), thereby accelerating delivery of revenue-generating products, services, and experiences.

- Realize the ideas coming from the technical marketing organization faster, so the business can deliver better experiences for customers while staying ahead of the competition and growing revenue.

To achieve these goals, AB Technologies needs a single infrastructure with unified pools of compute, storage, and network fabric. All of the resources need to be capable of being instantly configured according to the specific needs of each design project. Resources need to be provisioned together with their state (BIOS settings, firmware, drivers, protocols, and so on), and the operating system image should use repeatable templates. The infrastructure should accurately provision and de-provision logical infrastructures into any combination with minimal delay.

Additionally, IT team members are very progressive in their approach to satisfying the LOB managers. They are keen on integrating and automating infrastructure operations and applications through a unified application programming interface (API). The unified API must provide a single interface to discover, search, inventory, configure, provision, update, and diagnose the infrastructure.

Ultimately, the IT solution for this company must be cost-effective, highly available, and scalable. To optimize available bandwidth and maximize utilization, IT needs to precisely allocate bandwidth across diverse workloads such as data management, web infrastructure, product development, and high-performance computing.

AB Technologies has experienced a huge growth in the amount of generated data. The IT team needs to be able to store and share file, block, and object data with enterprise-class reliability and nondisruptive change management. This would allow IT to effortlessly provision the right capacity and performance.

Existing infrastructure

AB Technologies has been working with HPE for a number of years and has a large installed base of ProLiant G7, Gen8, and Gen9 server blades and rack-mount servers, alongside high-performance technical computing solutions from various vendors. The company also has storage solutions from HPE and EMC. Networking infrastructure is mainly provided by Cisco, with some HPE top-of-rack switches.

In summary, Alice is seeking an IT solution that will help ABTechnologies:

- Increase resource utilization to support design projects without increasing data center resources

- Implement a software-defined data center project to increase service delivery

- Find faster ways to deploy IT resources to meet the needs of the design teams

- Transform, simplify, and reduce costs in the data center

Questions

1. What products and services should you discuss with the customer and why?

   _____

   _____

   _____

   _____

   _____

_____

_____

_____

_____

_____

_____

_____

Answers

The following recommendations should not be considered to be the "correct" answers, but should give you an indication of where to begin.

1. What products and services should you discuss with the customer and why?

    HPE Synergy provides superior, enterprise-grade availability and allows IT to quickly and confidently implement infrastructure changes with minimal or no service interruption and significantly reduced operational costs. Its one-tool, one-step template-based operations allow an entire infrastructure to be quickly and accurately configured with zero impact on availability. Standardized resource templates can be configured, provisioned and released as development cycles demand. In fact, the whole process of resource needs can be requested, approved and provisioned on-demand. No longer does resource utilization become a topic of budget conversations.

    Synergy eliminates complexity and empowers IT to orchestrate a single infrastructure of physical and virtual compute, storage, and fabric pools with a single management interface to assemble and reassemble resources in any configuration. This maximizes resource utilization and eliminates the risk of stranded capacity to ensure right-sized resource allocation for development projects while reducing capital expenditures.

    With one automated tool for firmware, hardware and workloads, HPE Synergy automates lifecycle management to simplify operations and minimize operational expenses for developing new services. This means that the ideas coming out of your lines of business can be realized faster to deliver better experiences for customers while staying ahead of the competition and growing revenue.

    With Synergy, organizations can now free up resources to focus on value creation, enabling IT to operate like a cloud provider to the lines of business accelerating digital transformation and delivery of revenue-generating products, services, and experiences. This means the ideas coming out of the lines of business can be realized faster, so the business can deliver better experiences for customers while staying ahead of the competition and growing revenue. With HPE Synergy, IT can now break free from the ordinary and accelerate the extraordinary to become a value creation partner for all of enterprise.

The Synergy experience is designed around three core benefits:

- Run anything: Synergy optimizes any application and stores all data on a single infrastructure with fluid pools of compute, storage, and fabric. All the resources are now always available and can be instantly configured according to the specific needs of each application.

- Move faster. Work efficiently: Synergy enables IT to accelerate application and service delivery through a single interface that precisely composes and recomposes logical infrastructures into any combination at near-instant speeds. Composable resources are provisioned together with their state (BIOS settings, firmware, drivers, protocols, etc.) and the operating system image using repeatable templates.

- Unlock value: Synergy increases productivity and control across the data center by integrating and automating infrastructure operations and applications through a unified API. The unified API provides a single interface to discover, search, inventory, configure, provision, update, and diagnose the composable infrastructure.

HPE tools for selecting solution components

HPE helps partners and customers create the best solution for each IT problem by providing multiple sizing and planning tools. Access to some tools may require registration.

HPE Partner Ready Portal

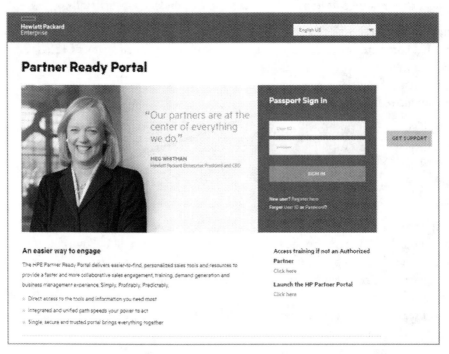

Figure 8-7 HPE Partner Ready Portal

The HPE Partner Ready Portal, shown in Figure 8-7, delivers easy-to-find, personalized sales tools and resources to provide a faster and more collaborative sales engagement, training, demand generation, and business management experience.

It is a secure and trusted portal for resources including information in the following areas:

- Storage sizing and assessment
- Software downloads and licenses
- Software quoting tools
- Accessories, supplies, and services

HPE Proposal Web

The Partner Ready Portal also includes access to Proposal Web. Proposal Web is an easy-to-use, web-based resource. You can create strong customer proposals quickly and easily using the HPE standard, global, online proposal platform. It offers a comprehensive library of up-to-date proposal boilerplate content along with a powerful tool suite for automated proposal assembly. This tool is used worldwide by sales representatives, proposal bid teams, and channel partners. It is structured around country- and region-specific portals with localized and translated content. Proposal Web content includes HPE products, services, solutions, and corporate and general information.

HPE Proposal Web offers the right resources to help with all your proposal-related sales activities. You can use this tool to:

- Prepare unsolicited proposals
- Respond to customer RFI or RFP requests
- Add boilerplate content to quotes or configurations

 Note

To access the Partner Ready Portal using an HPE Passport account, scan this QR code or enter the URL into your browser.

https://partner.hpe.com/

HPE configuration tools

HPE offers a variety of configurator tools to guide you in developing appropriate IT solutions for customers. These tools streamline the ability to select and configure HPE products and to create quotes for you and your customers.

HPE Sales Builder for Windows

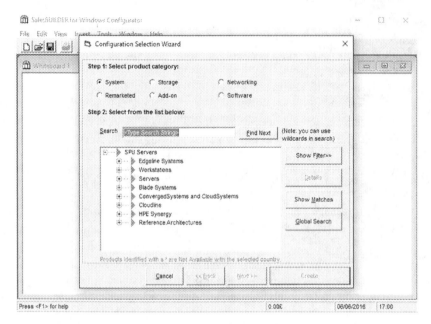

Figure 8-8 Sales Builder for Windows

Sales Builder for Windows (SBW), as shown in Figure 8-8, is a downloadable, stand-alone application that enables you to configure and quote enterprise servers, storage products, software, and workstations. SBW provides a configuration platform for sales and channel users who need a mobile product portfolio to meet their customers' needs. With SBW, you can configure associated software and services with current pricing and guaranteed validity. Configurations can be converted to budgetary quotes in SBW using list price and purchase agreements. SBW is accessible through the Partner Ready Portal.

Note

Pricing in SBW is updated weekly, so it provides approximate pricing. If there are discrepancies between SBW and other pricing sources, accurate pricing is always available in the Standard Pricing Viewer.

SBW creates complex, cross-enterprise solutions integrating multiple CTO and BTO configurations and exports schematic diagrams in Microsoft Visio format. For example, it is useful when configuring a new, custom CTO solution that involves multiple levels of integration or server partitioning.

The SBW Configurator displays the system diagram and modifications for HPE clusters, servers, and storage running HPE-UX, Windows, Linux, or mixed environments. You can use this tool to configure and customize technical solutions for new systems, upgrades, and add-ons. Components include:

- **Whiteboard**—Centerpiece that shows technical solutions and forms the unit of quoting and storing

- **Configurator worksheet**—Main configuration tool that enables you to configure complete technical solutions

- **System diagram**—A graphical view of your configuration that can be used for modifications

- **Quoter**—Prepares a budgetary quotation for the customer that shows numbers, descriptions, and prices of all the products in the solution

- **Price book**—Data files containing the latest product descriptions and prices that are updated every two weeks

- **Knowledge base**—Data file containing the rules and product modeling used by SBW to check configurations

HPE Power Advisor

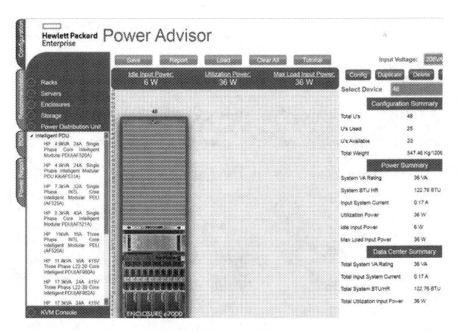

Figure 8-9 Power Advisor

When you are designing and expanding data centers or planning rack deployments, accurate estimates of power requirements are essential to ensure appropriate levels of power and cooling as well as to determine power-related operating costs for the customer's budgetary considerations.

Power Advisor, as shown in Figure 8-9, is an easy-to-use tool that estimates data center power requirements for server and storage configurations. Version 6.x includes ProLiant Gen9 servers and options. A downloadable Windows application and an online version are available. The Power Advisor online tool supports Google Chrome and Mozilla Firefox. This tool allows you to:

- Accurately estimate power consumption of HPE server and storage products

- Select the appropriate power supplies and other system components

- Configure and plan power usage at a system, rack, and multirack levels

- Access useful tools including a cost-of-ownership calculator, power report, and bill of materials (BOM)

Note

Scan this QR code or enter the URL into your browser for more information on the Power Advisor. After the webpage opens, scroll down to HPE Power Advisor.

https://www.hpe.com/us/en/integrated-systems/rack-power-cooling.html

---

HPE Server Memory Configurator

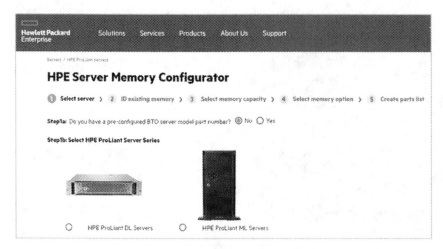

Figure 8-10 Server Memory Configurator

The Server Memory Configurator, shown in Figure 8-10, is a web-based tool used to assist with populating memory in ProLiant servers. This tool provides optimal configurations based on memory population guidelines. Non-optimal configurations are also shown for customers who require a specific memory configuration. Guidance provided is based on customer-provided information and does not guarantee specific performance.

This tool includes a five-step process that provides recommended memory configurations and RAM module installation locations:

1. Select the relevant ProLiant server.

2. Identify existing memory.

   – Auto-detect memory with HPE Insight Diagnostics or manually enter current memory.

3. Select memory capacity.

4. Select memory option.

5. Create parts list.

 Note

Scan this QR code or enter the URL into your browser to access the Server Memory Configurator.

http://h22195.www2.hp.com/DDR4memoryconfig/Home/Legal

HPE solution sizers

HPE offers several automated tools that assist with recommending a solution environment. The sizing information and algorithms in HPE solution sizers have been developed using testing and performance data on a wide range of HPE servers running solutions from partners such as Citrix, Microsoft, SAP, and VMware. These tools provide a consistent methodology to help determine a "best fit" server for the environment. Sizers are downloaded and run on the user's personal computer. Updates with the latest information on HPE hardware and solution software are available automatically when the user is connected to the Internet and may optionally be installed by the user.

There are several solutions sizers available through the HPE Information Library, including:

- **HPE Converged Infrastructure Solution Sizer Suite (CISSS)**—Solution sizers from HPE are conveniently available through the CISSS. This suite provides an easy way to install sizers, consolidate the BOM generated by multiple sizings, access reference architectures, and more. You can use the CISS to:

  - List the HPE solution sizers and select which ones to install through the Sizer Manager

  - Size an application solution using one of the installed solution sizers

  - Combine application solutions after two or more solutions have been sized and saved

  - Size an HPE ConvergedSystem solution

  - Calculate power requirements for a solution using Power Advisor

- **HPE Sizer for Server Virtualization**—This automated, downloadable tool provides quick and helpful sizing guidance for HPE server and storage configurations running in VMware vSphere 5.0 or Hyper-V R2 environments. The tool allows users to create new solutions, open existing solutions, or use other types of performance data collecting tools, such as the Microsoft Assessment and Planning (MAP) Toolkit, to build virtualized configurations based on HPE server and storage technologies. It enables the user to quickly compare different solution configurations and produces a customizable server and storage solution complete with a detailed BOM that includes part numbers and prices.

 Note

To access solution sizers, scan this QR code or enter the URL into your browser.

http://h17007.www1.hp.com/us/en/enterprise/converged-infrastructure/info-library/
index.aspx?type=20#.VvCf6PkrLlU

HPE Storage Sizer

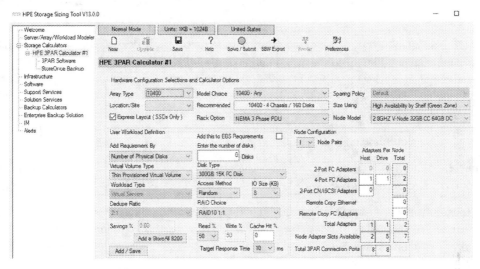

Figure 8-11 Storage Sizer

As shown in Figure 8-11, the Storage Sizer is a sizing tool that helps you design a storage infrastructure to meet the needs of a customer. The Storage Sizer can be downloaded from the HPE website. This is an important feature because it keeps the sizer current and any configuration prepared using this tool will be a valid, fully supported configuration.

The Storage Sizer supports the disk storage subsystem and other storage solutions such as backup systems, network-attached storage (NAS) solutions, and other storage components. The Storage Sizer requires a license.

Storage Sizer provides the following features and benefits:

- Simplifies the process of designing a storage solution

- Applies storage design, licensing, and services rules

- Provides output as a valid, supported configuration that can be imported directly into SBW for quotation

- Provides localized parts and pricing for different geographic regions

- Includes HPE Smart Update technology, which brings new products or functionality to you through an Internet connection

- Encompasses the HPE storage family

- Initiates an update for every product launch as part of the new product introduction process

- Includes new functionality, which was added based on user input, annual surveys, and quarterly focus groups

The Storage Sizer enables you to work with your customers to design a storage infrastructure that will meet their online and offline needs. You can define customer requirements, such as:

- Performance requirements with specific metrics

- Business requirements, such as server consolidation

- Pure capacity requirements

For example, additional requirements might include raw capacity, estimated IOPS, replication and backup criteria, and number of host ports.

Because the tool applies all the HPE SAN design rules, it provides a valid, supported storage infrastructure to meet the requirements of your customer. Use the Storage Sizer when you are not sure which combination of products will best address customer requirements. This tool lets you try different solutions.

A helpful wizard interface guides you through the process of sizing a SAN by asking a series of questions about the proposed configuration. This wizard is intended for those who have less experience using the Storage Sizer.

 Note

To access the Storage Sizer, scan this QR code or enter the URL into your browser.

https://sizersllb.itcs.hpe.com/swdsizerweb/

SAN Design Reference Guide

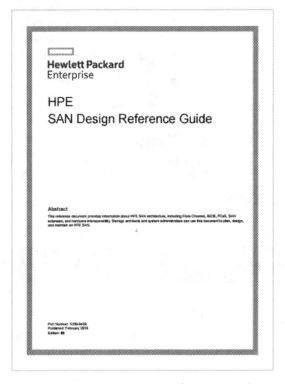

Figure 8-12 SAN Design Reference Guide

The SAN Design Reference Guide, as shown in Figure 8-12, is the single source for SAN configuration support, interoperability, and best practices. The guide provides access to HPE multi-vendor, end-to-end storage networking architectural information, including:

- SAN design rules

- SAN topologies and supported configurations

- SAN design philosophies, security, and management

- HPE best practices

- SAN components

    - Architecture

    - Configurations

    - Implementation

    - New technologies

Note

To access the SAN Design Reference Guide, scan this QR code or enter the URL into your browser.

http://h10032.www1.hp.com/ctg/Manual/c00403562.pdf

HPE iQuote Universal

Figure 8-13 iQuote Universal

iQuote Universal is a cloud-based service for channel partners that provides sales configuration and quoting features, as shown in Figure 8-13. This service simplifies the process of selling HPE products and helps users maximize revenue and margin on every sale. iQuote Universal generates quotes for products across the HPE portfolio, including ProLiant servers as well as HPE servers, storage, networking, commercial desktops, laptops, and workstations.

This subscription-based software guides HPE partners through each step of the configuration process, includes real-time information on promotional pricing and stock, and notifies users of technical errors. Resellers and IT providers can select a product, create a configuration, export the information, and send a validated BOM to the distributor or supplier.

 Note

To access iQuote Universal, scan this QR code or enter the URL into your browser.

http://split.hpiquote.net/aspx/Signin.aspx?Universal&mfr=hpe

HPE One Config Simple

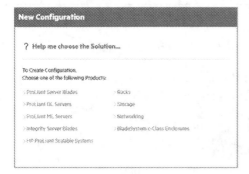

Figure 8-14 One Config Simple

Previously known as the Simplified Configuration Environment, the One Config Simple (OCS) online configuration tool, shown in Figure 8-14, actively gathers customer requirements and maps them to a set of products or service options. The guided selling capability directs users to an optimal solution for the customer's specific workloads or application needs. This capability makes OCS ideal for non-technical sales people, allowing them to quickly create simple CTO and BTO solutions. Final configuration files can be imported into partner quote and ordering systems for fast conversion to an order. It is most appropriate to use OCS early in the sales cycle for budgetary quotes and "what if" scenarios.

OCS is a web-based application that:

- Provides a self-service environment

- Generates initial solution configurations in three to five minutes

- Offers local list pricing

- Uses solution wizards based on applications

- Lists popular configurations that can be modified

- Can export configurations to Microsoft Excel file format

OCS includes two basic modes for creating configuration files: a guided or a platform-driven approach to solutions. With the guided solutions approach, the user selects a particular application (email server, web server, and so on) and uses a solution wizard to generate a configuration. In the platform-driven model, the user selects an HPE product family (ProLiant BL servers, storage, networking, and so on) and then completes a series of steps to add or confirm components (for example, disk, memory, and services).

OCS requires an Internet connection and access to the Partner Ready Portal.

 Note

To access OCS, scan this QR code or enter the URL into your browser.

https://h22174.www2.hp.com/SimplifiedConfig/Index

Additional HPE configuration tools

Additional HPE configuration tools include:

- **Elite**—Elite is an HPE-hosted web application with preapproved pricing. Elite is the updated version of ePrime. It functions as a direct-to-customer, B2B storefront with a customer-specific catalog. B2B customers with HPE Catalog access and HPE configuration specialists can use it for repeat catalog purchases.

- **Source**—Source is an HPE-hosted webstore for partners based on an order management system. It provides partners with business and e-commerce solutions that help fulfill business, servers, storage, networking, converged systems, services, and software objectives. Volume purchase agreement partners are eligible to use this tool, and a login is required.

- **HPE Switch Selector**—This selector tool helps with determining the correct HPE networking product based on specific requirements.

 Note

To access the Switch Selector, scan this QR code or enter the URL into your browser.

http://h17007.www1.hp.com/us/en/networking/products/switches/switch-selector.aspx#.VvFZufkrLlU

- **HPE Networking Online Configurator**—This configurator enables quick and easy creation of price quotations for HPE networking products. Users can create a complex networking solution with a network map, wiring guide, or inclusion of multi-vendor networks.

 Note

To access the configurator, scan this QR code or enter the URL into your browser.

http://h17007.www1.hpe.com/us/en/networking/products/configurator/index.aspx#.VvFZ7vkrLlU

You can use these tools to access information to support configurations:

- **HPE PartSurfer**—This tool provides fast, easy access to service parts information for a wide range of HPE products.

Note

To access PartSurfer, scan this QR code or enter the URL into your browser.

http://partsurfer.hpe.com/search.aspx

- **HPE Customer Self-Repair Services Media Library**—Users can find media for HPE products using this library.

 Note

To access this library, scan this QR code or enter the URL into your browser.

http://h20464.www2.hp.com/index.html

- **HPE Single Point of Configuration Knowledge (SPOCK)**—SPOCK is the primary portal used to obtain detailed information about supported HPE storage product configurations. In SPOCK, you will find details of supported configurations (hardware and software) as well as other useful technical documents.

 Note

To access SPOCK, scan this QR code or enter the URL into your browser.

https://h20272.www2.hpe.com/spock/

Using TCO and ROI tools

It is important to understand qualitative and quantitative data about the entire lifecycle cost of a technology project and the impact it will have on business processes. TCO measures cost, and ROI measures benefit. TCO data should feed ROI data, and ROI data should feed the overall business case for business technology decisions.

Total cost of ownership

TCO costs can be difficult to quantify. It is important to ask the customer meaningful questions about why a technology initiative exists and what impact it will have on the business.

TCO represents total direct and indirect costs over the entire lifecycle of a hardware or software product. The simple approach to TCO data collection and assessment is a template that requires collecting specific hard and soft data. Hard data is always preferred over soft data, but soft data should be analyzed if it can be monetized (for example, generating a premium for a company's stock price or enhancing the brand). Examples of soft costs include the cost of downtime, consulting from indirect sources, and costs connected with standardization.

Return on investment

ROI can be difficult to calculate. For example, companies developed websites for a variety of reasons in the mid-to-late 1990s. First-generation sites were essentially company marketing displays, but very few transactions took place. What was the ROI for these sites? They did not reduce costs; in fact, they increased them. They did not generate revenue. Companies built them to convince customers, Wall Street analysts, investors, and others that they understood that the World Wide Web was important. The ROI for this endeavor provided an intangible benefit, meaning the companies might have improved their credibility or reputation, which could have affected downstream profits.

There are many factors to consider when calculating ROI. Research suggests that although many ROI methods are used, the most popular approaches calculate cost reduction, customer satisfaction, productivity improvement, and contributions to profits and earnings. Most business technology executives consider two years to be a reasonable timeline for measuring ROI.

A few methods of calculating ROI are:

- Payback is one simple approach to ROI data collection and assessment. This method calculates the time it takes to offset the IT investment through increased revenues or reduced costs. If the payback period is short and the offsets are great, then the ROI is significant. Payback should be defined by using internal metrics. The payback for some projects will be a year, but others might take three years. Most IT projects should achieve a positive ROI within three years.

- Another way to determine ROI is based on a simple calculation that starts with the amount of money needed to purchase an IT solution (including TCO and other data). Then, the increased revenue or reduced costs that the investment would generate are projected. If a project costs $1 million but saves $2 million, then the ROI is healthy.

- Additional ROI methods are based on financial metrics such as economic value analysis (or economic value added), internal rate of return (IRR), net present value (NPV), total economic impact (TEI), rapid economic justification (REJ), information economic (IE), and real options valuation (ROV), among others.

Alinean ROI and TCO analysis

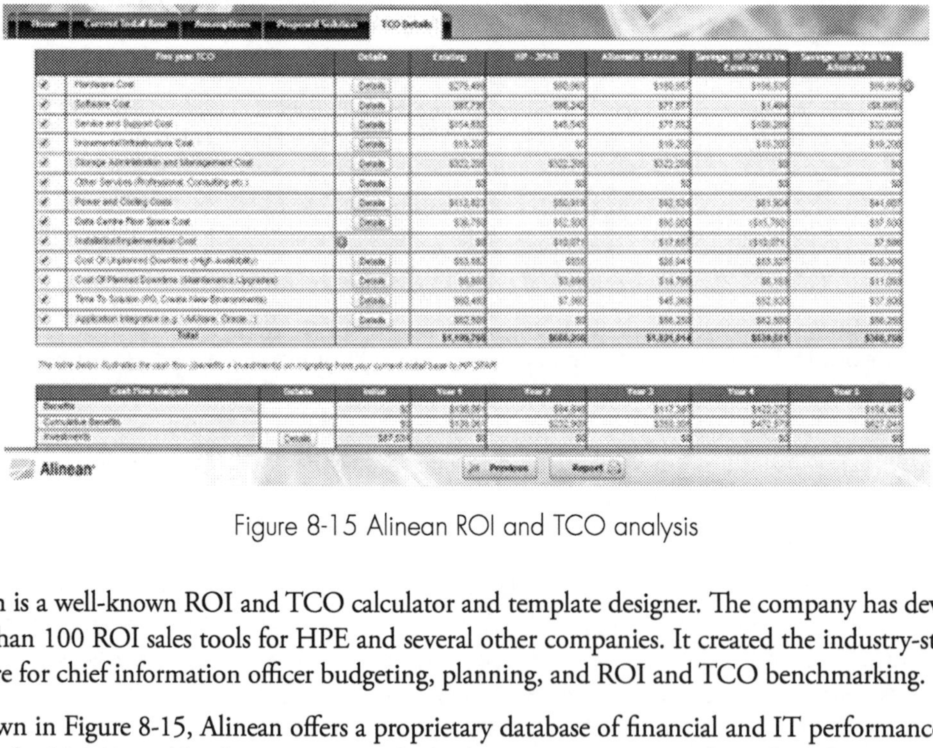

Figure 8-15 Alinean ROI and TCO analysis

Alinean is a well-known ROI and TCO calculator and template designer. The company has developed more than 100 ROI sales tools for HPE and several other companies. It created the industry-standard software for chief information officer budgeting, planning, and ROI and TCO benchmarking.

As shown in Figure 8-15, Alinean offers a proprietary database of financial and IT performance information for 20,000 worldwide corporations. It also has proprietary research methodologies to quantify the costs and benefits of IT projects. Alinean software helps users demonstrate ROI, TCO, and the overall value of IT solutions.

Note

To access the Alinean tools (login required), scan this QR code or enter the URL into your browser.

http://www8.hp.com/us/en/business-solutions/tco-calculators.html

Alinean analysis tools use this data to perform calculations and determine the benefits of migrations to HPE solutions:

- Server, power and cooling, and operating system and database license costs

- Operations and administration costs

HPE Storage Quick ROI Calculator

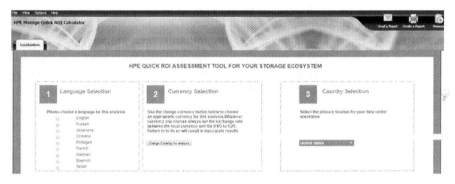

Figure 8-16 Storage Quick ROI Calculator

The Storage Quick ROI Calculator (Figure 8-16) can make it very easy to build the business case. The Alinean tool set contains several important HPE templates and calculators.

Alinean tools are available for these products:

- HPE 3PAR StoreServ

- HPE StoreOnce Backup

- HPE StoreVirtual

When you are working with decision makers, it is important to provide financial justification. That means presenting a business case that includes the value proposition as well as essential financial metrics. The Storage Quick ROI Calculator provides these metrics along with the NPV and IRR. Sales skills combined with Alinean tools and financial metrics will help you develop a compelling business case.

Financial metrics

ROI Analysis (Solution B) (Probable Case)	Initial	Year 1	Year 2	Year 3
Benefits (to Solution B from Current (AS IS))	$0	$585,020	$663,566	$713,397
Cumulative Benefits		$585,020	$1,248,586	$1,961,983
Investment (Solution B)	$308,181	$157,327	$163,501	$169,770
Cumulative Investment	$308,181	$465,508	$629,009	$798,779
Cash Flow	($308,181)	$427,693	$500,065	$543,627
Cumulative Cash Flow	($308,181)	$119,512	$619,577	$1,163,204
ROI	146%			
Risk Adjusted ROI	127%			
NPV Savings	$913,522			
IRR	138%			
Payback period (including deployment period)	6 month(s)			
Risk Adjusted Discount Rate	9.5%			

Figure 8-17 Financial metrics

Figure 8-17 shows financial metrics generated by the Storage Quick ROI Calculator. The top half of the chart shows the year-by-year cash flow and the cumulative cash flow. The rows highlighted in orange show the ROI, NPV, IRR, and the payback period. Remember that the output charts within the Alinean tool are based on the customer's information, which is important to know when developing or proposing an HPE solution.

Alinean customer deliverables

The Storage Quick ROI Calculator allows you to create a complete business value report. Using key inputs from your customer, you can generate a report that has third-party credibility along with the ability to dig deeper when necessary.

PowerPoint presentations, interactive surveys, assessment summaries, and blueprints can be generated using Alinean tools. These customer deliverables include information such as customer data, financial metrics, and competitive analysis.

Developing the proposal

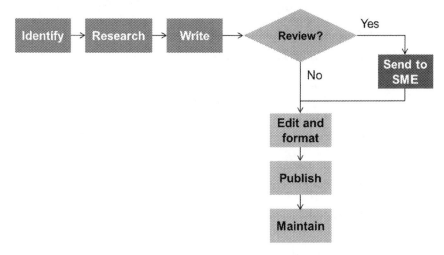

Figure 8-18 Developing the proposal

After assembling the business and technical information from the customer, work can begin on the solution proposal. The creation of timely, accurate, high-quality, proposal-ready content requires a rigorous development and maintenance process as shown in Figure 8-18. At this point, you should focus on various aspects of the solution, including addressing architectural and transitional issues such as functional and technical design, organizational design, technology governance, and change management.

Matching the challenge with the opportunity

When designing a solution, it is not enough to present the technical aspects. Remember that you want to present the value of the solution in business terms that matter to the customer. Frame the technology in terms of how it addresses business drivers and initiatives, how it overcomes obstacles, and how it meets the customer's goals. You will architect the solution to meet the customer's business, technical, and financial needs by:

- Developing a logical architecture that will host the solution, including:
    - Network layout
    - Server requirements
    - Application services
    - Storage space requirements

- Incorporating licensing options based on current QuickSpecs

- Outlining how to integrate your solution into the customer's IT infrastructure

- Describing the business value for the customer

From a content perspective, proposal-ready documents typically include:

- Key benefits and differentiators

- Latest HPE marketing messages

- Customer and analysis quotes

- High-level technical information

- Proof points

The order in which you present information might be dictated by points of focus in the request for information (RFI) or request for proposal (RFP). Depending on the solution you are proposing and the resources required, you might also include support information from channel partners, program managers, special interest groups, and others.

The majority of proposal-ready content must be reviewed before publication to ensure accuracy. Reviewers can include:

- Product managers

- Program managers

- Marketing specialists

- Special interest groups

    - HPE Solution Architect (SA) community

    - Ambassador program

 Note

Proposal-ready content that is based on nontechnical, external sources such as the HPE website does not require review by a subject matter expert (SME).

After review, changes need to be incorporated from the reviewers to ensure that the content passes a final editing process.

Writing a scope of work

A *scope of work* is a preproject overview you prepare for the proposal. This document captures the plan, time frame, required resources, and completion milestones of a project. It is crucial for ensuring a mutual understanding with the customer. Executive support is essential. Without leadership support, it can be difficult to implement an IT project.

 Note

A scope of work should not be confused with a statement of work, which is a final project overview prepared for billing.

The scope of work should provide a summary of the plan you create for the solution, including:

- Overall time frame

- Completion milestones for each aspect of the solution

- Resources required

    - Channel partners, HPE sales representatives, HPE services, and any other parties involved in delivering the solution

    - Executive support for the project (name, position, and so on)

Learning check

1. Which factors should you consider when conducting a site survey? (Select three.)

   a. Backup requirements

   b. Site services/utilities

   c. High-availability requirements

   d. Projected growth

   e. Projected role of server solution

   f. Physical security

2. When designing a VM oversubscription ratio, what are the four aspects to consider?

   _____

   _____

   _____

   _____

3. Which tool is a downloadable, stand-alone application that enables you to configure and quote enterprise servers, storage products, software, and workstations?

   a. Server Memory Configurator

   b. HPE Storage Sizer

   c. HPE Proposal Web

   d. HPE Power Advisor

   e. SBW

4. How can the HPE Storage Quick ROI Calculator help develop a business case for customers?

   _____

   _____

   _____

   _____

5. Proposal-ready content that is based on nontechnical, external sources such as the HPE website does not require review by a SME.

   ☐ True

   ☐ False

Learning check answers

1. Which factors should you consider when conducting a site survey? (Select three.)

   a. Backup requirements

   b. Site services/utilities

   c. High-availability requirements

   d. Projected growth

   e. Projected role of server solution

   f. Physical security

2. When designing a VM oversubscription ratio, what are the four aspects to consider?

   - **Processing capacity per VM**

   - **Memory per VM**

   - **IO per VM**

   - **Licensing cost**

3. Which tool is a downloadable, stand-alone application that enables you to configure and quote enterprise servers, storage products, software, and workstations?

   a. Server Memory Configurator

   b. HPE Storage Sizer

   c. HPE Proposal Web

   d. HPE Power Advisor

   e. SBW

4. How can the HPE Storage Quick ROI Calculator help develop a business case for customers?

   - **When you are working with decision makers, it is important to provide financial justification**

   - **The business case should include the value proposition as well as essential financial metrics**

   - **The Storage Quick ROI Calculator provides these metrics along with the NPV and IRR**

   - **This calculator also provides customer deliverables**

5. Proposal-ready content that is based on nontechnical, external sources such as the HPE website does not require review by a SME.

☐ **True**

☐ False

Summary

- When you are planning an HPE server solution, it is crucial to thoroughly evaluate the customer's requirements. Factors that need to be examined include the number of users, existing SLAs, future plans, businesses requirements, obstacles, site specifications, and so on. This information can be used to determine the best HPE platform for each customer.

- When designing a solution, specific considerations range from basic server, storage, networking, and operating system requirements to VM design goals and guidelines. These factors are important to building a compelling business case.

- HPE helps partners and customers create the best solution for each IT problem by providing multiple sizing and planning tools. These tools include the Partner Ready Portal and HPE configuration tools such as SBW and the Storage Sizer.

- The HPE Storage Quick ROI Calculator generates qualitative and quantitative data about the entire lifecycle cost of a technology project and demonstrates the impact it will have on business processes.

- After assembling the business and technical information from the customer, work can begin on the solution proposal. The creation of timely, accurate, high-quality, proposal-ready content requires a rigorous development and maintenance process.

Hands-on exercises

You now have the opportunity to get some hands-on experience with some tools that would be used to size a customer solution. There are two exercises to complete:

1. Using HPE QuickSpecs

2. Using the HPE Power Advisor Tool

Exercise 1—Using HPE QuickSpecs

This exercise uses the QuickSpecs document of a ProLiant DL380 Gen9 server to identify supported operating temperatures, and support for extended ambient operating temperatures.

1. On the HPE.com website, search for the ProLiant DL380 Gen9 QuickSpecs.

 Note

    This specific QuickSpecs document is available online at: http://www8.hp.com/h20195/ v2/GetDocument.aspx?docname=c04346247

2. Locate the storage section in the QuickSpecs document. What is the maximum amount of disk drives that a ProLiant DL380 Gen9 server can support?

_____

3. On the HPE website, search for the extended ambient temperature guidelines for the ProLiant DL380 Gen9.

 Note

    The HPE Extended Ambient Operating Support documents can be found at: http://www. hpe.com/servers/ashrae

4. Specifically for the ProLiant DL380 Gen9, what is the maximum supported extended ambient operating temperature?

_____

5. For the ProLiant DL380 Gen9, what is the maximum number of supported disk drives when operating at the highest supported temperature?

_____

6. Which network adapter is not supported when using HPE Extended Ambient Operating Support on a ProLiant DL380 Gen9?

_____

Exercise 2—Using the HPE Power Advisor Tool

Objectives

After completing this exercise you should be able to:

- Install the Hewlett Packard Enterprise (HPE) Power Advisor tool and select the input voltage

- Select a rack and an enclosure, and then configure the enclosure

- Select and install server blades in the enclosure, and then configure the server blades

- Create, review, and save a power report

Requirements

To complete this exercise you need:

- Internet access
- A supported Internet browser

**Important**

Microsoft Internet Explorer 11 is supported only with Microsoft Windows 10.

Introduction

Power Advisor is an online and a downloadable tool provided by HPE (http://www.hp.com/go/hppoweradvisor) to assist in the estimation of power consumption and in the proper selection of components, such as power supplies at the system, rack, and multi-rack levels. Additional features of the tool include a condensed bill of materials (BOM), a cost of ownership calculator, and a power report.

Note

Screenshots in this document are for your reference and might differ slightly from the ones you will see on your screen during these exercises.

This exercise demonstrates how to use the online HPE Power Advisor tool to identify the amount of power and cooling is required for a solution proposal. This lab is designed to be performed from your laptop or workstation.

1. From your local computer, navigate to the URL: http://www.hpe.com/info/poweradvisor/online

2. To accept the license agreement, click **I Agree** (Figure 8-19).

Figure 8-19 Accept licensing agreement

3. On the Input Voltage screen, select **220VAC** from the drop-down list, and then click **Go** (Figure 8-20).

Figure 8-20 Select input voltage

4. To insert a rack into the utility, click the rack size from the Racks menu. For this exercise, click **22 U Rack** (Figure 8-21).

Figure 8-21 Insert rack

5. In the Rack name field, enter LGxRack1, and then click **OK** (Figure 8-22).

Figure 8-22 Enter rack name

6. To add a BladeSystem enclosure to the rack, expand the **Enclosures** menu, click **BladeSystem,** then click **BladeSystem c7000** (Figure 8-23).

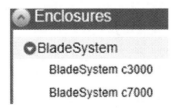

Figure 8-23 Add BladeSystem enclosure

7. The enclosure is added to the rack (Figure 8-24).

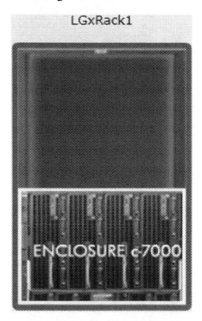

Figure 8-24 Enclosure added

8. Expand the Servers menu, click DL/RX, and then click **ProLiant DL380 Gen9** twice to add two ProLiant DL380 Gen9 servers to the rack (Figure 8-25).

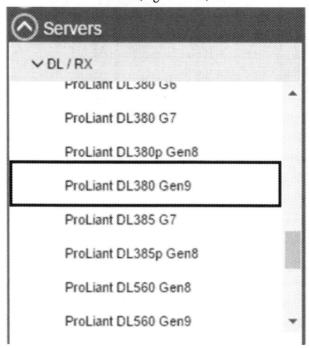

Figure 8-25 Add ProLiant DL380 Gen9 servers

9. In the rack topology, left-click to select and then right-click the c7000 enclosure, and then click **Config** (Figure 8-26).

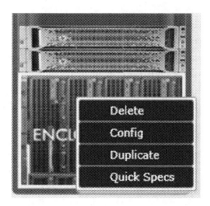

Figure 8-26 Rack topology

10. On the BladeSystem c7000 screen, click the **BL460c Gen9** label 16 times to add 16 server blades to the enclosure, and then click **Save** (Figure 8-27).

Figure 8-27 Add BL460c Gen9 servers

11. In the rack topology, select and then right-click a ProLiant DL380 Gen9 server, and then click **Config**. Use the following information to add the components to the server (Figure 8-28) and then click **Save**:

   – Processor: One ProLiant DL380 Gen9 E5-2623v3 Kit (779556-B21)

   – Memory: Two 8 GB 1Rx4 PC4-2133P-R Kit (726718-B21)

   – Storage: Eight SFF 600GB 6G SAS 10K 2.5in SC ENT HDD (652583-B21)

   – Expansion: 350GB PCIe Workload Accelerator (708088-B21)

   – Power Supply: Two 800W FS Plat Ht Plg Pwr Supply Kit (720479-B21). You will receive a warning that two different power supplies cannot be mixed. Click **OK** to replace the existing power supply.

Model(s)	Quantity
Processor	
HP DL380 Gen9 E5-2623v3 Kit(779556-B21)	1
Memory	
HP 8GB 1Rx4 PC4-2133P-R Kit(726718-B21)	2
Storage	
HP 600GB 6G SAS 10K 2.5in SC ENT HDD(652583-B21)	8
Expansion	
HP 350GB HE PCIe Workload Accelerator(708088-B21)	1
Micro UPS	
Power Supply	
HP 800W FS Plat Ht Plg Pwr Supply Kit(720479-B21)	2

Configuration is Power Redundant

Figure 8-28 Add components to DL380 Gen9 server

12. Repeat this process for the second ProLiant DL380 Gen9 server that was added to the topology. After completing the configurations, answer the following questions.

What is the total input system current for what is configured?

_____

What is the total wattage of the configuration?

_____

What is the total system BTU per hour?

_____

13. On the configuration screen, click **Report** (Figure 8-29).

Figure 8-29 Click **Report**

14. In the cost per kWh field, enter 0.15. In the Server Lifecycle field, enter 4 and then click **Generate Report** (Figure 8-30).

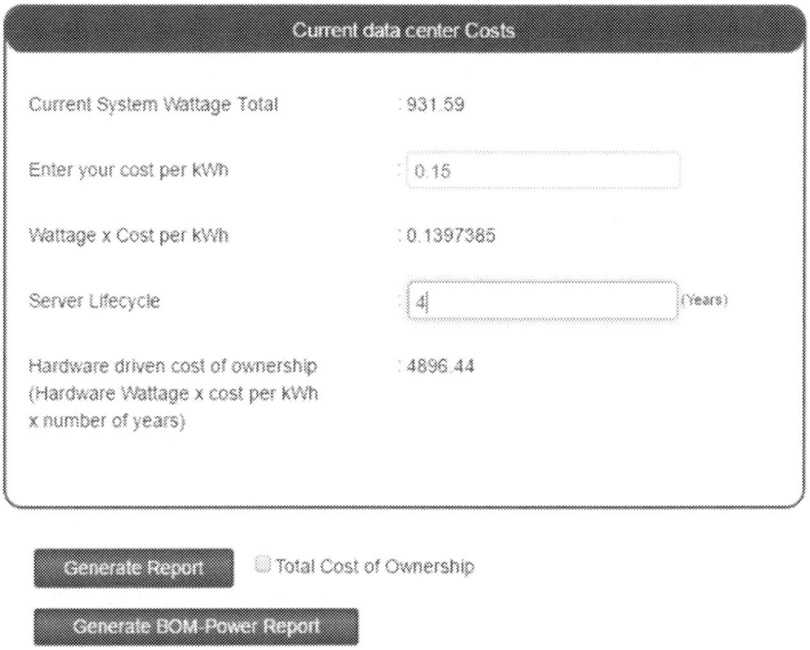

Current data center Costs	
Current System Wattage Total	: 931.59
Enter your cost per kWh	: 0.15
Wattage x Cost per kWh	: 0.1397385
Server Lifecycle	: 4 (Years)
Hardware driven cost of ownership (Hardware Wattage x cost per kWh x number of years)	: 4896.44

Generate Report ☐ Total Cost of Ownership

Generate BOM-Power Report

Figure 8-30 Generate report

15. On the popup window, click **HTML**. Save the file to the desktop of your PC, and then click **OK**. Depending on the browser that you are using, you might receive a popup message asking to save or open the file. If necessary, save the file to your desktop to open it (Figure 8-31).

Export file as ?

HTML Excel Cancel

Figure 8-31 Export report

16. Open the generated report (Figure 8-32).

Data Center Summary	
Line Voltage	220 VAC
BTU HR	1424.48 BTU
System Current	1.96 A
Total Utilization Input Power	417.74 W
VA Rating	431.06 VA
Total Idle Input Power	166.78 W
Total Max Load Input Power	417.74 W

Figure 8-32 Open report

17. Read through the report and close the browser tab when you have finished.

9 Managing the Server Infrastructure

WHAT IS IN THIS CHAPTER FOR YOU?

After completing this chapter, you should be able to:

✓ Describe features of Hewlett Packard Enterprise (HPE) Insight Control server provisioning (ICsp)

✓ Explain the options available for remote support from HPE

✓ Describe the functions of HPE Insight Online in the HPE Support Center

  – Insight Online: My IT Environment

  – Insight Online: My Customer

OPENING CONSIDERATIONS

Before proceeding with this section, answer the following questions to assess your existing knowledge of the topics covered in this chapter. Record your answers in the space provided here.

1. How familiar are you with the process of server provisioning?

_____

_____

_____

_____

_____

2. What is your experience working with HPE Insight Control server provisioning and Insight Remote Support?

_____

_____

_____

_____

_____

_____

3. Have you ever solved software issues using assistance from the HPE Support Center? If so, what was your experience?

_____

_____

_____

_____

_____

_____

_____

_____

HPE Insight Control server provisioning

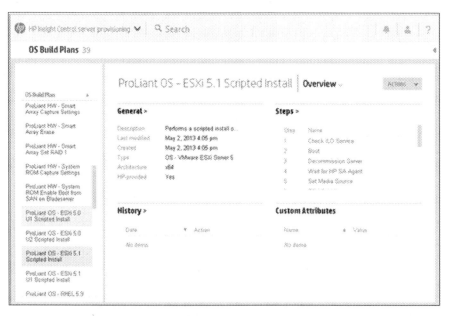

Figure 9-1 Insight Control server provisioning (ICsp)

With support for the latest HPE ProLiant servers, ICsp, shown in Figure 9-1, features updated server provisioning capabilities designed to help customers reduce the time associated with server provisioning tasks.

 Note

HP Insight Control server deployment has been replaced by ICsp. For support information, scan this QR code or enter the URL into your browser.

http://www.hp.com/go/insightcontrol/docs

ICsp is a complete provisioning solution for HPE servers with capabilities for multi-server operating system and firmware provisioning to tower, rack-mount, and BladeSystem servers. ICsp automates the process of deploying and provisioning server software, enabling an IT team to adapt to changing business demands quickly and easily. It increases server provisioning speed significantly, reduces unplanned downtime, optimizes data center capacity, and reduces system admin expenses and travel costs with complete remote control.

ICsp is optimized for Microsoft Windows and Hyper-V, Red Hat and SUSE Linux, and VMware ESXi on ProLiant Gen8 and Gen9 servers (server blades, tower, and rack-mount servers).

 Note

You can run ICsp on VMware vSphere/ESXi 5.5U3 managed mode, and VMware vSphere 6.0 CMS. For the latest Insight Management support matrix, scan this QR code or enter the URL into your browser.

https://h20565.www2.hpe.com/hpsc/doc/public/display?sp4ts.
oid=3312156&docId=emr_na-c04762085&docLocale=en_US

With ICsp, you can:

- Install Windows, Linux, and ESXi on ProLiant servers

- Deploy operating systems to virtual machines (VMs)

- Update drivers, utilities, and firmware on ProLiant servers using the HPE Service Packs for ProLiant (SPPs)

- Configure ProLiant system hardware, iLOs, BIOS, HPE Smart Array controllers, and Fibre Channel host bus adapters (HBAs)

- Configure network settings on one or more servers at a time during operating system installation or postinstallation

- Deploy to target servers without using PXE (for ProLiant Gen8 servers or later). PXE-less deployment can be done on these servers without any special configuration because they include built-in service Boss as part of the embedded HPE Intelligent Provisioning feature.

- Customize ProLiant deployments by using a browser-based interface

- Create and run customized build plans to perform additional configuration tasks before or after operating system deployment

- Use REST API calls to perform all of the functions available from the ICsp UI

- Migrate from HPE Insight Control server deployment to ICsp

 Important

HPE recommends that customers running ICsp versions 7.2, 7.3.x, and 7.4.x upgrade to the latest version.

ICsp components

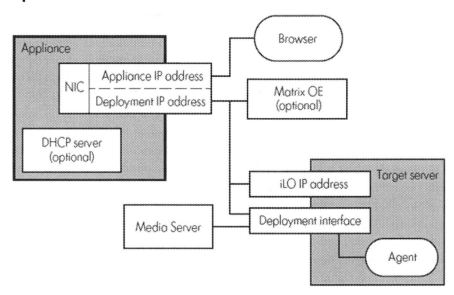

Figure 9-2 ICsp components in a VMware hypervisor environment

The ICsp appliance is a virtual appliance residing on an ESXi or Hyper-V hypervisor. The appliance is accessed through a web browser. As illustrated by Figure 9-2, ICsp includes the following components:

• The **appliance** is the ICsp product that is delivered as a virtual machine optimized to run the application.

• The ICsp appliance ships with an **embedded DHCP server**. Depending on your environment, you might configure this server for use or disable it using the appliance UI Settings screen.

• The **appliance IP address** is the IP address assigned to the appliance. Use this IP address to browse to the appliance using a supported browser, or when making REST calls to perform specialized functions.

• The **deployment IP address** is the IP address used for all deployment operations and target server communications. If you are using ICsp with the Matrix Operating Environment, this is also the IP address that the CMS will communicate with.

When configuring the appliance, you have the option to have the appliance IP and the deployment IP addresses sharing one network interface (single NIC), or to use a separate network interface for each IP address (multi-NIC).

 Note

For details, see the section on "Appliance networking considerations" in the *HP Insight Control Server Provisioning Installation Guide.* To access the guide, scan this QR code or enter the URL into your browser.

http://h20564.www2.hpe.com/hpsc/doc/public/display?docId=c04686196

- The **target server** represents a server managed by IC server provisioning. Each managed server runs an agent, which is software used to make changes to the server. The agent is used for software installation and removal, software and hardware configuration, and server status reporting.

- The **media server** contains vendor-supplied operating system media used during provisioning of the operating system. It might also contain media for other purposes, such as firmware and driver updates, and is also where captured images are stored. The media server is a separate server from the ICsp appliance and is not included as part of the appliance backup and restore actions that you set up and configure.

ICsp OS build plans

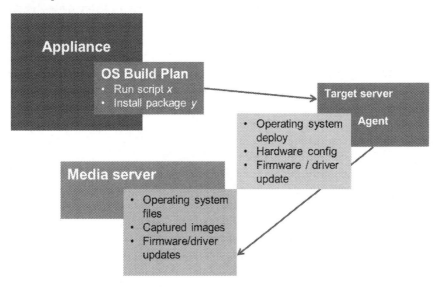

Figure 9-3 ICsp OS build plans

ICsp provides OS build plans, scripts, packages, and configuration files that are used to deploy operating systems, configure hardware, and update firmware, as illustrated by Figure 9-3. OS build plans are the way tasks are accomplished in ICsp. These are the items you run to cause actions such as installing a server or updating firmware to happen.

OS build plans are a collection of ordered steps and parameters associated with those steps that when placed together, in the proper order, can perform just about any action you require. Anything that can be done by a script can be done in a build plan.

ICsp ships ready to run with sample build plans and build plan steps that are designed to work right out of the box. These sample build plans are important because they demonstrate the steps needed to perform the most common deployment related operations. Although you can create your own build plans from scratch, it is expected that most users will start with one of the provided samples and modify it to perform the functions they need. ICsp provides four types of build plans:

- **ProLiant HW**—Build plans labeled with *ProLiant HW* perform hardware-related functions on target servers such as booting the target server to the operating system or capturing or configuring hardware settings.

- **ProLiant OS**—Build plans labeled with *ProLiant OS* deploy an operating system to target servers either through scripted or image installation.

- **ProLiant SW**—Build plans labeled with *ProLiant SW* perform functions on target servers to update the firmware or install/update software on target servers running a production operating system.

- **ProLiant Combination**—Build plans labeled with *ProLiant Combo* perform a combination of functions on target servers, such as hardware-related configurations, deploying an operating system, and installing software.

These build plans do more than operating system deployment. For example, they can also be used to:

- Configure a target server's hardware.

- Capture a target server's hardware configuration so that the same configuration can later be applied to other servers.

- Update the firmware on a target server.

- Install software on a target server with a running operating system.

- Add scripts to perform additional tasks on the target server after it has been installed.

Updated OS build plan features

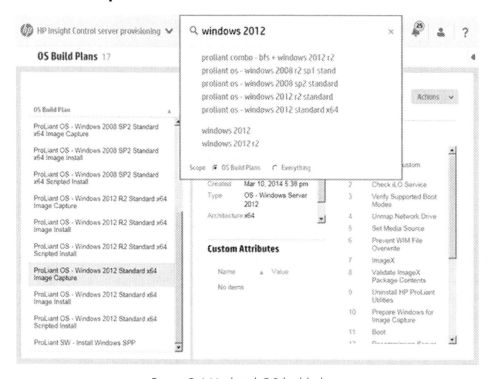

Figure 9-4 Updated OS build plans

With each new release of ICsp, the build plans provided by HPE, some of which are shown in Figure 9-4, are updated to provide more functionality, support more platforms, be more robust, and correct defects. Any build plans you may have customized in a previous release might not have the latest changes because those build plans are copies of the ones HPE provided in the older release. The appendix in the *Insight Control Server Provisioning 7.5 Build Plan Reference Guide* provides a description of all the build plan changes made for each release so that you can update your customized build plan to take advantage of the latest features. Some of the changes listed here are optional, and others are mandatory.

Depending on the customizations you make to your build plans and the changes for each release, you might find it easier to re-customize the latest HPE provided build plans rather than go back and change old ones. For example, if all you do is change the unattend file, it is easier to copy a new Windows build plan and replace the unattend file than it would be to modify the old build plan to incorporate all the latest changes.

 Note

If you are upgrading from a previous release, it is important to read the *Insight Control Server Provisioning 7.5 Build Plans Reference Guide, Appendix* for a complete list of all build plan changes that might affect existing build plans. To reference the guide, scan this QR code or enter the URL into your browser.

http://h20564.www2.hpe.com/hpsc/doc/public/display?docId=c04686185

Running build plans against device groups

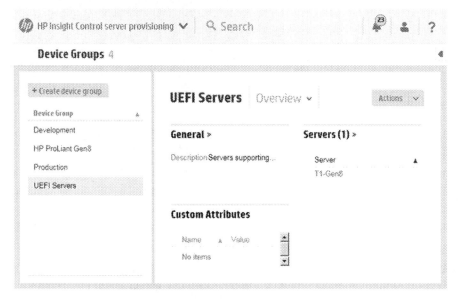

Figure 9-5 Build plans and device groups

You can create and manipulate device groups, which are simple, user-defined groups of servers used for organizing servers in meaningful ways and acting on them together. As shown in Figure 9-5, you can run a build plan against a device group or assign custom attributes to the group, which are inherited by the servers in that group. A server can be in multiple device groups at the same time. After a server is added to a group, it remains in that group until it is removed from the group or deleted from the appliance.

You can add servers to a device group from either the Servers screen or the Device Groups screen. For more information on device groups, see the online help.

With device groups, you can:

- Run OS build plans against all the servers in a group
- Display a list of all the groups of which a server is a member
- Add a server to more than one group
- Add a server to a group from the group view or the server view

- Run a build plan against all the servers in a group

- Assign to device groups custom attributes that are inherited by servers in the group and are available when running build plans

Build plan interoperability with HPE OneView

ICsp can work with HPE OneView appliances operating in your environment. HPE OneView can perform the following functions that can also be performed using ICsp OS build plans:

- Configure BIOS settings

- Manage firmware updates

- Configure SAN boot

If an HPE OneView appliance is managing these aspects of your servers, then you might not want ICsp to run build plans that perform those same actions. ICsp needs to be aware of the HPE OneView appliances. You can identify the HPE OneView appliances by using the OneView Appliances screen available from the ICsp main menu. The HPE OneView user account that is configured in the ICsp appliance requires only read-only access. An account with more privileges is unnecessary and should not be used.

ICsp build plans use the plan *Type* field to identify what services they manage. The Type field is used when integrating with HPE OneView to identify where the services managed by a build plan might overlap services of HPE OneView appliances.

When ICsp runs an OS build plan, it checks to see if you have identified HPE OneView appliances and if the OS build plan type is one of the following three types:

- HW—SAN Configuration

- HW—BIOS Configuration

- SW—Firmware

ICsp then compares the list of servers and build plans selected with the HPE OneView appliances listed to see if there is a conflict. If a conflict is found, a notification is displayed where you can cancel and make corrections or choose to force run the build plan. For details on these notifications, see the online help.

Build plans provided by HPE already have the plan type field set so no additional configuration is required. Examples of some HPE provided OS build plans that already have the plan type field set:

- **HW**—SAN Configuration plan Type is set in the HPE provided ProLiant HW—Fibre Channel

- **HBA Configure Boot Device and ProLiant HW**—Fibre Channel HBA Display Configuration Build Plans.

- **HW**—BIOS Configuration plan Type is set in the HP provided System ROM capture, System ROM enable boot from SAN (BFS) and Erase server Build Plans.

- **SW**—Firmware plan Type is set in the HPE provided Firmware Upgrade Build Plan.

If you are integrating with HPE OneView and you make custom build plans, you will need to set the plan type correctly so that ICsp will identify HPE OneView conflicts as it does with build plans provided by HPE. For build plans you create that configure BIOS settings, manage firmware updates, or configure SAN boot, specify one of these plan types listed. For information on editing a build plan, see the online help.

If one or more of the HPE OneView appliances is unavailable at the time an OS build plan is run, it is treated as a conflict with HPE OneView and the same options will be provided.

You can check the Appliances screen in the ICsp UI to see the status of the HPE OneView appliances. The status indicates whether ICsp could communicate with the HPE OneView appliance.

HPE OneView with ICsp Installation and Configuration Service

HPE OneView with ICsp Installation and Configuration Service is a basic fixed-price, fixed-scope installation and startup service. The service includes the installation of HPE OneView on a supported VMware vSphere (ESXi) or Microsoft Hyper-V host VM, the first-time setup of the appliance, and configuration and setup of all managed devices within a single BladeSystem c7000 enclosure or a single ProLiant DL rack-mounted server environment. In a blade environment, it provides configuration and setup of all managed devices within a single enclosure. In a rackmount environment, it provides configuration and setup of all managed devices within a single rack. This includes bringing the enclosure or DL server under management, updating the enclosure or server to a specified firmware baseline, defining the network configuration (networks, network sets, SAN connectivity, and so forth), and creating and assigning server profiles for the server blades or DL servers. This service also includes:

- Installation and startup of the ICsp appliance on another separate supported VMware ESXi host VM

- Installation and startup of the ProLiant media server on a separate supported ProLiant server running Windows Server

As part of this service, HPE provides the customer's organization with a test and verification session to help ensure that everything has been configured and set up properly, along with a brief customer orientation session. Service benefits are:

- Availability of an HPE service specialist to answer basic questions during the delivery of this service

- Delivery of the service at a mutually scheduled time convenient to the customer's organization

- Verification before installation that all service prerequisites are met

- Installation and configuration of HPE OneView

- A customer orientation session

Service feature highlights include:

- Service planning

- Service deployment

  - HPE OneView installation and startup

  - ICsp installation and startup

    - Installation verification tests (IVT)

    - Customer orientation session

 Note

Scan this QR code or enter the URL into your browser to read the technical paper titled *HPE OneView with Insight Control server provisioning Installation and Startup.*

http://www8.hp.com/h20195/v2/GetPDF.aspx%2F4AA5-0792ENW.pdf

Video demonstration—using ICsp for on-premises management

Watch this six-minute demonstration that shows how to use ICsp to erase and deploy an operating system to a physical server blade.

Note

To view this demonstration, scan this QR code or enter the URL into your browser.

https://youtu.be/MTXsOfx11Gw

Remote management

HPE iLO, Insight Online, and related remote support tools are available as part of an HPE warranty or support agreement. iLO is an embedded management technology that supports the complete lifecycle of all ProLiant Gen8 and Gen9 servers, from initial deployment to remote management and service alerting.

Remote management with iLO Advanced

All HPE OneView SKUs currently ship with an iLO Advanced license (except for the upgrade SKU). iLO Advanced provides remote access to server power control and event logs.

 Note

iLO Advanced remote management is only accessible to HPE OneView customers running HPE OneView Advanced.

Features used to help solve complex IT problems include:

- **Graphical remote console** turns a supported browser into a virtual desktop, giving the user full control over the display, keyboard, and mouse of the host server. The console is independent from the operating system; the console displays remote host server activities (such as shutdown or startup operations) and can be launched from the HPE OneView server profile page.

- **Shared console and console replay functions** allow up to six team members to view and share control of a single virtual KVM session. They can also capture and save screen video for later review.

- **USB-based virtual media** allows an IT administrator to boot the remote server from the client machine (or anywhere on the client's network) and execute functions remotely.

- **Integration with Microsoft Terminal Services** provides a graphical remote console when the operating system is fully loaded and available on the host system. It also offers a secure, hardware-based, lights-out console for remote access to the host server when the operating system is not operational.

- **Serial record and play back functions** save the text-based output data for later access.

- **Remote System** keeps a log of activity for later troubleshooting or reference.

 Note

After the release of HPE OneView 2.0, OneView for ProLiant DL and stand-alone SKUs were introduced. These SKUs do not include the iLO Advanced license. Customers can purchase an HPE OneView SKU, iLO Advanced SKU, or both. This provides customers more flexibility and choice of software features. The blade SKUs are not affected.

Right to use Insight Control

The purchase of HPE OneView Advanced licenses provides integrated license capability with the right to use Insight Control and aids users in their transition to HPE OneView. License keys for both HPE OneView Advanced and Insight Control are provided for use on the same system, in the same purchased quantity, but not for use at the same time. For example, Insight Control can be licensed for use for a year, and HPE OneView Advanced can be licensed for use afterward on the same system.

This integrated license capability applies to all HPE OneView Advanced license purchases except for upgrades. It is also available retroactively to previous purchasers of HPE OneView. New customers must take their entitlement order number to the HPE software licensing portal to receive their license keys. Existing HPE OneView customers will receive an email when their licenses are ready, and then they can take their Support Agreement ID (SAID) to the software updates portal to receive their license keys. An HPE Passport account is required, but a new account can be created when using the portal.

HPE OneView managed device support

The devices that HPE OneView supports depend on the license:

- With HPE **OneView Advanced** support:

  - **BladeSystem c7000 enclosures**—BladeSystem c7000 enclosures are supported but BladeSystem c3000 enclosures are not. The c7000 enclosure must be populated with at least one pair of Virtual Connect Flex-10, Flex-10/10D, FlexFabric 10 Gb, or FlexFabric-20/40 F8 modules. Virtual Connect Fibre Channel modules are also supported, but one of these aforementioned Virtual Connect module types must also be installed. In addition, Cisco Fabric Extender modules for BladeSystem are supported in the enclosure, for which HPE OneView will provide a monitoring-only service.

> **▶ Note**
>
> Other interconnect bays can be populated with modules that are not Virtual Connect modules, but these are not manageable by HPE OneView. These types of modules include:
>
> - HPE Pass-Thru Modules
>
> - Layer 2 and Layer 3 switches such as the HPE 6125 (Comware) and HPE 6120 (ProVision)
>
> - Fibre Channel switches such as Cisco MDS 9000 or Brocade

- **ProLiant BL Gen8 or Gen9 server blades and WS Gen8 or Gen9 workstation blades**—Only the Flex-10, Flex-20, and FlexFabric (LAN on motherboard [LOM], FlexibleLOM Blade [FLB], or mezzanine) server adapters in the ProLiant BL and workstation blades are supported by HPE OneView.

- **ProLiant BL G7 servers**—ProLiant BL G7 servers that are supported allow a Virtual Connect type of server profile to be assigned. These types of server profiles do not include boot order, BIOS, or firmware settings, and this is due to the lack of iLO 4 and Intelligent Provisioning features.

- **ProLiant DL Gen8 and Gen9 rack servers**—ProLiant DL580 Gen9 and DL560 Gen9 servers are supported.

- With HPE **OneView Standard** support:

  - **ProLiant DL Gen8 and Gen9 rack servers**—These rack servers are supported for HPE OneView Standard health and alert management functions. Server profile management is supported, but the connection management functions are not. The management of BIOS settings is supported only for ProLiant DL360 and DL380 models.

  - **ProLiant DL G7 servers**—These servers are supported for HPE OneView Standard monitoring only. The features supported include health and alert management functions; power, cooling, and utilization (CPU and RAM) data; launching the iLO console along with single sign-on capability; and the ability to power the server on and off.

HPE Insight Online

Figure 9-6 Insight Online

Insight Online, shown in Figure 9-6, is a cloud-based infrastructure management and support portal available through the HPE Support Center. Powered by HPE remote support technology, it provides a personalized dashboard to simplify tracking of IT operations and to view support information from anywhere at any time. It is an addition to the HPE Support Center portal for IT staff who deploy, manage, and support systems, plus HPE Authorized Channel partners who support an IT Infrastructure.

Insight Online is complementary to HPE OneView. Insight Online provides access to device and support information in a cloud-based personalized dashboard, so you can stay informed while in the office or on the go. Use the Insight Online dashboard to track service events and support cases, view device configurations, and proactively monitor HPE contracts and warranties for all devices monitored by HPE remote support tools, as well as HPE Proactive service credit balances.

No installation is required to use Insight Online. From http://hp.com/go/insightonline, enter your HPE Passport username and password for secure access, or go to the HPE Support Center and click the **Insight Online My IT Environment** tab. You can also view Insight Online from any PC, tablet, or mobile phone browser.

Choice of HPE remote support tools with Insight Online

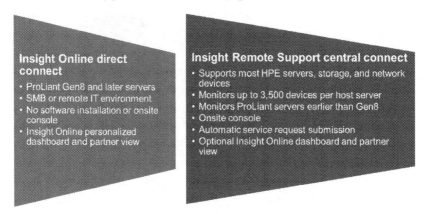

Insight Online direct connect
- ProLiant Gen8 and later servers
- SMB or remote IT environment
- No software installation or onsite console
- Insight Online personalized dashboard and partner view

Insight Remote Support central connect
- Supports most HPE servers, storage, and network devices
- Monitors up to 3,500 devices per host server
- Monitors ProLiant servers earlier than Gen8
- Onsite console
- Automatic service request submission
- Optional Insight Online dashboard and partner view

Figure 9-7 HPE remote support connection methods

To take full advantage of Insight Online, you need to install HPE remote support technologies. Remote support with Insight Online is provided by two connection methods as shown in Figure 9-7:

- **Insight Online direct connect**—Available for ProLiant Gen8 and Gen9 servers and BladeSystem enclosures. Direct connect enables these devices to automatically submit hardware failure and diagnostic information directly to HPE for analysis, case generation, and automated parts replacement. No centralized hosting device is required; instead, you use Insight Online as the online console. ProLiant Gen8 and Gen9 servers benefit from agentless remote support monitoring provided with iLO management. This method is ideal for SMB and remote sites with ProLiant Gen8 and Gen9 servers.

- **Insight Remote Support central connect**—Available for servers, storage, and networks. Using central connect, you register the device to communicate with HPE through an Insight Remote Support centralized hosting device in your local environment. This method is ideal for HPE Converged Infrastructure IT environments with multiple device types. Insight Remote Support is optimized to support up to 3500 monitored devices per hosting device and can be installed on a Windows ProLiant hosting device or a Windows virtual guest.

These solutions automatically send hardware failures and configuration information to HPE for fast, accurate diagnosis and repair. With Insight Online, all devices monitored by Insight Remote Support central connect or Insight Online direct connect can be auto-populated to the Insight Online personalized dashboard to provide 24 x 7 access to product and support information regardless of location.

Note

To download the Insight Remote Support software, scan this QR code or enter the URL into your browser.

http://www.hp.com/services/getconnected

Insight Online configuration options

Insight Online direct connect

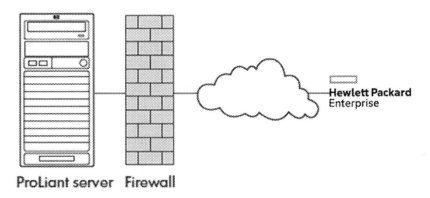

ProLiant server   Firewall

Figure 9-8 Insight Online direct connect

Figure 9-8 shows a server configured to communicate directly with Insight Online without the need to set up an Insight Remote Support centralized hosting device in the local environment. Insight Online is the primary interface for remote support information.

Insight Remote Support central connect

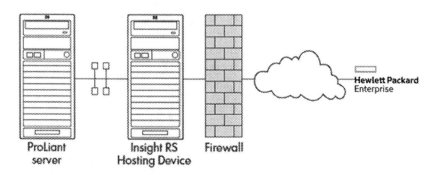

Figure 9-9 Insight Remote Support central connect

Figure 9-9 shows a server configured to communicate with HPE through an Insight Remote Support centralized hosting device in your local environment. All configuration and service event information is routed through the hosting device. This information can be viewed using the local Insight Remote Support Console or the web-based view in Insight Online (if it is enabled in Insight Remote Support).

For some IT environments, Insight Remote Support central connect might be more suitable because it offers specific capabilities that Insight Online direct connect does not:

- Monitoring of ProLiant servers earlier than Gen8

- Monitoring of HPE storage and networking products

- Monitoring of large IT environments with up to 3500 IT devices. Additional hosting servers can be deployed to monitor larger environments.

Embedded Remote Support

Embedded Insight Remote Support is included in Intelligent Provisioning, which allows you to register supported servers for HPE remote support. Insight Remote Support provides automatic submission of hardware events to HPE to prevent downtime and enable faster issue resolution.

Embedded Remote Support enables you to easily and securely export the Active Health file to an HPE Support professional to help resolve issues faster and more accurately. When you use the embedded remote support feature, choose from the two configuration options: Insight Online direct connect and Insight Remote Support central connect.

The following Insight Online direct connect device types support direct connect registration:

- ProLiant Gen8 servers

- ProLiant Gen9 servers

The following Insight Remote Support central connect types support central connect registration:

- ProLiant Gen8 servers
- ProLiant Gen9 servers

Software support for Insight Online includes:

- iLO firmware
  - To address third-party software vulnerabilities, HPE recommends using iLO 4 2.03 or later.
- ProLiant Gen8 servers
  - Version 1.40 or later is required for Insight Online direct connect registration.
  - Version 1.10 or later is required for Insight Remote Support central connect registration.
- ProLiant Gen9 servers
  - Version 2.00 or later is required for remote support registration.

Reducing time to resolution with HPE support

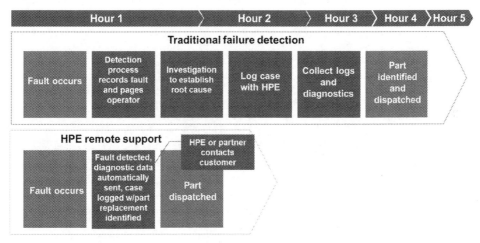

Figure 9-10 An illustration of the advantages of HPE remote support

In the first (top) scenario shown in Figure 9-10, an IT issue triggered by traditional failure detection involves a minimum sequence of six steps and as long as five hours to reach the point when the faulty part is identified and dispatched.

In the second (bottom) scenario, leveraging standard Internet security protocols and principles, each customer has a process that includes detecting the fault and alerting an operator (Short Message Service [SMS] and other messaging services are often used). HPE remote support technology automatically forwards all actionable events to HPE or an HPE partner and a case is logged. This reduces the sequence to three steps and enables the part to be identified and dispatched in approximately two hours.

When using HPE remote support technology, it is not unusual for HPE or an HPE partner to know about an IT event and begin addressing the issue before the customer IT staff is aware that the event occurred.

Target users for remote support

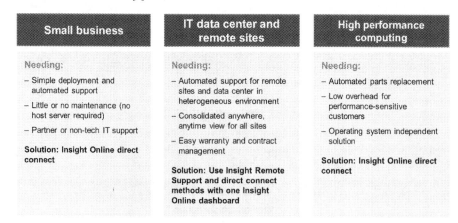

Figure 9-11 Target users for remote support

Figure 9-11 illustrates the target users for each remote support method.

Direct connect for remote support provides automated support, faster resolution, simplified contract, and warranty management. It is ideal for the following types of customers:

- Small business—Insight Online direct connect

  – Simple deployment and automated support

  – Little or no maintenance (no host server required)

  – Partner or nontech IT support

  – ProLiant Gen8 and Gen9 servers

- IT data center and remote sites—Insight Remote Support and direct connect methods with one Insight Online dashboard

  – Automated support for remote sites and data center in heterogeneous environments

  – Consolidated anywhere, anytime view for all sites

  – Easy warranty and contract management

- High performance computing (HPC)—Insight Online direct connect

    - Automated parts replacement for ProLiant Gen8 and Gen9 servers

    - Low overhead for performance-sensitive customers

    - Operating system independent solution

Insight Online with Insight Remote Support use case

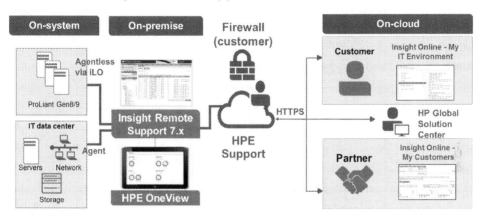

Figure 9-12 Insight Online with Insight Remote Support use case

Customers with a main data center and several remote sites can use the Insight Online dashboard in combination with Insight Remote Support. This provides automated support for remote sites and the data center in a heterogeneous environment, easy-to-use warranty and contract management, and consolidated anywhere, anytime view for all sites.

This integrates management and support software technologies into one seamless experience across the server lifecycle—from server deployment to ongoing support and continual improvement. Figure 9-12 shows how three key software products integrate:

- iLO Management provides onboard intelligence with ProLiant Gen8 and Gen9 servers

- Insight Remote Support provides remote monitoring, diagnosis, and problem resolution

- Insight Online allows customers to access the support and management information they need anytime, anywhere

Data is transmitted via a secure HTTPS connection from the customer's IT environment through Insight Remote Support and then to HPE. This data populates the Insight Online portal information. If a device requires service, the solution sends diagnostic information that automatically opens a support ticket with HPE. Customers also have the option of sharing Insight Online information with their HPE partner when appropriate.

Insight Online complements HPE OneView; together they offer a modern and integrated workspace. Using Insight Online with HPE OneView allows you to view your device and support information in a cloud-based personalized dashboard.

Activity: Touring Insight Online

In this activity, you will watch a video about Insight Online to gain a better understanding of its features and functions.

 Note

To access the video, scan this QR code or enter the URL into your browser.

http://h20621.www2.hp.com/video-gallery/us/en/products/blades/48DEA92C-
C72F-4BE0-A09A-92F071849FD5/r/video/

Be prepared to pause the video as needed to answer these questions:

1. Using Insight Online, what type of information can users access about their IT environments?

   _____

   _____

   _____

2. By clicking on a specific device, what type of additional details can users access about that device?

   _____

   _____

   _____

3. What type of information does the contracts and warranties page show?

   _____

   _____

   _____

   _____

Getting started

To get started with Insight Online and Insight Remote Support, first make sure that the following steps have been taken:

1. The Preparing for registration steps are complete.

2. The server you want to register is not in use as an Insight Remote Support hosting device—HPE does not support Insight Online direct connect registration of a server that is used as an Insight Remote Support hosting device. If you register an active hosting device for Insight Online direct connect, all of the devices that are monitored by that hosting device will be unable to communicate with HPE to receive remote support.

3. The device you want to register meets the Insight Online direct connect network requirements.

After you have confirmed that the previous steps have been completed, there are two configuration options available as previously mentioned:

- **Insight Online direct connect**—Register a server or an enclosure to communicate directly with Insight Online without the need to setup an Insight Remote Support centralized hosting device in your local environment. Insight Online is your primary interface for remote support information.

  Insight Online direct connect relies on communication between the environment and HPE to deliver support services. Be sure that the network meets the port requirements.

 Note

For information about port requirements, refer to the *HP Insight Remote Support and Insight Online Setup Guide for HP ProLiant Servers and HP BladeSystem c-Class Enclosures.* To access the guide, scan this QR code or enter the URL into your browser.

http://h20564.www2.hpe.com/hpsc/doc/public/
display?docId=c03508827&lang=en-us&cc=us

- **Insight Remote Support central connect**—Register a server or an enclosure to communicate with HPE through an Insight Remote Support centralized hosting device in your local environment. All configuration and service event information is routed through the hosting device. This information can be viewed using the local Insight Remote Support Console or the web-based view in Insight Online (if it is enabled in Insight Remote Support).

If you do not want to use the embedded remote support feature to register your servers or enclosures, you can register them in the Insight Remote Support Console. If you have many servers or enclosures to register, it is faster to discover them from the Insight Remote Support Console.

 Note

To access additional Insight Remote Support documentation, scan this QR code or enter the URL into your browser.

http://www.hpe.com/info/insightremotesupport/docs

Registering for remote support

Remote support is available at no additional cost as part of your warranty or contractual support agreement with HPE. You can configure automatic integration into the Insight Online portal when installing Insight Remote Support or subsequently by entering your HPE Passport ID and password. On login to HPE Support Center, within the My IT Environment view, you will automatically see all devices that are registered with Insight Remote Support 7.x.

If all Insight Remote Support registered devices do not appear automatically in Insight Online, it might be possible that the device was already registered to another person in your company. To see this device in your view, you need them to share or transfer ownership to yourself.

 Important

When you disable your connection to Insight Online or change to a different Passport account, the change takes time to be reflected in Insight Online. Wait until all of your devices have been removed from your current Insight Online account before connecting to a new account. For details and prerequisites, see the Getting Started with HPE Insight Online Guide for details. To access the guide, scan this QR code or enter the URL into your browser.

http://www.hp.com/go/insightremotesupport/docs

Registering for remote support direct connect

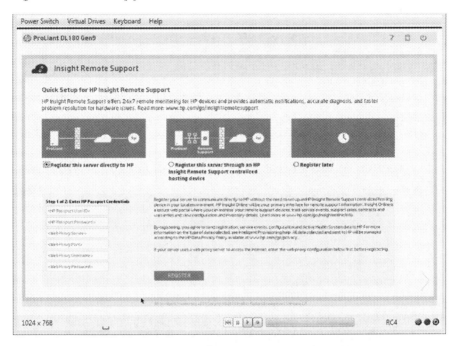

Figure 9-13 Registering for remote support direct connect

Use the option shown in Figure 9-13 to register a ProLiant server for direct connect remote support. By registering, you agree to send registration, service events, configuration, and Active Health System data to HPE. When you register for Insight Remote Support direct connect, you must complete steps in both Intelligent Provisioning and the Insight Online portal.

1. First, verify that the server meets the prerequisites for using the Insight Remote Support solution.

2. Select **Register this server directly to HPE**. The page refreshes to show the Insight Online direct connect registration options.

3. Enter your HPE Passport credentials in the HPE Passport User ID and HPE Passport Password boxes. In most cases, your Passport user ID is the same as the email address you used during the Passport registration process.

4. Optional: Enter information about the web proxy server if the ProLiant server uses a web proxy server to access the Internet.

5. Click **Register**. Clicking **Register** is Step 1 of a two-step registration process. Step 2 is completed in Insight Online. Allow up to five minutes for your registration request to be fully processed.

6. Navigate to the Insight Online website at http://www.hp.com/go/InsightOnline and then log in with your Passport credentials.

7. Follow the instructions and provide your site, contact, and partner information so that HPE can deliver service for the server. If you have multiple servers to register, complete Step 1 for all of the servers and then complete Step 2 for all of the servers during one Insight Online session.

8. Return to the Insight Remote Support page in Intelligent Provisioning, and then click **Confirm**.

9. Click the **Continue** right arrow to proceed to the Intelligent Provisioning home page (Overview).

Registering for remote support through a centralized hosting device

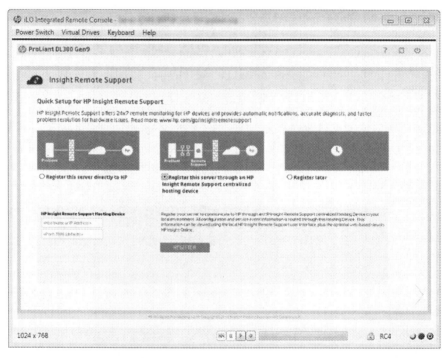

Figure 9-14 Registering for remote support through a centralized hosting device

Use the option shown in Figure 9-14 to register a ProLiant server for Insight Remote Support central connect.

1. Verify that the server meets the prerequisites for using Insight Remote Support.

2. Select **Register this server through an HPE Insight Remote Support centralized hosting device**. The page refreshes to show the Insight Remote Support central connect registration options.

3. Enter the Insight Remote Support hosting device host name or IP address and port number. The default port is 7906. Click **Register**.

4. Click the **Continue** right arrow to proceed to the Intelligent Provisioning home page (Overview).

Agentless Management Service

HPE recommends installing Agentless Management Service (AMS). AMS is one way in which iLO can obtain the server name. If iLO cannot obtain the server name, the displayed server name in Insight Online and Insight Remote Support is derived from the server serial number.

AMS is installed automatically if you use the Intelligent Provisioning Recommended installation method for Windows installation.

If you did not install AMS, ensure that the server name is displayed correctly in Insight Online and Insight Remote Support.

 Note

For more information, refer to the HPE iLO 4 User Guide, which is available by scanning the following QR code or entering the URL into your browser.

http://h10032.www1.hp.com/ctg/Manual/c03334051

Using Insight Remote Support with HPE Proactive Care service

After installing HPE remote support tools and registering with Insight Online, take advantage of the HPE Proactive Insight Experience by adopting HPE Proactive Care Support.

Proactive Care service customers must register their ProLiant servers for Insight Remote Support central connect or Insight Online direct connect in order to receive Proactive Scan and Firmware/Software Version Report and recommendations.

- The direct connect option requires the installation of AMS.

- The central connect option requires the installation of AMS or the SNMP/WBEM agents.

AMS installation is required in order to receive Proactive Scan and Firmware/Software Version Report and recommendations.

 Note

For more information about the Proactive Care service, visit the following website by scanning the QR code or entering the URL into your browser.

http://www.hp.com/services/proactivecarecentral

Data collected by Insight Remote Support

By registering for Insight Remote Support, you agree to send registration, service events, configuration, and Active Health System data to HPE.

During server registration, iLO collects data (including server model, serial number, and iLO NIC address) to uniquely identify the server hardware. When service events are recorded, iLO collects data to uniquely identify the relevant hardware component and to enable proactive advice and consulting.

This data is sent to the Insight Remote Support hosting device (Insight Remote Support central connect) or directly to HPE (Insight Online direct connect). iLO or the Insight Remote Support hosting device sends Active Health System information to HPE every seven days and sends configuration information every 30 days. All data collected and sent to HPE is used to provide remote support and quality improvement.

 Note

For information about the data that is collected, refer to the iLO 4 User Guide, which is available by scanning the following QR code or entering the URL into your browser.

http://h10032.www1.hp.com/ctg/Manual/c03334051

Insight Remote Support configuration settings

There are three communication paths within Insight Remote Support that require configuration settings for the firewall. Make sure the correct firewall settings are configured to allow communication between the components. The three paths are:

- **Configuring communication from a web browser to Insight Remote Support**—Port 7906 (HTTPS): Open port 7906 on the hosting device to access the Insight Remote Support user interface on the hosting device from other systems inside of the same network. Also make sure that Windows firewall allows the connection to port 7906.

- **Configuring communication from the hosting device to HPE**—Insight Remote Support communicates directly with the HPE data center through the firewall or web proxy server (if a web proxy server is in use). Insight Remote Support supports connecting directly to the Internet or connecting through a proxy server. Insight Remote Support does not support proxies using proxy auto-configuration scripts, NTLM (NT LAN Manager) authentication, or Kerberos authentication.

  The configuration might fail if the firewall or security software filters network communication between the hosting device and the HPE data center. For example, some firewall software, such as WatchGuard firewall, filters some HTTP protocols by default. It might block HTTP redirection, HTTP download of compressed files, and so forth. In those cases, change the firewall settings so that it does not block any HTTP communication between the hosting device and the HPE data center. Verify that it passes any HTTP standard protocol between the hosting device and the HPE data center so that it meets the communication requirement (TCP 443 outbound with established return).

  Configure the following port and alias in the firewall:

  - **Port 443 (HTTPS)**—Insight Remote Support communicates over HTTPS/443 to submit incidents to and retrieve warranty and contract information from HPE. HTTPS provides encryption for confidentiality of software configuration data collected from the hosting device and transferred to HPE. HPE recommends you configure your firewall before installing Insight Remote Support.

  - **services.isee.hp.com**—Set the firewall rules to allow access to HPE using the alias services. isee.hp.com. All data sent to HPE is through an HTTPS connection to the destination services.isee.hp.com. This destination is a virtual IP address that is automatically routed to an active server in one of the HPE Data Centers. HPE strongly recommends configuring only the alias.

- **Configuring the hosting device to use DNS**—Insight Remote Support uses redundant data centers to provide resiliency and load balancing. Global Server Load Balancing (GSLB) redirects traffic based on server load and availability. GSLB uses DNS to return the IP address of an available server. The hosting device must use DNS to communicate with the redundant sites. If you configure IP addresses in the hosts file instead of using DNS, you will lose service when the GSLB redirects traffic.

 Note

For more information, refer to the *Insight Remote Support Installation and Configuration Guide*. To access the guide, scan this QR code or enter the URL into your browser.

http://h20564.www2.hpe.com/hpsc/doc/public/display?docId=c04743806

Insight Online in the HPE Support Center

Insight Online adds two sections to HPE Support Center:

- My IT Environment, which is a custom view of the IT environment. Within My IT Environment, users can view the following:

  - Personalized dashboard

  - Device status and configurations

  - Contracts and warranty status

  - Auto-generated events tracking

  - Support cases

  - Check service credit balance

  - Proactive reports

- My Customers, where one or multiple HPE authorized partners may be allowed to see or manage designated remote support devices, as the customer chooses from Insight Remote Support. In this view, they will see:

  - Customer Grid

  - Customer dashboard

  - Contracts and warranty status

  - Monitor service events

  - Proactive reports

Insight Online: My IT Environment (customer view)

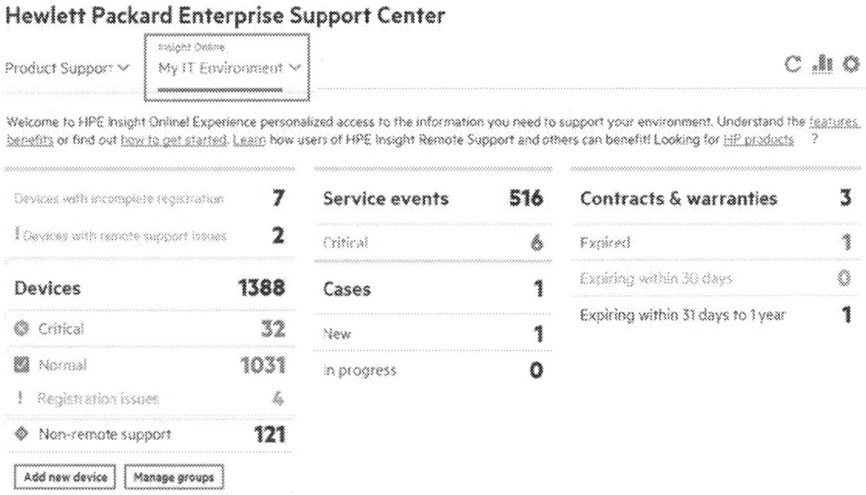

Hewlett Packard Enterprise Support Center

Product Support ∨ | Insight Online
My IT Environment ∨ | C ▁▁▪ ⚙

Welcome to HPE Insight Online! Experience personalized access to the information you need to support your environment. Understand the features, benefits or find out how to get started. Learn how users of HPE Insight Remote Support and others can benefit! Looking for HP products ?

		Service events	**516**	Contracts & warranties	**3**
Devices with incomplete registration	**7**				
Devices with remote support issues	**2**	Critical	**6**	Expired	**1**
Devices	**1388**	**Cases**	**1**	Expiring within 30 days	**0**
⚙ Critical	**32**	New	**1**	Expiring within 31 days to 1 year	**1**
▣ Normal	**1031**	In progress	**0**		
! Registration issues	**4**				
◆ Non-remote support	**121**				

Add new device | Manage groups

Figure 9-15 Insight Online login

Insight Online is an offering in the HPE Support Center portal for IT staff who deploy, manage, and support systems, as well as HPE Authorized Resellers who support IT infrastructure. Insight Online provides IT staff with the option of viewing device configuration and hardware event information in the Support Center to better support IT infrastructure.

Through the Support Center, Insight Online can automatically display devices remotely monitored by HPE. It enables IT staff to track service events and support cases easily, view device configurations, and proactively monitor HPE contracts and warranties—from anywhere at any time.

As shown in Figure 9-15, the main functional areas in Insight Online My IT Environment are:

- The personalized dashboard to monitor device health, hardware events, and contract and warranty status. Even without remote support, the dashboard can be used to view support contract status and navigate to support details.

- The devices feature, which provides a visual status of individual devices and device groups. Devices can be organized for easier management and information can be shared with other IT administrators.

- The contracts and warranties, support cases, and service credit feature areas, where HPE support can be proactively managed. Contract and warranty status can be monitored by a device as well as by a user, and HPE Proactive Select contract customers can view their service credit balances online.

Managing and organizing devices

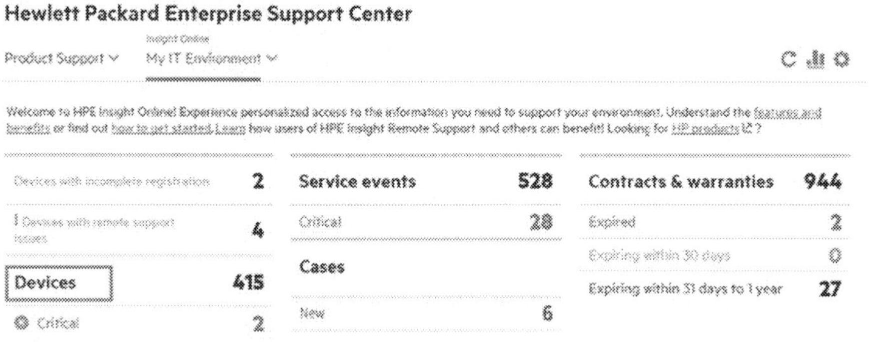

Hewlett Packard Enterprise Support Center

Figure 9-16 Devices dashboard

When you first sign in to My IT Environment, the dashboard is the initial display. The personalized device dashboard provides status and details on a specific device.

Click **Devices** on the left as highlighted in Figure 9-16 or select **Devices** from the main menu. The All Devices view is a mixed view, containing:

- Device groups shared with you by other users. These device groups may be top-level groups, child groups, or lowest-level groups, in the environment of the user who shared the group with you.

- Second-level (child) device groups you created

- Individual devices registered using the direct connect method, whose registration must be completed

- Individual devices in your default device group that you have not put into a child group

- Individual devices that you have put into a child group that you have not removed from your default device group

- Solutions

Hover your cursor over an object to display summary information. Click a summary to view details. For a device whose registration is complete, click the device to go to its individual screen.

To simplify what you see in the All Devices view, you can create child device groups, put devices into those child groups, and then remove the devices from the default device group. Then only groups consisting of child groups and groups shared with you by other users appear in this view.

Benefits include:

- Personalized, secure device information

- Easy navigation to details

- Rich configuration and service event details for devices remotely monitored by HPE

- Contract and warranty details by device

- Information about support cases

- Integrated product and support view

Managing contracts and warranties

Figure 9-17 Contracts and warranties menu in My IT Environment

Managing and tracking warranties and support agreements for the IT environment are often time-consuming tasks, especially as devices are added. With Insight Online, customers and HPE partners can more easily and efficiently manage HPE warranty and support agreements, monitor support status, view contract and service-level details, share entitlements, and change ownership. Figure 9-17 shows the Insight Online contracts and warranties menu and illustrates the range of activities that can be performed.

Contracts and warranties for devices monitored by HPE remote support are automatically added to Insight Online through remote support registration. In addition, contracts and warranties can be manually added individually or in a bulk upload from a spreadsheet.

Linking a contract or warranty

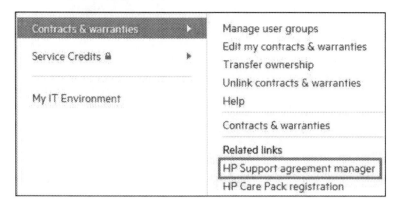

Figure 9-18 Linking a contract or warranty

To enable devices to be registered for remote support in Insight Online, you must link a contract or a warranty that covers those devices to your Passport ID. If you sign in to Insight Online using the same Passport ID used to register the device, you do not have to link your ID to the contracts and warranties that cover the remote support devices. They are linked to your ID automatically and appear as a derived contract and warranty share. Figure 9-18 highlights the link to the HPE Support Agreement manager, which can be used to manage support agreements.

Because contracts and warranties are associated with the devices they cover, the link between your HPE Passport ID and a contract or a warranty provides access to information about the devices covered by the contract or the warranty. This information is available to you from the Insight Online devices area in the GUI.

When you sign in to Insight Online using your HPE Passport ID, devices that you registered remotely using that ID and their associated contracts and warranties are visible to you. These contracts and warranties were derived and linked to your HPE Passport ID automatically.

 Note

Contracts and warranties include support agreements, Care Packs, and warranties. Contracts and warranties are managed through the HPE Support Case Manager website. To access the website, scan this QR code or enter the URL into your browser.

http://h20566.www2.hpe.com/portal/site/hpsc/public/scm/home?ac.
admitted=1458963893801.125225703.1938120508

Automated case management and parts replacement

Figure 9-19 Automated case management and parts replacement

Support cases are associated with devices and with your HPE Passport Account. Figure 9-19 shows the Cases link in the My IT Environment section of Insight Online. The following are actions that you can take when creating, updating, or investigating a support case:

- **View support cases**—Sign in to Insight Online to view support cases from several places. You can search for an active case with its case ID.

- **View a case details page**—Click a case ID in the list to reach the case details page, which is part of the Support Case Manager (SCM). A case details page lists a description of each detail of the case:

  - Case ID

  - Case title

  - HPE Support Contract

  - Source

  - Type

  - Case Status

  - Submitted

  - Customer Tracking Number

  - Schedule and Parts Information

 Note

On the case details page you can find information by clicking the case details and history or service and parts tabs.

- **Update a case**—From the list of support cases, click the ID of the case you want to update to bring up the case details page. Click the **Case details and history** tab. You can add comments, email attachments, or request to close the case. Click **Update** after you have made your changes. A request to close a case does not automatically close the case. An HPE support agent will close the case upon review of the request.

- **Submit a case**—Submit a new case using the **Submit case** button in the case section of the dashboard. If not already populated, enter a serial number or contract ID in the *contract or* the *warranty* field: Serial number. If the serial number is valid and the warranty is active, a case details page appears. (The serial number does not have to be linked to your profile.) Enter the case information and submit.

  If you submit a case within the contract's coverage hours, the first available support agent will respond. For support outside of the contract's coverage hours, submit the case by phone and say that you need an "uplift" to your support contract. This will result in an additional charge. Uplift is not available in some countries. For onsite support, contract type determines onsite response time. Onsite response time refers to the time when an engineer will be onsite to address the problem, not problem resolution time.

Contract IDs (HPE Support Agreement ID, Service Agreement ID, Support Account Reference, nickname) might be used. If the contract identifier is valid, a list of products associated with the contract appears. Select the product and submit to reach the case details page.

Additional actions you can take regarding a support case are:

- Provide case information

- Provide troubleshooting information

- Add comments to a case

- Close a case

- View case report

Service events

The service events feature provides online monitoring and management of devices using Insight Remote Support. Users can view all key hardware event information, such as severity, problem description, date and time generated, status, and related support case ID.

This feature lets IT staff monitor hardware service events and related support case details from any-where, anytime. Hardware service events and support cases are automatically generated and sent to HPE or an Authorized Partner for faster problem resolution. The service event remains in an active state until the corresponding support case is closed or is closed if a support case is not required.

Insight Online reports

Figure 9-20 Insight Online New report screen

You can click the **Reports** icon on any page in the My IT Environment, and you can create a new report by clicking **New report** as shown in Figure 9-20. The reporting feature in Insight Online allows you to:

- Create new reports and save templates for re-use

- Save the selections you made when creating the reports into a template for reuse

- View and download generated reports

- Maintain a list of your existing reports

Report types in Insight Online My IT Environment include:

- HPE Care Pack–Proactive Care (available with a Proactive Care Service contract)

- Contract

- Contracts and warranties

- Device configuration

- Firmware

- Service event

- Support case

Viewing reports

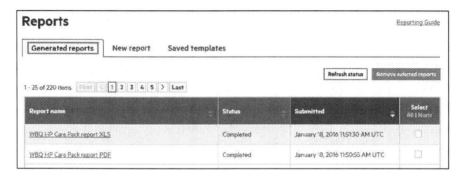

Figure 9-21 Viewing reports

Click the **Reports** icon on any My IT Environment page. **Generated reports** is the default tab and lists your reports, an example of which is shown in Figure 9-21. The reports list consists of these fields:

- Report name

- Status: In Progress, Completed, Failed, or Retry

- Submitted—Date and time you generated the report

- Select check box

Use the buttons on the page to refresh the report status or remove all checked reports from the list.

Sort the list by clicking any of the headings. Click the heading again to sort in reverse order.

Click a report name to view the report. You can view the report directly or download it. You can only view completed reports.

You can also access reports from your email. When creating a report, enter your email address to receive a notification when the report completes, and then click the link in the email notification.

 Important

Report data is not in real time and has a 24-hour lag to complete. Report data might appear blank due to daily updates to the database

For best results, use PDF format for printing:

1. Click the report name in the list. The File Download message box appears.

2. Click **Open** to view the report.

3. In the PDF viewer, click **Print**.

To remove a report, complete the following steps:

1. Click **Generated** reports.

2. In the *All generated reports* section, locate the report you want to remove.

3. In the Remove column, select the check box for the report.

4. Click **Remove all checked.**

The report is permanently removed from the All generated reports section.

HPE Proactive Care

Proactive Care Service is a flexible, comprehensive, and cost-effective service that combines smart technology and support to boost performance. HPE works in partnership with the customer to provide proactive consultation, recommendations, and reporting from HPE technology experts—as well as rapid expert support if needed.

Gain access to advanced HPE technical experts and get connected to anticipate change and increase agility. The HPE support solution is structured on three guiding principles for service delivery:

- **Personalized**—Customers have access to advanced technical expertise for rapid problem diagnosis and resolution with a premium call experience, where a Technical Solution Specialist acts as a single point of contact for end-to-end case ownership.

- **Proactive**—Proactive Care leverages industry-leading remote support technology for real-time monitoring, alerts to diagnose and fix issues early, and call logging for reactive support—all of which help avoid downtime. This also facilitates analysis, which HPE uses to deliver proactive reports with firmware and patch analysis and recommendations, as well as proactive scan health check reporting on the customer's infrastructure.

- **Simplified**—Proactive Care offers a single point of contact to manage support cases from end to end and an integrated set of support deliverables at one of three service support levels selected by the customer. Proactive Care can be purchased at any time (when customer purchases HPE products or contractually afterwards) and can cover the entire stack.

Proactive Care helps prevent problems and stabilize IT, by using products connected to HPE, for secure, real-time, analytics, tailored reports with analysis and recommendations.

My IT Environment mobile dashboard

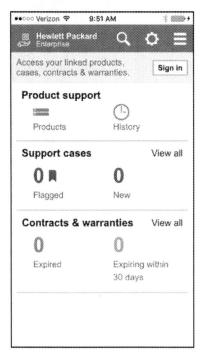

Figure 9-22 My IT Environment mobile dashboard

As shown in Figure 9-22, Insight Online provides a mobile dashboard for monitoring when you are traveling or away from a computer. The HPE Support Center Mobile App is geared toward consuming or providing information that requires attention. The app is:

- **Simplified**—Easy to install and use while on the go, anytime, anywhere, in a secure manner

- **Personalized**—Dashboard with click-down details for Contracts, Warranties, Configuration and Support Case status, and Customizable IT hardware configuration views

- **Connected**—24 x 7 remote monitoring with proactive notification; automated fault detection, case status and break fix service (with Proactive Care Service)

  – Automated case creation and problem resolution for HPE servers, storage and networking

- **Available**—For Apple iOS (version 6.1 or later) and Google Android (version 2.3 or later) devices

- **Available**—From iTunes and Android Play Store

Note

For more information on the HPE Support Center Mobile App, scan the QR code or enter the URL into your browser.

http://www8.hp.com/us/en/business-services/it-services.html?
compURI=1078323#.VunwUxhRYpl

Insight Online: My Customer (partner view)

Figure 9-23 Personalized partner dashboard

HPE Authorized Channel partners can view asset information shared to them by their customers in a separate "My Customers" section in Insight Online. The My Customer view of Insight Online for HPE Authorized Partners, as shown in Figure 9-23, provides:

- A personalized dashboard to monitor device health, hardware events, and contract and warranty status for customer's devices, with easy-to-use navigation to detailed information.

- The devices feature, which presents a visual status of individual devices and device groups. Devices can be organized for easier management, and information can be shared with colleagues.

- The service event, which includes details regarding hardware failure and part replacement information for faster and accurate problem resolution.

- A reporting feature for device lifecycle management to assist with identifying product and support upsell opportunities within the customer account.

The My Customer dashboard enables you to save time and resources with automation and simplified information access. Benefits of the My Customer dashboard include:

- Round-the-clock monitoring
- Easy all-in-one access; secure, personalized dashboard
- Automated cases and spare parts sent to the site
- Simplified warranty, contract, and case management

Learning check

1. What are some of the benefits of using ICsp? (List four benefits.)

_____

_____

_____

_____

_____

2. What are some of the key functions of OS build plans? (List four functions.)

_____

_____

_____

_____

_____

3. What are some of the features of the My IT Environment dashboard in Insight Online? (List at least four features.)

_____

_____

_____

_____

_____

4. Who are target users of Insight Online direct connect for remote support? (Select two.)

    A. SMBs needing simple deployment and automated support

    B. IT data center and remote sites needing automated support for remote sites and data centers in heterogeneous environments

    C. HPC customers needing automated parts replacement for ProLiant Gen8 and Gen9 servers

    D. SMBs needing site-to-site data replication services

    E. IT data center and remote sites needing high-availability clustering

5. Which key features does the HPE integrated management and support experience rely on?

_____

_____

_____

_____

Learning check answers

1. What are some of the benefits of using ICsp? (List four benefits.)

   – **Helps customers simplify and reduce the time associated with server provisioning tasks**

   – **Enables an IT team to adapt to changing business demands quickly and easily**

   – **Increases server provisioning speed**

   – **Reduces unplanned downtime**

   – **Optimizes data center capacity**

   – **Reduces system admin expenses and travel costs**

2. What are some of the key functions of Insight Control OS build plans? (List four.)

   – **Deploy operating systems**

   – **Configure a target server's hardware**

   – **Capture a target server's hardware configuration so that the same configuration can later be applied to other servers**

   – **Update the firmware on a target server**

   – **Install software on a target server with a running operating system**

   – **Add scripts to perform additional tasks on the target server after it has been installed**

3. What are some of the features of the My IT Environment dashboard in Insight Online? (List at least four.)

   – **Personalized dashboard**

   – **Device status, configurations**

   – **Contracts and warranty status**

- **Auto-generated events tracking**

- **Support cases**

- **Check service credit balance**

- **Proactive reports**

4. Who are target users of Insight Online direct connect for remote support? (Select two.)

    A. SMBs needing simple deployment and automated support

    B. IT data center and remote sites needing automated support for remote sites and data centers in heterogeneous environments

    C. HPC customers needing automated parts replacement for ProLiant Gen8 and Gen9 servers

    D. SMBs needing site-to-site data replication services

    E. IT data center and remote sites needing high-availability clustering

5. Which key features does the HPE integrated management and support experience rely on?

    - **OS build plans are used to deploy operating systems, configure hardware, and update firmware**

    - **iLO provides onboard intelligence**

    - **Insight Remote Support provides remote monitoring, diagnostics, and problem resolution**

    - **Insight Online allows customers to access the support and management information they need anytime, anywhere**

Summary

- ICsp enables multi-server operating system and firmware provisioning to rack-mount and BladeSystem servers. ICsp automates the process of deploying and provisioning server software, enabling an IT team to adapt to changing business demands quickly and easily.

- HPE iLO, Insight Online, and related remote support tools are available as part of an HPE warranty or support agreement. HPE iLO is an embedded management technology that supports the complete lifecycle of all ProLiant Gen8 and Gen9 servers, from initial deployment to remote management and service alerting.

- HPE Insight Online is a cloud-based infrastructure management and support portal available through the HPE Support Center. Powered by HPE remote support technology, it provides a personalized dashboard to simplify tracking of IT operations and to view support information from anywhere at any time. Remote support with Insight Online is provided by two connection methods:

- Insight Online direct connect

- Insight Remote Support central connect

- Insight Online adds two sections to HPE Support Center:

  - My IT Environment, which is a custom view of the IT environment.

  - My Customers, where one or multiple HPE authorized partners may be allowed to see or manage designated remote support devices, as the customer chooses from Insight Remote Support.

10 Practice Test

INTRODUCTION

The HPE ASE Architecting Server Solutions certification is considered an advanced-level certification.

This certification targets solution architects, including the following job titles:

- ✓ Presales architects
- ✓ Consultants
- ✓ Presales engineers
- ✓ Systems engineers

The intent of this book is to set expectations about the context of the exam and to help candidates prepare for it. Recommended training to prepare for this exam can be found at the HPE Certification and Learning website (http://certification-learning.hpe.com), as well as in books like this one. It is important to note that although training is recommended for exam preparation, successful completion of the training alone does not guarantee that you will pass the exam. In addition to training, exam items are based on knowledge gained from on-the-job experience and application, as well as other supplemental reference material that might be specified in this guide.

MINIMUM QUALIFICATIONS

The ASE Architecting Server Solutions certification is targeted at the ASE skill level and is the next progressive step after a candidate has achieved HPE ATP—Building Server Solutions certification.

Although anyone may take the exam, it is recommended that candidates have a minimum of two years' experience with architecting HPE server solutions. Candidates are expected to have industry-standard server technology knowledge from training or hands-on experience.

CANDIDATE PROFILE

The ASE Architecting Server Solutions certification is ideal for individuals who perform planning and design tasks for HPE ProLiant servers. Architects use strong business skills and technical skills to identify customer requirements and recommend appropriate technical solutions. The architect performs demonstrations of equipment and tools for fellow employees and customers. The architect advises integration staff on design proposals. Architect responsibilities include:

- Design and consult on HPE solutions

- Use business and technical skills to assess customer requirements

- Design tangible metrics-driven IT business solutions based on a customer's needs, leveraging the HPE broad portfolio of products and solutions

- Explain how technology implementation will meet achievable business metrics and objectives

- Provide a competent working understanding of HPE's broad portfolio of products with the ability to demonstrate solutions to fellow employees and customers

Who should take this exam?

This certification exam is designed for candidates with at least two years' experience in a presales consultative role who want to acquire the HPE ASE-Architecting HPE Server Solutions certification.

Alternatively, if eligible, you can take the HPE0-S47 delta exam instead of HPE0-S46 (HPE Server Solutions).

Exam details

The following are details about the exam:

- **Exam ID:** HPE0-S46

- **Exam Type:** Proctored exam taken at dedicated testing center

- **Number of items**: 60

- **Item types**: Input text, input numbers, matching, multiple choice (single-response), multiple choice (multiple-response), point and click

- **Exam time**: 1 hour 30 minutes

- **Passing score**: 70%

- **Reference material**: No online or hard copy reference material will be allowed at the testing site.

HPE0-S46 testing objectives

24%—Foundational server architectures and technologies

- Differentiate between processor classes and types to provide design guidance based on customer needs.

- – Describe IO accelerator technologies.

- – Describe and explain networking technologies.

- – Identify storage technologies.

- – Explain server management technology features and their functionality.

- – Propose high availability and disaster recovery solutions to meet the customer's business requirements.

- – Differentiate between scale-out and scale-up benefits and purpose.

- – Differentiate current server operating system and virtualization solutions.

- – Determine an appropriate plan for data center components based on industry best practices and standards.

33%—Functions, features, and benefits of HPE server products and solutions

- – Differentiate and explain the HPE server product offerings, architectures, and options.

- – Locate and describe HPE health and fault technologies.

- – Propose HPE data center rack and power infrastructure solutions based on site conditions and requirements.

- – Given a use case, propose appropriate HPE server IO connectivity options.

- – Given a customer environment scenario, propose which HPE management tools optimize administrative operations.

- – Describe the HPE standard warranties for server solutions and options.

16%—Analyzing the server market and positioning HPE server solutions to customers

- – Compare and contrast the HPE Server solution marketplace.

- – Compare and contrast how HPE server solutions provide competitive advantage and add value.

27%—Planning and designing HPE server solutions

- – Given customer requirements and constraints, determine information needed to understand the customer's needs.

- – Explain concepts of designing, sizing, and validating the solution.

- – Interpret customer requirements and integrate them into an HPE solution.

Test preparation questions and answers

The following questions will help you measure your understanding of the material presented in this book. Read all of the choices carefully, because there might be more than one correct answer. Choose all correct answers for each question.

Questions

1. Which applications can take full advantage of Intel Xeon Phi coprocessors? (Select two.)

   a. Those that take advantage of CPU cores running at extreme speeds

   b. Those that scale well to more than one hundred threads

   c. Those that are single-threaded

   d. Those that take advantage of large CPU cores

   e. Those that make extensive use of vectors

2. What is a feature of HPE Advanced Memory Error Detection?

   a. Analysis of correctable errors to determine which ones are likely to lead to uncorrectable errors in the future

   b. Counting and storing correctable errors in a database for subsequent transmission to an HPE support center

   c. Detection of potential data corruption as a result of L1 and L2 cache errors

   d. Monitoring of data at rest and in transit in storage systems using SSD technologies

3. Which memory technology is ideal for customers who require maximum memory capacity in ProLiant Gen9 servers?

   a. RDIMM

   b. LRDIMM

   c. UDIMM

   d. NVDIMM

4. Which storage technology provides the minimum latency?

   a. PCie SSD

   b. CPU cache

   c. DRAM

   d. MRAM

5. How does a Nonvolatile Memory Express (NVMe) interface reduce the latency and power consumption of flash-based storage devices?

   a. It eliminates the need for an HBA to connect directly to devices through the PCIe bus.

   b. It uses a dedicated low-power SAS HBA in an x16 PCIe slot.

   c. It bypasses the need for bad block relocation by automatically storing two copies of each data block.

   d. It uses advanced double chip-sparing technology to correct multi-bit errors.

6. How does HPE DDR4 Smart Memory provide lower latency than DDR3 RAM?

   a. It uses 16 banks of memory per rank.

   b. It uses rank multiplication to enable four ranks of memory on LRDIMMs.

   c. It operates at a maximum data rate of 1866 MT/s.

   d. It uses a linear error-correcting code that encodes four bits of data into seven bits.

7. A customer has a need for storage devices with latency of less than one millisecond. What should you recommend? (Select two.)

   a. SAS SSDs

   b. SATA SSDs

   c. PCIe SSDs

   d. NVDIMMs

   e. SAS NAND flash

8. How does NVDIMM technology protect data during a power outage?

   a. A controller moves data from the DRAM to a workload accelerator.

   b. Power is maintained using an array of supercaps.

   c. A controller moves data from the DRAM to its own onboard flash memory.

   d. NVDIMMs only use persistent memory and are immune to loss of power.

9. How does the HPE Dynamic Smart Array controller reduce the cost of protecting data?

   a. It uses low cost flash-based cache memory.

   b. It only requires a single supercap to protect data.

   c. It provides the maximum performance at the lowest cost for IO intensive workloads.

   d. It replaces many hardware RAID functions with software.

10. What are elements of HPE SmartCache technology? (Select three.)

    a. Bulk storage

    b. Battery backup

    c. NVDIMM

    d. Accelerator

    e. Metadata

11. What are advantages of HPE FlexFabric RoCE-capable adapters? (Select two.)

    a. Latency is reduced by bypassing the adapter's RAID engine.

    b. Data is protected with write-through and write-back caching.

    c. Latency is reduced by bypassing the host TCP/IP stack.

    d. Data at rest is protected using secure encryption.

    e. Server CPU utilization is reduced.

12. An SMB customer has a need for an entry-level rack server that supports two Intel Xeon processors in a 1U form factor. Which ProLiant Gen9 server should you recommend?

    a. DL20

    b. DL60

    c. DL80

    d. DL120

13. An enterprise customer has a need for a rack server in a 1U form factor with support for NVDIMM persistent memory. Which ProLiant Gen9 server should you recommend?

    a. DL160

    b. DL180

    c. DL360

    d. DL380

14. An enterprise customer has a need for a server that supports four processors and four power supplies to maximize redundancy and reduce unplanned downtime. Which ProLiant Gen9 server should you recommend?

    a. DL360

    b. DL380

    c. DL560

    d. DL580

15. A customer has a need for a premium two-port 12Gb SAS storage controller to protect the internal hard drives in a ProLiant Gen9 rack server. Which Smart Array controller should you recommend?

    a. P240

    b. P741m

    c. P840ar

    d. P841

16. A customer needs to protect internal SATA storage in ProLiant Gen9 rack servers and has requested an embedded controller. Which Smart Array controller should you recommend?

    a. B140i

    b. P240

    c. P840ar

    d. P841

17. How does RAID 10 ADM protect data?

    a. It uses three drives to create redundant copies of data.

    b. It stripes data across two or more sets of RAID 1 ADM volumes.

    c. It stripes data across two or more sets of RAID 0 volumes.

    d. It stripes data across two or more sets of RAID 6 volumes.

18. What is the maximum server inlet air temperature for select ProLiant Gen9 servers with support for ASHRAE A4?

    a. 30°C

    b. 35°C

    c. 40°C

    d. 45°C

19. What are functions of HPE Virtual Connect? (Select two.)

    a. Acts as an edge port aggregator

    b. Server changes affect the network

    c. Uplink ports are termination ports

    d. Is part of the LAN

    e. Is managed as a switch

20. What is an IRF domain?

    a. A high-availability server clustering solution

    b. A method for encapsulating OSI Layer 2 Ethernet frames within Layer 4 UDP packets

    c. A virtualized switch consisting of multiple physical switches

    d. A server system in which memory access time depends on the memory location relative to the processor

21. Which uplinks and downlinks does the Virtual Connect FlexFabric 20/40 F8 module provide?

    a. 20 Gb downlinks and 4 x 8 Gb uplinks

    b. 40 Gb downlinks and 8 Gb uplinks

    c. 20 Gb downlinks and 40 Gb uplinks

    d. 40 Gb downlinks and 20 Gb uplinks

22. What is a FlexNIC?

    a. A virtualized switch consisting of multiple physical switches

    b. A virtual NIC contained in a software layer

    c. A PCIe function with its own NIC driver instance

    d. A pair of NICs bonded within an operating system

23. Which technology provides the dual-hop FCoE feature in Virtual Connect?

    a. FIP snooping

    b. IGMP snooping

    c. DHCP snooping

    d. TCP snooping

24. A customer is using HPE FlexFabric 20Gb 2-port 650FLB adapters in their server blades and needs a switching solution to support the RoCE capabilities of the adapters. Which solution should you recommend?

    a. FlexFabric 10/10D

    b. FlexFabric 20/40 F8

    c. HPE 6125XLG

    d. HPE 6125G/XG

25. Which physical function (PF) of a FlexFabric adapter can be configured as Ethernet, iSCSI, or FCoE?

    a. PF1

    b. PF2

    c. PF3

    d. PF4

26. What are advantages of active optical cables (AOCs) when compared with copper cables? (Select three.)

    a. Increased immunity to EMI

    b. QSFP+ connectors are purchased separately

    c. Low cost

    d. Large size for increased physical durability

    e. Increased cable lengths

27. An HPE BladeSystem PCI Expansion Blade installed in Bay 1 of a c7000 chassis provides PCI expansion slots to which server blades?

    a. Bay 2

    b. Bay 3

    c. Bay 9

    d. All even-numbered bays

28. What is the maximum number of Active Cool fans that can be installed in a BladeSystem c7000 enclosure?

    a. 6

    b. 8

c. 10

d. 12

29. What is the maximum number of HPE ProLiant XL190r Gen9 server nodes that can be installed in a single HPE Apollo 2000 system?

a. One

b. Two

c. Four

d. Eight

30. A customer is looking for a new two-server clustered storage solution for their Lustre parallel file system workload. Which HPE Apollo system should you recommend?

a. Apollo 4510

b. Apollo 4520

c. Apollo 4530

d. Apollo 4540

31. What is the purpose of the Apollo 8000 primary water loop?

a. It channels water from the gravity drain collection system to the water collection pan.

b. It isolates the heat exchanger from the water collection pan using a bleeder valve.

c. It distributes water from the iCDU to the f8000 rack where it flows through the heat exchanger.

d. It circulates facility water through the cooling distribution unit located within the iCDU rack.

32. What are key elements of a composable infrastructure? (Select three.)

a. Software-defined intelligence

b. Hardware-defined intelligence

c. Autonomic storage systems

d. Fluid resource pools

e. High-availability clustering

f. A unified API

33. Which HPE Synergy component provides a resource information control point and reports asset and inventory information about all the devices in the frame?

    a. Composer

    b. Frame link module

    c. Image streamer

    d. 12000 frame

34. Put the five CI-CM stages in the correct order starting with stage one.

    a. Service enabled

    b. Silo

    c. Dynamic reconfigurable services

    d. hybrid delivery

    e. Optimized

35. What IO bandwidth does HPE recommend for each VM in a virtualized environment?

    a. 50 Mb/s IP and 100 Mb/s storage

    b. 100 Mb/s IP and 100 Mb/s storage

    c. 200 Mb/s IP and 100 Mb/s storage

    d. 100 Mb/s IP and 200 Mb/s storage

36. How does ROI measure the value of a particular proposed solution?

    a. It demonstrates TCO over a one-, three-, or five-year period.

    b. It highlights the costs associated with unplanned downtime.

    c. It compares CAPEX and OPEX savings over a particular time period.

    d. It compares the investment required with the resulting cost reduction or increased revenue.

37. What is the primary purpose of a scope of work?

    a. It documents the final project overview and is used for billing purposes.

    b. It captures the plan, time frame, required resources, and completion milestones of a project.

c. It outlines the support services required to achieve the customer-agreed SLAs.

d. It is sent by customers interested in acquiring a solution to potential suppliers who then submit business proposals.

38. When used to provision systems, ICsp compares the selected build plans and servers with the known HPE OneView appliances to see if there is a conflict. Which OS build plans does this apply to? (Select three.)

a. HW—SAN Configuration

b. HW—Clear UEFI Boot Menu

c. HW—Erase Server

d. HW—BIOS Configuration

e. SW—Firmware

f. SW—Validate Media Server Settings

39. The HPE OneView with ICsp Installation and Configuration Service includes the installation of HPE OneView on which hypervisor hosts? (Select two).

a. VMware vSphere (ESXi)

b. Citrix Xen

c. Canonical Ubuntu

d. KVM

e. Microsoft Hyper-V

40. What does HPE Insight Online provide?

a. A virtual desktop, giving the user full control over the display, keyboard, and mouse of the host server

b. Access to device and support information in a cloud-based personalized dashboard

c. Shared console and console replay functions

d. Access to a tool for configuring and pricing HPE servers and storage systems

41. Which environment is ideal for Insight Remote Support central connect?

a. Small to mid-sized businesses

b. Remote sites with ProLiant Gen8 and Gen9 servers

c. Customers who do not want a centralized hosting device in the local environment

d. HPE Converged Infrastructure environments with multiple device types

42. Which report types are included in Insight Online My IT Environment? (Select three.)

a. Contracts and warranties

b. Device configuration

c. Service events

d. Compliance

e. Application availability

Answers

1. ☑ **B** and **E** are correct. To take full advantage of Phi coprocessors, an application must scale well to more than one hundred threads and make extensive use of vectors.

   ☒ **A, C,** and **D** are incorrect. The speed of Phi coprocessors is lower than that of Xeon processors, and the high degree of parallelism compensates for the lower speed of each individual core. Single-threaded applications and those that take advantage of large CPU cores do not take full advantage of Phi coprocessors.

   For more information, see Chapter 3.

2. ☑ **A** is correct. HPE Advanced Memory Error Detection analyzes correctable errors to determine which ones are likely to lead to uncorrectable errors in the future.

   ☒ **B, C,** and **D** are incorrect. HPE Advanced Memory Error Detection does not count and store errors in a database for transmission to HPE, detect cache errors, or monitor data in SSD storage.

   For more information, see Chapter 3.

3. ☑ **B** is correct. LRDIMM memory technology is ideal for customers who require maximum memory capacity in ProLiant Gen9 servers.

   ☒ **A, C,** and **D** are incorrect. RDIMM, UDIMM, and NVDIMM do not enable the maximum amount of memory to be installed in ProLiant Gen9 servers.

   For more information, see Chapter 3.

4. ☑ **B** is correct. CPU cache shares processor real estate with the CPU cores and has the lowest latency of all memory technologies (except processor registers).

   ☒ **A, C,** and **D** are incorrect. PCi-e SSD, DRAM, and MRAM do not provide the minimum latency.

   For more information, see Chapter 3.

5. ☑ **A** is correct. NVMe interfaces eliminate the need for an HBA to connect directly to devices through the PCIe bus. Not needing an HBA means less consumed power and lower latency.

   ☒ **B, C,** and **D** are incorrect. NVMe interfaces do not use a dedicated low-power SAS HBA in an x16 PCIe slot, bypass the need for bad block relocation by automatically storing two copies of each data block, or use advanced double chip-sparing technology to correct multibit errors.

   For more information, see Chapter 3.

6. ☑ **A** is correct. DDR4 memory has 16 banks of memory in a DRAM chip compared with the eight banks in DDR3; this helps to reduce memory latency.

   ☒ **B, C,** and **D** are incorrect. Rank multiplication is used by DDR3, not DDR4 memory. DDR4 memory operates at up to 2133 MT/s bandwidth. DDR4 memory does not use a linear error-correcting code that encodes four bits of data into seven bits.

   For more information, see Chapter 3.

7. ☑ **C** and **D** are correct. The latency of PCIe SSDs is 100s of microseconds; the latency of NVDIMMs is in the region of nanoseconds.

   ☒ **A, B,** and **E** are incorrect. SAS/SATA SSDs and SAS NAND flash have latency of tens of milliseconds.

   For more information, see Chapter 3.

8. ☑ **C** is correct. A controller on the NVDIMM moves data from the DRAM to its own onboard flash memory.

   ☒ **A, B,** and **D** are incorrect. Data is not moved from DRAM to a workload accelerator, supercaps are not used, and NVDIMMs use DRAM and persistent memory.

   For more information, see Chapter 3.

9. ☑ **D** is correct. The Dynamic Smart Array controller eliminates most of the hardware RAID controller components and relocates the RAID algorithms from a hardware-based controller into device driver software.

   ☒ **A, B,** and **C** are incorrect. The Dynamic Smart Array controller does not use flash-based cache memory or supercaps. It is ideal for applications that do not generate massive IO workload.

   For more information, see Chapter 3.

10. ☑ **A**, **D,** and **E** are correct. The three elements of SmartCache technology are bulk storage, accelerator, and metadata.

    ☒ **B** and **C** are incorrect. SmartCache technology does not include battery backup or NVDIMM.

    For more information, see Chapter 3.

11. ☑ **C** and **E** are correct. HPE FlexFabric RoCE-capable adapters reduce latency by bypassing the host TCP/IP stack and reduce server CPU utilization by using the direct memory access engine on the adapter.

    ☒ **A, B,** and **D** are incorrect. HPE FlexFabric RoCE-capable adapters do not bypass the adapter's RAID engine, use write-through or write-back caching, or protect data using encryption.

    For more information, see Chapter 3.

12. ☑ **B** is correct. The ProLiant DL60 Gen9 is a 1U rack server that supports two Xeon processors.

     ☒ **A, C,** and **D** are incorrect. The ProLiant DL20 Gen9 supports only one processor. The ProLiant DL80 Gen9 is a 2U server. The ProLiant DL120 Gen9 supports only one processor.

     For more information, see Chapter 4.

13. ☑ **C** is correct. The ProLiant DL360 Gen9 is a 1U rack server with support for NVDIMM persistent memory.

     ☒ **A, B,** and **D** are incorrect. The ProLiant DL160 Gen9 and DL180 Gen9 do not support NVDIMM memory. The ProLiant DL380 Gen9 supports NVDIMM memory, but it is a 2U rack server.

     For more information, see Chapter 4.

14. ☑ **D** is correct. The ProLiant DL580 Gen9 server supports four processors and four power supplies.

     ☒ **A, B,** and **C** are incorrect. The ProLiant DL360 Gen9 and DL380 Gen9 support up to two processors. The ProLiant DL560 Gen9 supports up to two power supplies.

     For more information, see Chapter 4.

15. ☑ **C** is correct. The Smart Array P854ar is a premium two-port internal 12 Gb SAS controller for ProLiant rack servers.

     ☒ **A, B,** and **D** are incorrect. The Smart Array P240 is a base/entry-level 12 Gb SAS controller. The Smart Array P741m is a mezzanine controller for server blades. The Smart Array P841 supports external hard drives.

     For more information, see Chapter 4.

16. ☑ **A** is correct. The Smart Array B140i is an embedded SATA controller.

     ☒ **B, C,** and **D** are incorrect. The Smart Array P240, P840ar, and P840 are not embedded controllers.

     For more information, see Chapter 4.

17. ☑ **B** is correct. RAID 10 ADM stripes data across two or more sets of RAID 1 ADM volumes.

     ☒ **A, C,** and **D** are incorrect. RAID 1 ADM creates redundant copies of data using three drives. RAID 10 ADM does not stripe data across two or more sets of RAID 0 or RAID 6 volumes.

     For more information, see Chapter 4.

18. ☑ **D** is correct. For select ProLiant Gen9 platforms, 45°C (ASHRAE A4) support is available with configuration limitations.

    ☒ **A, B,** and **C** are incorrect. ASHRAE A4 supports a maximum air inlet temperature of 45°C, not 30°C, 35°C, or 40°C.

    For more information, see Chapter 4.

19. ☑ **A** and **C** are correct. Virtual connect is an edge port aggregator and its uplink ports are termination ports.

    ☒ **B, D,** and **E** are incorrect. With Virtual Connect, server changes do not affect the network. It is part of the server system, not the LAN, and it is not a switch.

    For more information, see Chapter 5.

20. ☑ **C** is correct. With IRF, multiple physical switches can be combined into one virtualized switch known as an IRF domain.

    ☒ **A, B,** and **D** are incorrect. AN IRF domain is not a high-availability server clustering solution, a method for encapsulating OSI Layer 2 Ethernet frames within Layer 4 UDP packets, or a server system in which memory access time depends on the memory location relative to the processor.

    For more information, see Chapter 5.

21. ☑ **C** is correct. The Virtual Connect FlexFabric 20/40 F8 module provides 20 Gb downlinks and 40 Gb uplinks.

    ☒ **A, B,** and **D** are incorrect. The Virtual Connect FlexFabric 20/40 F8 module provides 20 Gb downlinks and 40 Gb uplinks.

    For more information, see Chapter 5.

22. ☑ **C** is correct. A FlexNIC is a PCIe function that appears to the system ROM, operating system, or hypervisor as a discrete physical NIC with its own driver instance.

    ☒ **A, B**, and **D** are incorrect. A FlexNIC is not a virtualized switch consisting of multiple physical switches, a virtual NIC contained in a software layer, or a pair of NICs bonded within an operating system.

    For more information, see Chapter 5.

23. ☑ **A** is correct. The dual-hop FCoE feature in Virtual Connect employs a technology known as FCoE Initialization Protocol (FIP) snooping.

    ☒ **B, C,** and **D** are incorrect. The dual-hop FCoE feature in Virtual Connect does not use IGMP, DHCP, or TCP snooping.

    For more information, see Chapter 5.

24. ☑ **C** is correct. The HPE 6125XLG Ethernet blade switch supports RoCE.

    ☒ **A, B,** and **D** are incorrect. The FlexFabric 10/10D, FlexFabric 20/40 F8, and HPE 6125G/XG do not support RoCE.

    For more information, see Chapter 5.

25. ☑ **B** is correct. PF2 can be configured as Ethernet, iSCSI, or FCoE.

    ☒ **A, C,** and **D** are incorrect. PF1, PF3, and PF4 cannot be configured as iSCSI or FCoE.

    For more information, see Chapter 5.

26. ☑ **A, C,** and **E** are correct. Active optical cables provide increased immunity to EMI, lower cost, and support for increased cable lengths when compared with copper cables.

    ☒ **B** and **D** are incorrect. QSFP+ connectors are included with AOCs, and AOCs are physically smaller than traditional copper cables.

    For more information, see Chapter 5.

27. ☑ **A** is correct. The HPE BladeSystem PCI Expansion Blade provides PCI card expansion slots to an adjacent server blade, and if an expansion blade is installed in Bay 1, the expansion slots will be available to the server blade in Bay 2.

    ☒ **B, C,** and **D** are incorrect. A BladeSystem PCI Expansion Blade installed in Bay 1 will not provide PCI expansion slots to Bays 3, 9, or all even numbered bays.

    For more information, see Chapter 5.

28. ☑ **C** is correct. The c7000 enclosure can be configured with a maximum of 10 Active Cool fans.

    ☒ **A, B,** and **D** are incorrect. The c7000 enclosure can be configured with a maximum of 10 Active Cool fans.

    For more information, see Chapter 5.

29. ☑ **B** is correct. The maximum number of HPE ProLiant XL190r Gen9 server nodes that can be installed in a single HPE Apollo 2000 system is two.

    ☒ **A, C,** and **D** are incorrect. The maximum number of HPE ProLiant XL190r Gen9 server nodes that can be installed in a single HPE Apollo 2000 system is two.

    For more information, see Chapter 6.

30. ☑ **B** is correct. The Apollo 4520 is a two-server system and is ideal for clustered storage environments. It is an optimal fit for workloads such as Lustre parallel file system.

    ☒ **A, C,** and **D** are incorrect. The Apollo 4510 is a single-server system and is ideal for object storage solutions. The Apollo 4530 is a three-server system purpose-built for the wide variety of big data analytics solutions based on parallel Hadoop-based data mining, as well as NoSQL-based big data analytics solutions. The Apollo 4540 does not exist.

For more information, see Chapter 6.

31. ☑ **D** is correct. The Apollo 8000 primary water loop circulates facility water through the cooling distribution unit (CDU) located within the iCDU rack.

    ☒ **A**, **B**, and **C** are incorrect. The Apollo 8000 primary water loop does not channel water from the gravity drain collection system to the water collection pan, isolate the heat exchanger from the water collection pan using a bleeder valve, or distribute water from the iCDU to the f8000 rack where it flows through the heat exchanger.

For more information, see Chapter 6.

32. ☑ **A**, **D**, and **F** are correct. The three key elements of a composable infrastructure are software-defined intelligence, fluid resource pools, and a unified API.

    ☒ **B**, **C**, and **E** are incorrect. Hardware-defined intelligence, autonomic storage systems, and high-availability clustering are not key elements of a composable infrastructure.

For more information, see Chapter 7.

33. ☑ **B** is correct. The Synergy frame link module provides a resource information control point and reports asset and inventory information about all the devices in the frame.

    ☒ **A**, **C**, and **D** are incorrect. The Synergy composer provides a single interface for assembling and re-assembling flexible compute, storage, and fabric resources. The Synergy image streamer is a physical appliance repository of bootable (golden) images. The Synergy 12000 frame is the base infrastructure that pools resources of compute, storage, fabric, cooling, power, and scalability.

For more information, see Chapter 7.

34. ☑ **B**—Silo. **E**—Optimized. **A**—Service enabled. **D**—Hybrid delivery. **C**—Dynamic reconfigurable services.

For more information, see Chapter 8.

35. ☑ **D** is correct. HPE recommends 100 Mb/s IP and 200 Mb/s storage IO for each VM.

    ☒ **A**, **B**, and **C** are incorrect. HPE recommends 100 Mb/s IP and 200 Mb/s storage IO for each VM.

For more information, see Chapter 8.

36. ☑ **D** is correct. ROI compares the investment required with the resulting cost reduction or increased revenue.

    ☒ **A**, **B**, and **C** are incorrect. ROI does not demonstrate TCO over a one, three, or five-year period, highlight the costs associated with unplanned downtime, or compare CAPEX and OPEX savings over a particular time period.

For more information, see Chapter 8.

37. ☑ **B** is correct. A scope of work captures the plan, time frame, required resources, and completion milestones of a project.

 ☒ **A** is incorrect; it describes a statement of work. **C** is incorrect; documenting required support services is not the primary purpose of a scope of work. **D** is incorrect; it describes a request for proposal (RFP).

 For more information, see Chapter 8.

38. ☑ **A, D,** and **E** are correct. When ICsp runs an OS build plan, it checks to see if you have identified HPE OneView appliances and if the OS build plan type is one of the following three types: HW—SAN Configuration, HW—BIOS Configuration, or SW—Firmware, ICsp checks to see if there is a conflict.

 ☒ **B, C,** and **F** are incorrect. ICsp does not check for conflicts for build plans HW—Clear UEFI Boot Menu, HW—Erase Server, or SW—Validate Media Server Settings.

 For more information, see Chapter 9.

39. ☑ **A** and **E** are correct. The HPE OneView with ICsp Installation and Configuration Service includes the installation of HPE OneView on a supported VMware vSphere (ESXi) or Microsoft Hyper-V host VM.

 ☒ **B, C,** and **D** are incorrect. The HPE OneView with ICsp Installation and Configuration Service does not include the installation of HPE OneView on a Citrix Xen, Canonical Ubuntu, or KVM hypervisor host.

 For more information, see Chapter 9.

40. ☑ **B** is correct. Insight Online provides access to device and support information in a cloud-based personalized dashboard.

 ☒ **A, C,** and **D** are incorrect. Insight Online does not provide a virtual desktop, giving the user full control over the display, keyboard, and mouse of the host server, shared console and console replay functions, or access to a tool for configuring and pricing HPE servers and storage systems.

 For more information, see Chapter 9.

41. ☑ **D** is correct. Insight Remote Support central connect is ideal for HPE Converged Infrastructure environments with multiple device types.

 ☒ **A, B,** and **C** are incorrect. Insight Remote Support central connect is not ideal for SMB, remote sites, or customers who do not want a centralized hosting device in the local environment.

 For more information, see Chapter 9.

42. ☑ **A**, **B,** and **C** are correct. Insight Online My IT Environment includes reports for contracts and warranties, device configuration, and service events.

    ☒ **D** and **E** are incorrect. Insight Online My IT Environment does not include reports for compliance or application availability.

    For more information, see Chapter 9.

Index